Alexander Dallas Bache

Axel Jansen is assistant professor of American studies at Frankfurt University. He currently teaches in the Department of History, Heidelberg University.

Axel Jansen

Alexander Dallas Bache

Building the American Nation through Science and Education in the Nineteenth Century

Campus Verlag
Frankfurt/New York

Bibliographic Information published by the Deutsche Nationalbibliothek.
Die Deutsche Nationalbibliothek lists this publication in the Deutsche Nationalbibliografie.
Detailed bibliographic data are available in the Internet at http://dnb.d-nb.de.
ISBN 978-3-593-39355-1

Cover illustration: © "Navigation Chart of the City of San Francisco and Its Vicinity, 1853" (Image Archives of the Historical Map & Chart Collection, CP278C); photograph of Alexander Dallas Bache, ca. 1858 (NOAA's People Collection, theb3392). Both images courtesy of the Office of Coast Survey/National Ocean Service/NOAA.
Cover design: Campus Verlag, Frankfurt-on-Main

Printed on acid free paper.
Printed in Germany

For further information:
www.campus.de
www.press.uchicago.edu

"Axel Jansen's writing is exemplary in its deployment of close readings and literary perspectives to rethink large issues in the history of American science and the American state. His focus on Alexander Dallas Bache, self-conscious descendant of Benjamin Franklin, supports a new understanding of professionalizing campaigns. Bache did not mainly look to the nation to build up science, but labored to harness science to the task of lifting the burden of sectionalism and forming a genuine nation."

—*Theodore M. Porter, UCLA*

"In this bold reassessment of Alexander Dallas Bache, Jansen employs new theories of professional development to re-conceptualize the relationship between the rise of science and the project of nation-building in the antebellum United States. A critical reappraisal of Bache's early career is augmented with close textual analyses of key documents to reveal an institutional realist with a powerful—though previously misconstrued—vision for his nation's future. Jansen's intriguing perspective transforms the founding of the National Academy of Sciences in 1863 from a war-time happenstance of minimal lasting importance into the culmination of a project designed to bond science and the state on special terms. While the Academy itself may never have functioned as Bache hoped it would, the projects that Bache influenced would continue apace. He had laid the groundwork for one kind of tie between science and the state, which would have huge implications and profound lessons for later American history. This challenging new perspective is a must-read for anyone interested in the various roots of modern science and professionalism."

—*James C. Mohr, College of Arts and Sciences Professor of History and Philip H. Knight Professor of Social Sciences, University of Oregon*

"American historians have long debated the timing of the National Academy's founding without reaching consensus. Now Axel Jansen sets forth a new and insightful analysis that revolves around the extraordinary life and character of Alexander Dallas Bache, the first president of the academy. He and the most prominent scientists of his generation embarked on what Jansen calls "state building," establishing authority and encouraging greater consolidation even in the face of national disintegration. This is a book historians will not fail to read."

—*Thomas Haskell, Rice University*

Table of Contents

Acknowledgements

The Institut für England- und Amerikastudien, Johann-Wolfgang-Goethe-Universität Frankfurt, and the Department of History at the University of California, Los Angeles have been my institutional homes for this project. The German Research Foundation (DFG) as well as the Fulbright Commission provided the funds to write this book. A fellowship at the German Historical Institute in Washington D.C. as well as travel grants by the Thyssen Stiftung and the Center for North America Studies (ZENAF, Universität Frankfurt) made initial research possible. I am also grateful for the support for the publication of this book by the DFG and the Dr. Bodo-Sponholz-Stiftung.

I am indebted to many colleagues who have discussed with me aspects of this project and drafts of some chapters. Ulrich Oevermann was an invaluable intellectual resource and catalyst, and I am also grateful for the many conversations and the critical assessment by Andreas Franzmann, Peter Münte, and Lorenz Rumpf. Ian Copestake read the entire manuscript and I am greatly obliged to him for his valuable comments and suggestions. I wish to thank Roland Burkholz, Marcus Gräser, Johannes Hätscher, Olaf Hansen, Thomas Haskell, James Mohr, Michael Osborne, Cynthia Ott, Theodore Porter, Michael Reidy, Peter Scholz, Billy G. Smith, Mary Terrall, Jessica Wang, Norton Wise, and Katharina Worch for their support, advice, and critical observations at various stages of this project. Edward Hogan and Marc Rothenberg kindly offered important insights. Robert and Margaret Bicknell, Kevin Roth and Bettina Hansel, as well as Michael and Ines Jindra hosted me on trips related to this project, and I am grateful for their kind hospitality.

I would also like to thank students in seminars I taught in Frankfurt, at UCLA, in Tübingen, and in Heidelberg for their questions and ideas. I presented drafts of chapters at the American Studies colloquium in Frankfurt, the History of Science colloquium at UCLA, at the Department of History,

University of California, Santa Barbara, at the Department of History and Philosophy, Montana State University in Bozeman, at the German Historical Institute in Washington D.C., and at the Center for the Study of Religion and Society, University of Notre Dame. I thank all participants for their perceptive commentary. I also wish to thank reviewers of an earlier draft of this book (*Habilitationsschrift*) for their detailed comments and suggestions: Christa Buschendorf, Moritz Epple, Olaf Hansen, Michael Hochgeschwender, Ulrich Oevermann, and Theodore Porter. Publication of this book was remarkably facilitated by Dorothéa Frangopoulos-Blank, Judith Wilke-Primavesi, and Joachim Fischer at the Campus Verlag.

I gratefully acknowledge permission to quote from documents and to reproduce images from collections in the American Philosophical Society, the Historical Society of Pennsylvania, the Historical Society of Washington D.C., the Huntington Library, the Library Company of Philadelphia, the Office of Coast Survey/National Ocean Service (NOAA), and the Smithsonian Institution Archives. I wish to thank the staff in all of these institutions, and would also like to mention the friendly help by the staff at Girard College, at the Library of Congress, the National Archives and in the Houghton Library, Harvard University.

In concluding these acknowledgments, I wish to express my gratefulness to my wife Wencke for all of her warm support and caring.

Chapter 1

Introduction

The Curious Case of Alexander Dallas Bache

In the history of American science, Alexander Dallas Bache (1806–1867), great-grandson of Benjamin Franklin, occupies a singular and unparalleled position. More than anyone else in his generation and in perhaps any generation before or since, he embodied the American scientific profession, directed its development, and shaped its institutions. Most major national scientific institutions and organizations between 1830 and 1865 relied on his support or leadership: In the 1830s, Bache was the principal organizer of Philadelphia's Franklin Institute, then the most prominent research organization in the United States. In 1843, he became the superintendent of the U.S. Coast Survey, the country's largest government-run scientific enterprise with more scientific employees than any other contemporary science-related institution including Harvard University. From 1847, Bache helped instigate and direct the American Association for the Advancement of Science (AAAS), the country's first national platform for science. He was one of the regents of the Smithsonian Institution and helped secure the post of secretary (i.e. director) for his colleague Joseph Henry in 1846. Finally, Bache helped found the National Academy of Sciences and became its first president in 1863. In view of this ubiquitous role, A. Hunter Dupree considers him (with physicist Joseph Henry and geologist John Wesley Powell) among the three "great hierarchs of federal science" in the nineteenth century, and Robert V. Bruce has concluded that Bache spoke "more authoritatively for antebellum science than anyone else."[1]

---

1 Quotes from Robert V. Bruce, *The Launching of Modern American Science, 1846–1876* (New York: Knopf, 1987), 255, and Nathan Reingold, *Science in Nineteenth-Century America, a Documentary History* (London: Macmillan, 1966), 8, respectively. This assessment dates back to Bache's own time. In his eulogy of Bache, astronomer Benjamin Apthorp Gould

Fig. 1. Alexander Dallas Bache

(From American Philosophical Society, Proceedings 84:2, 1941)

While Bache was the acknowledged leader of mid-nineteenth century American science, however, the authority for his leadership remains enigmatic. One problem is that Bache was less pioneering in his research than in his institutional efforts. In a symposium in honor of Bache's legacy, organized by the American Philosophical Society in 1941, Frank B. Jewett conceded that while Bache's contributions to science "dealt largely with … [scientific problems] of recognized fundamental importance," they nevertheless concerned "departments of physics which neither then nor later

---

suggested in 1868 that to his colleague, "the scientific progress of the nation is indebted, more than to any other man who has trod her soil." Benjamin Apthorp Gould, "An Address in Commemoration of Alexander Dallas Bache," American Association for the Advancement of Science, *Proceedings* 17 (1868): 35.

could be regarded as spectacular or especially productive."[2] In his more recent assessment, Bruce perhaps overemphasizes this point by arguing that as "a scientist, Bache fell far short of both his famous ancestor [Benjamin Franklin] and his friend Professor [Joseph] Henry."[3] These observations reflect the fact that while Bache plays a prominent role in accounts of the institutional development of American science in the nineteenth century, he is less prominent in accounts of the development of the cognitive content of science in that period.[4] This has left Bache with a somewhat ambivalent reputation. Bache was well connected through relatives in Pennsylvania and in federal politics. Was he not much more than an apt administrator, an institutional booster with good connections and a knack for federal fundraising?

Another aspect of Bache's career complicates matters, and that is his involvement in education before 1842. While historians of American science have focused on his institutional role and his leadership in the professional community, historians of education have focused on Bache's role as president of the Girard College for Orphans and as first principal of Central High School in Philadelphia.[5] In 1836, Bache gave up his professorship at the University of Pennsylvania in order to assume these and other educational activities. How do such efforts fit into the pattern? Was Bache interested in cultural control, a Whiggish interest in "moral and intellectual discipline" both in his educational and in his professional leader-

2 Frank B. Jewett, "Alexander Dallas Bache. A Founder, First President and Benefactor of the National Academy of Sciences," American Philosophical Society, *Proceedings* 84, no. 2 (1941): 181.

3 Bruce, *Launching of Modern American Science*, 17. Similar comments abound. Another example is Mary Ann James, *Elites in Conflict: The Antebellum Clash over the Dudley Observatory* (New Brunswick: Rutgers Univ. Press, 1987), 26.

4 Geodesists of course remember Bache, as attested by the American Philosophical Society's symposium in 1941. *Commemoration of the Life and Work of Alexander Dallas Bache and Symposium on Geomagnetism*, American Philosophical Society, *Proceedings* 84, no. 2 (1941). He is mentioned in Mark Littmann, *The Heavens on Fire: The Great Leonid Meteor Storms* (Cambridge, UK: Cambridge Univ. Press, 1998). My argument here pertains to Bache's research record in relation to his institutional role. For more on this, see chap. 4 below.

5 These include: David F. Labaree, *Making of an American High School* (New Haven: Yale Univ. Press, 1992) and David B. Tyack and Elisabeth Hansot, *Managers of Virtue: Public School Leadership in America, 1820–1980* (New York: Basic Books, 1982).

ship, or was his educational involvement an extension of his administrative interests?[6]

In the absence of a comprehensive biography of Alexander Dallas Bache, and considering his extensive involvement and leadership in mid-nineteenth century American science, any attempt to clarify such issues will provide insights relevant well beyond the immediate task of identifying the motivational coordinates of his career. Bache's singular role in American science is of particular significance when considered in the context of recent developments in theories of the professions.

2. The Revised Theory of Professionalization

Historians have most commonly discussed Bache's career in the context of the emergence of the American scientific community.[7] In his pioneering work on the history of American science, A. Hunter Dupree had focused on the history of science as a development leading to the federal support

6 Hugh R. Slotten, "The Dilemmas of Science in the Unites States. Alexander Dallas Bache and the U.S. Coast Survey," *Isis*, no. 84 (1993): 47. Slotten does not discuss Bache's career overall but focuses on his Coast Survey work. He does use similar ideas for explaining Bache's educational work in his essay on "Science, Education, and Antebellum Reform: The Case of Alexander Dallas Bache," *History of Education Quarterly* 31, no. 3 (Autumn 1991): 323–42. For more on this, see chap. 5.

7 An earlier generation of historians focused on the cognitive content of science. To them, American achievements in the nineteenth century seemed negligible when compared to European science. See, for example, Richard H. Shryock, "American Indifference to Basic Research," *Archives internationales d'histoire des sciences* XXVII (1948): 50–65. See also I. Bernhard Cohen, "Science in America: The 19th Century," in *Paths of American Thought*, ed. Arthur M. Schlesinger Jr. and Morton White (Boston: Houghton Mifflin, 1963); Ronald L. Numbers and Charles E. Rosenberg, eds., "Science in American Society: A Generation of Historical Debate," in *The Scientific Enterprise in America: Readings from Isis* (Chicago: Univ. of Chicago Press, 1996). This focus on the cognitive development of science was modified by Nathan Reingold, "American Indifference to Basic Research: A Reappraisal," in *Nineteenth-Century American Science: A Reappraisal*, ed. George H. Daniels (Evanston: Northwestern Univ. Press, 1972), 38–62, and George H. Daniels, *American Science in the Age of Jackson* (New York: Columbia Univ. Press, 1968). An important book of the early phase, in which the professionalization of the scientific community was discussed, is Sally Kohlstedt, *The Formation of the American Scientific Community: The American Association for the Advancement of Science, 1848–1860* (Urbana: Univ. of Illinois Press, 1976). The most recent overview of the development of American science in the nineteenth century is Bruce, *Launching of Modern American Science*.

of research by the twentieth-century activist state.[8] In the 1970s, historians shifted their emphasis toward explaining the emergence of professional institutions in the United States. George Daniels suggested that the American scientific profession got started between 1820 and 1840 as it moved from gathering facts to developing "esoteric" knowledge, a process that culminated in the public acceptance of science before the Civil War.[9] Sally Kohlstedt's classic work on the *Formation of the American Scientific Community* views the founding of the AAAS in 1848 as a decisive moment. She provides a detailed account of the struggles that led to the organization's founding and of conflicts within the profession.[10] The historiographic focus altered slightly in the 1980s with authors such as Hugh R. Slotten who stressed "boundary work," and that scientists used a particular ethos to facilitate social and cultural control. His work was receptive to views that stressed the role of individual and group interests.[11]

The historical evidence suggested that as a profession, science was somehow distinct from other occupations, and sociological theories seemed to offer the best mode for explaining what it was that scientists were doing and how it was similar to and different from other activities.

In historical writing about the professions, it has proven to be of little benefit to use the term "profession" as one found it at large, because adopting the term from historic sources was to associate it with any occupation claiming professional status.[12] This is why more recent theories have tried to explain the peculiar characteristics of some occupations, such as

---

8 A. Hunter Dupree, *Science in the Federal Government: A History of Policies and Activities to 1940* (Baltimore: Johns Hopkins Univ. Press, 1986).

9 George H. Daniels, "The Process of Professionalization in American Science: The Emergent Period, 1820–1860," *Isis* 58, no. 2 (Summer 1967): 150–66. Important literature also includes Nathan Reingold, "Definitions and Speculations: The Professionalization of Science in America in the Nineteenth Century," in *The Pursuit of Knowledge in the Early American Republic: American Learned and Scientific Societies from Colonial Times to the Civil War*, ed. Alexandra Oleson and Sanborn C. Brown (Baltimore: Johns Hopkins Univ. Press, 1976), 33–69.

10 Kohlstedt, *Formation of the American Scientific Community*.

11 Bruce, *Launching of Modern American Science*, 263. Slotten, "Dilemmas of Science," 43; see also his *Patronage, Practice, and the Culture of American Science: Alexander Dallas Bache and the U.S. Coast Survey* (Cambridge: Cambridge Univ. Press, 1994).

12 Burton J. Bledstein, *The Culture of Professionalism: The Middle Class and the Development of Higher Education in America* (New York: W. W. Norton, 1976) uses the term too broadly. Laurence Veysey ("Who's a Professional? Who Cares?," *Reviews in American History* 3 (December 1975): 419–23) has criticized inflationary uses of the term but has also questioned the relevance of trying to define it.

the tendency by professions to invoke autonomy from outside social and political interference and to organize their own affairs. Very broadly speaking, there have been two sociological positions relevant for historians. A structural-functionalist approach (Talcott Parsons, William J. Goode) stressed the profession's role in developing, preserving, and using esoteric knowledge considered to be an important cultural value. One problem with this idea was that it could not explain why the professions successfully insisted on autonomy and how they had averted control by outside experts or administrators. Another approach focused on the profession as an interest group (Terence J. Johnson, Magali Sarfatti Larson). It considered the profession's claims of representing esoteric knowledge as an ideological tool for establishing market control in order to protect pecuniary interests and advantages. Neither of these two theoretical perspectives addressed the issue of whether professions pursue a specific type of activity different from other activities that do not require autonomy and exclusive organization.[13]

In his revised theory of professionalization, Ulrich Oevermann does not restrict "professionalization" to the emergence of organizations or successful claims for autonomy by occupational groups. He argues that professions are distinct from other types of vocations because of the peculiar type of activity in which they are engaged. He suggests that professions seek to restore a client's autonomy with reference to the client's particular autonomy potential and that they are responsible for a "vicarious crisis management" (or "vicarious problem solving"). In considering a therapy for a given disease, a medical doctor, for example, will have to take into consideration a patient's specific health and living situation. This requires a particular "habitus," a readiness to become aware of the particularities of unforeseen patterns as well as a readiness to intervene to the best of one's ability even in cases where available knowledge provides no answer. This

13 I am following Peter Münte's overview: Peter Münte, *Die Autonomisierung der Erfahrungswissenschaften im Kontext frühneuzeitlicher Herrschaft: Fallrekonstruktive Analysen zur Gründung der Royal Society* (Frankfurt: Humanities Online, 2004), 1:21 ff. For the different positions mentioned here, see Talcott Parsons, *The Social System* (Glencoe, IL: Free Press, 1951); Talcott Parsons, "The Professions and Social Structure," Social Forces, no. 4 (May 1939): 457–61; William J. Goode, "Community within the Community: The Professions," *American Sociological Review* 22 (1957): 194–200; Terence James Johnson, *Professions and Power* (London: Macmillan, 1972); Magali Sarfatti Larson, *The Rise of Professionalism: A Sociological Analysis* (Berkeley: Univ. of California Press, 1977).

makes the physician's relationship with a client both diffuse and specific.[14] Unlike the expert, therefore, the physician's role is not restricted to making use of available medical knowledge, and predefined checklists are useless for establishing a medical doctor's "efficiency." An evaluation of a physician's work will have to address the case-specific adequacy of intervention, which is based on a diagnosis and on a consideration of a given patient's living situation. This precludes assessment through market or administration and necessitates collegiate criticism shielded from outside (political etc.) interference. Oevermann distinguishes between three areas of professional activity that correspond to the three foci of human sociality: (1) therapy aimed at the constitution and preservation of autonomy by individuals, communities, etc.; (2) judicature aimed at the preservation of a community's normative order; and (3) science and the arts as the justification and development of knowledge. Among these three foci, science and the arts are of particular relevance because they provide the basis for the other two.[15]

In contrast to the physician's patient, of course, the "client" of science is abstract. Science and the arts represent an analytical logic that is also part

14 Regarding this observation and argument, see also Parsons, "The Professions and Social Structure."

15 For the revised theory of the professionalization, see Ulrich Oevermann, "Theoretische Skizze einer revidierten Theorie professionalisierten Handelns," in *Pädagogische Professionalität. Untersuchungen zum Typus pädagogischen Handelns*, ed. Arno Combe and Werner Helsper (Frankfurt: Suhrkamp, 1996), 70–182. For science, see Peter Münte and Ulrich Oevermann, "Die Institutionalisierung der Erfahrungswissenschaften und die Professionalisierung der Forschungspraxis im 17. Jahrhundert. Eine Fallstudie zur Gründung der Royal Society," in *Wissen und soziale Konstruktion*, ed. Claus Zittel (Berlin: Akademie Verlag, 2002), 165–230; Ulrich Oevermann, "Wissenschaft als Beruf—Die Professionalisierung wissenschaftlichen Handelns und die gegenwärtige Universitätsentwicklung," *Die Hochschule—Journal für Wissenschaft und Bildung* 14, no. 1 (2005): 307–18, Peter Münte, "Institutionalisierung der Erfahrungswissenschaften in unterschiedlichen Herrschaftskontexten. Zur Erschließung historischer Konstellationen anhand bildlicher Darstellungen," *Sozialer Sinn* 1 (2005): 3–44. For the perception of the role of science in the French context, see Andreas Franzmann, "Die Krise Frankreichs von 1870 und ihre Ausdeutung durch den Wissenschaftler Louis Pasteur—Eine Deutungsmusteranalyse," in *Wissen in der Krise*, ed. Carsten Kretschmann, Henning Pahl, and Peter Scholz (Berlin: Akademie Verlag, 2004), 117–56. For the case of art, see Ulrich Oevermann, "Für ein neues Modell von Kulturpatronage," *Die Kunst der Mächtigen und die Macht der Kunst*, ed. Ulrich Oevermann, Johannes Süßmann, and Christine Tauber (Berlin: Akademie Verlag, 2007), 13–23.

of other professions.[16] The "crisis" to be resolved here is the development and testing of the validity of cognitive and aesthetic truth claims, and the authoritative establishment of interpretations. In principle, truth claims are universal. For science, therefore, the structural equivalent to the physician's client is humanity and this includes future generations. Practically speaking, however, humanity has no political or institutional equivalent. The United Nations represents member countries and derives its legitimacy from them. It does not represent a community that would coincide with humanity. The role of "client" thus devolves to the nation-state as the most comprehensive legitimate political entity. On the basis of his study of the seventeenth-century founding of the Royal Society of London for Improving of Natural Knowledge, Peter Münte has suggested that national (or royal) academies assume the important role of legitimizing science, its radical questioning of recognized ideas, and of stabilizing investigative coherence by providing a common institutional and cognitive focus and monopoly. A nation-state, by accepting science in this particular way, acknowledges the universality and the rationality represented by the scientific discourse.[17] The revised theory of professionalization provides a foil for assessing the history of American science and of Alexander Dallas Bache's role within it.

Science as a Profession and the American Nation-State

The political context for science in nineteenth-century America differed radically from the situation in European nations. Even though the American states agreed on a Constitution in 1789, a regional and state-centered perspective carried over into the emerging federal arena. The Constitution established the idea of dual citizenship in both the individual states and in the federal state, but national citizenship was a political project rather than the social and cultural status quo. The country lacked a coherent national public and a capital city that would serve as a cultural center. In the eight-

16 For an earlier formulation of similar ideas outside of a theory of professionalization, see Alvin W. Gouldner, *The Future of Intellectuals and the Rise of the New Class: A Frame of Reference, Theses, Conjectures, Arguments, and an Historical Perspective on the Role of Intellectuals and Intelligentsia in the International Class Contest of the Modern Era* (New York: Seabury Press, 1979).

17 Münte, "Institutionalisierung der Erfahrungswissenschaften … Erschließung anhand bildlicher Darstellungen," and Münte and Oevermann, "Institutionalisierung der Erfahrungswissenschaften."

eenth century, the expansion of the settlement area in the American colonies had run counter to British interests but the United States actively pursued continental ambitions. In 1787, the Northwest Ordinance established a system of converting settled areas into territories and states, and the Louisiana Purchase in 1803 added a huge and largely unknown area to the country's settlement plans. In all of these ways, the country looked to the future rather than the past, and the prominent national perspective was not to have one. "Americans undertook their grand experiment in nation-making without a distinctive national history and culture," Peter Onuf and Leonard J. Sadosky have observed. "As republicans, who acknowledged no superior authority, they looked to each other; as provincials, who aspired to higher levels of refinement and civilization, they continued to look to the European metropolis."[18] While scholars have traditionally discussed the history of the early republic in terms of political ideology, Onuf and Sadosky have stressed that in the wake of an agreement on a formal state structure, the political basis for this structure, American nationhood, remained fragile.[19] The Civil War attests to the fact that the American states, in 1861, had not grown into a national political community, which would have made such bloodshed impossible.[20]

Historians of science have of course been aware of the infrastructural and political difficulties for the emergence of science in the United States but they have usually shared an interpretation of American politics that underemphasized the lack of a consolidated national perspective. They considered the national political framework and living conditions in the

18 Peter S. Onuf and Leonard J. Sadosky, *Jeffersonian America* (Oxford: Blackwell Publishing, 2002), 120.

19 About this earlier generation of historians, Onuf and Sadowsky write that "they focused on ideological appeal and popular response … [and] tended to underestimate the fragility of the union, and therefore the possibility of violence." Ibid., 225. See also James Roger Sharp, *American Politics in the Early Republic: The New Nation in Crisis* (New Haven: Yale Univ. Press, 1993).

20 Norman K. Risjord has pointed to difficulties in establishing coherent national symbols in the early nineteenth century because "the elimination of the monarchy meant that Americans could not look to a crown as a symbol of nationhood, and the Constitution, subject as it was to conflicting interpretations, did not serve as a valid replacement until after the Civil War. The flag … was a natural rallying point, but the makeup of stars and bars was subject to constant fluctuation with the admission of new states. The flag, which Francis Scott Key saw at dawn on September 14, 1814, contained fifteen stripes and fifteen stars, and his poem (put to the music of an English drinking song) did not become the national anthem until 1931." Norman K Risjord, *Jefferson's America, 1760–1815*, 1st ed. (Madison: Madison House, 1991), 205.

United States to be a circumstance, not an aspect intrinsic to the development of American science. If the revised theory of professionalization carries any weight in the American case, however, the development of the scientific profession in the United States requires reexamination, and with it the role of Alexander Dallas Bache.

Bache was certainly not the most innovative American scientist of his generation, but he stood out in other ways: He came from a prominent Philadelphia family that had long been affiliated with building the American national state. His maternal grandfather Alexander James Dallas was one of the instigators and leaders of the Republican-Democratic movement in Pennsylvania, and during the War of 1812, he was U.S. secretary of the treasury and, for a time, secretary of war as well. Dallas was an immigrant from England who had decided to join the emerging nation right after the Treaty of Paris had confirmed American independence in 1783. In 1805, his oldest daughter Sophia Dallas married Richard Bache, a descendant of Benjamin Franklin, an icon of American ingenuity and political independence. Alexander Dallas Bache was Sophia and Richard Bache's oldest son. His background was confirmed by his education at the United States Military Academy at West Point, then the only school founded by the federal government. None of Bache's immediate colleagues within the leadership of the American scientific community had a similar background. Joseph Henry came from more humble circumstances; Louis Agassiz, the Harvard biologist, immigrated from Switzerland in 1847; Benjamin Peirce, the Harvard mathematician, came from Cambridge, Massachusetts; William Barton Rogers, the founder of the Massachusetts Institute of Technology (MIT), had an Irish background not associated with national leadership; and so on. Bache stands out as a figure who could represent a national perspective for all areas of American culture including science. It is in line with these observations that Bache helped found the National Academy of Sciences, becoming its first president in 1863.

In this study, I will test the hypothesis that Bache's career and leadership, as well as the history of nineteenth-century American science, cannot be explained without a better understanding of the unconsolidated state of the United States as a political nation. In the absence of mature nationhood, the scientific profession could not expect to have its work confirmed by the federal government even though it required such focus and stabilization for its work. Bache's background and his role in founding the Academy suggest that prior to 1863, and in lieu of an institutional arrangement, he represented science to the prospective nation and a political legitimacy for American science.

Approach and Methodology

In analyzing Alexander Dallas Bache's career, his scientific work, and his institutional objectives, I make use of a methodological approach known as objective hermeneutics.[21] While large sections of this study will be devoted to an analysis of sequences of historical decisions based on a broad sample of sources, I will occasionally switch to a detailed interpretation of individual documents such as a particular letter or speech. Even where I proceed in a more general mode, I try to contrast what was in fact done at a given time or what was in fact written in a given document with what could have been done or could have been written. This counterfactual comparison serves as a tool to go beyond a mere description of texts and events, and to identify in them conscious and unconscious motives by analyzing sequences and tracing solidified decision patterns.

The difference between decisions and the traces of decisions is crucial. It represents a basic hiatus in the social sciences (and in the humanities) between actions that are fleeting and without use in a research setting, and the traces of these actions, "texts" sufficiently permanent to become available for research. This refers to the basic notion that research requires accountability for the deduction of conclusions from evidence. The durability and availability of this evidence is a prerequisite for a methodologically controlled approach to the analysis of such evidence. The term "text" is used broadly here and taken to mean all traces of human action including letters, publications, interviews, machines, paintings, landscapes, and so on. In view of this distinction, the differences among the fields within the

21 Throughout my methodological remarks, I am drawing on Ulrich Oevermann, *Strukturprobleme supervisorischer Praxis. Eine objektiv hermeneutische Sequenzanalyse zur Überprüfung der Professionalisierungstheorie* (Frankfurt: Humanities Online, 2001), 27–42. See also Ulrich Oevermann, "Regelgeleitetes Handeln, Normativität und Lebenspraxis. Zur Konstruktionstheorie der Sozialwissenschaften," in *'Normalität' im Diskursnetz soziologischer Begriffe*, ed. Jürgen Link, Hartmut Neuendorf, and Thomas Loer (Heidelberg: Synchron, 2003), 183–219. I have sketched some of the theoretical assumptions underling this approach in Axel Jansen, "Die objektive Hermeneutik als Instrument der historischen Fallrekonstruktion," *traverse—Zeitschrift für Geschichte/Revue d'Histoire* 13, no. 2 (2006): 43–56. For a brief overview in English, see Ewald Terhart, "The Adventures of Interpretation: Approaches to Validity," *Curriculum Inquiry* 15, no. 4 (Winter 1985): 451–64.

social sciences/humanities are reduced to interpretive traditions. Such seemingly divergent fields as American studies, history, and sociology differ from the natural sciences by their common task of deciphering meaningful traces of human activity.

The analysis of such traces is possible with reference to cultural and linguistic rules. According to Ulrich Oevermann's theoretical model of "rule-governed action," autonomy manifests itself through decisions made by an individual, a community, a nation-state, a company, or any other subject ("agent", "actor" etc.), and these decisions become possible, and are indeed forced upon us, through cultural and linguistic rules. Rules come in different shapes and sizes and they have common qualities, but they are not mere agreements. Universal rules provide the means for communication across cultural borders. They enable us to identify houses and clothes in other cultures that look different from our own, and to understand, for example, that the death of a relative will be important in any culture. These universal rules are distinct from rules that are specific to a particular culture, rules that include mores and rituals that frequently provide particular answers to a universal problem.

In analyzing a given sequence of text or a sequence of events in reference to underlying rules, we can draw on what John Searle has called "Background," i.e. "a certain sort of knowledge about how the world works" and "a certain set of abilities for coping with the world." This "Background" enables us to comprehend a text because it relies on rules that are independent of it.[22] This does not mean that we are familiar with every rule and any culture and its language, traditions, etc., but we can familiarize ourselves with them because of more general, underlying rules that make translation possible. These abilities are more general than language and they enable us to understand metaphors or learn a foreign tongue. We take for granted that intention, sincerity, deception, etc. exist in any culture, as otherwise we would be going in circles.[23] Oevermann

22 This cancels out the relevance of the infamous "hermeneutic circle." Interpretation takes place with reference to rules, not through an advance understanding of a particular text, even if this interpretation will relate to a particular research question.

23 John Searle uses examples such as the following to make the point that in everyday situations we rely on all kinds of tacit expectations (Background): "If you consider the sentence 'Cut the grass!' you know that this is to be interpreted differently from 'Cut the cake!' If somebody tells me to cut the cake and I run over it with a lawnmower or they tell me to cut the grass and I rush and stab it with a knife, there is a very ordinary sense in which I did not do what I was told to do. Yet nothing in the literal meaning of those

stresses that rules provide the raw material through which autonomy becomes possible and manifests itself because rules imply "sequentiality." At any given moment, options present themselves to us, and even if we choose not to decide, we will in fact do so. We are presented with options through rules, and by selecting an option we close a branch of options while we open another. In his theoretical work, Oevermann spells out the consequences of such observations for the sociology of religion and for other areas.[24] In his methodological papers, he has developed tools for tracing decisions and their patterns and to reconstruct autonomy.

What I am trying to do in this study is to employ an approach based on these considerations both in discussing sequences of events, and in analyzing sequences of text in documents. In both cases, I will try to contrast a decision with relevant alternatives, i.e. options that were not chosen at the time. This is what I mean by "counterfactual comparison": I will make use of an unusual perspective in which I consciously introduce relevant hypothetical options to which I compare the (biographical, lexical, etc.) decisions that were in fact made. The purpose is to lay out rule-driven options that presented themselves to a historic actor, and this serves as a foil for charting the meaning of the selection that was made. In order to get a hand on the initial choices, I am consciously using a strategy of "artificial naiveté." The aim is not to paraphrase a decision or a particular piece of text but to deduce its implicit assumptions and contrast them with alternatives—not just *any* alternatives but alternatives that are relevant in the respective "text" situation. This involves asking questions such as: Given a particular family background, what does it mean to choose a certain name for one's child? Why was a particular career chosen instead of another? What difference does it make to use a particular verb instead of another verb that would also have been an appropriate choice? What is the (perhaps unconscious) objective benefit of (mistakenly) leaving out a word or

sentences blocks those wrong interpretations. In each case we understand the verb differently, even though its literal meaning is constant, because in each case our interpretation depends on our Background abilities." John R. Searle, *The Construction of Social Reality* (New York: Free Press, 1995), 130 f. For an extrapolation of John Searle's theoretical observations and of Oevermann's theoretical and methodological work for the case of literary studies, see Lorenz Rumpf, *Naturerkenntnis und Naturerfahrung: Zur Reflexion epikureischer Theorie bei Lukrez* (München: Beck, 2003).

24 Ulrich Oevermann, "Ein Modell der Struktur von Religiosität. Zugleich ein Strukturmodell von Lebenspraxis und von sozialer Zeit," in *Biographie und Religion. Zwischen Ritual und Selbstsuche*, ed. Monika Wohlrab-Sahr (Frankfurt: Campus, 1995), 27–102.

using the wrong one?[25] The aim is to deduce implications of a given text or historic sequence in order to identify motives (related to "decision patterns") that may or may not have been conscious to the historic actor.[26]

In discussing a particular decision (such as the decision to accept a particular post, to assume a particular attitude towards the idea of a national scientific organization, or to use a particular phrase for characterizing a colleague) I will not take the validity of my hypothesis for granted but compare it to alternative explanations.[27] This allows for a repeated testing of my hypothesis. For example, I will check my analysis of Alexander Dallas Bache's early career against the results of my analysis of his educational work, which I will in turn check against the results of my analysis of his later speeches and letters. This provides both a means for identifying and testing the underlying logic of his career, and for refining its historic development and variation. The following chapters are designed as a sequence for trying out the idea that Bache's prominence in mid-nineteenth century America somehow reflects and explains the political setting of science and that the profession's support of Bache indicates that the history of science as a profession must be explained by taking seriously the unconsolidated state of the American nation.

The systematic procedure for analyzing documents and sequences of events allows for a successive testing and sharpening of the emerging in-

25 It makes no difference whether the decision had been made consciously or unconsciously as this approach aims at the text's implicit logic rather than the author's intention. It is frequently possible to deduce the latter but it is important to distinguish it from the former. While discussing a given document or a detail from a document, I will sometimes use the present tense which reflects the presence of the evidence rather than the historic moment accessible through it. By using phrases such as "Bache infers" or "Bache implies" I do not suggest that Bache (or whoever the author of a given document may have been) was conscious of what he (or she) wrote or that he (or she) did so intentionally. Rather, the aim is to deduce the meaning of a given text regardless of whether the author was conscious of its implications.

26 In this way, this approach differs from other close-reading strategies such as the one used by Alexandre Koyré who focuses on details of the writings of scientists but does not analyze them in the way suggested here, namely, as a sequence. Alexandre Koyré, *From the Closed World to the Infinite Universe* (New York: Harper & Row, 1957).

27 This mode of investigation, an analysis through "counterfactual comparison," must not be mistaken for ignorance of available historical sources. Every effort has been made to look at all relevant material pertaining to Alexander Dallas Bache's career and motives. But for the reasons spelled out in this introduction, I will occasionally select individual documents and interpret them step by step and in detail, consciously using the interpretive strategy of "artificial naiveté" as a methodological tool for analyzing the text.

terpretation. I will use the interpretation of a given sequence as a hypothesis to be tested against an analysis of the following sequence. Falsification (in Karl Popper's sense of the term) is thus attempted both on a microscopic level of analyzing a particular document, and on a macroscopic level of testing these findings against the analysis of other documents or phases of Bache's career. The overall intent is to find specific patterns sufficiently general to explain the diversity of phenomena to which the initial question had pointed.

Investigative Agenda

For the reasons explained above, Alexander Dallas Bache will be our single "sample case." While biographical elements play a role in my analysis of this important figure in American science, this is not a comprehensive treatment of his life. The first section of this book focuses on Bache's motives for pursuing a career in science against the backdrop of a range of studies that have oscillated between a depiction of Bache as a scientist, an educator, and a manager. In chapter two, I discuss the history of both the Bache and the Dallas families that brings into focus parallels between it and Bache's own career decisions. Following a discussion of Bache's initial career choice (in chapter three), I investigate Bache's pre-Washington career that is less known and accessible than his later work. In chapters four and five, I discuss his early efforts in institution building and his educational work, respectively. These four initial chapters serve as a basis for delineating the specific advantages, abilities, and perspectives that Bache brought to a post in the federal administration in 1843. As will be seen, when Bache became superintendent of the U.S. Coast Survey, a post he would hold until his death in 1867, his program for developing science in America was settled and he laid it out in speeches he gave in 1842, 1844, and 1851.

In this second section of the book, I will slightly adjust my mode of investigation by focusing on these speeches in chapters six, seven, and eight. While I discuss individual documents in the first section of this book as well, I will concentrate on these speeches rather than historic events. In chapter nine, I continue my investigation of Bache's Washington role by analyzing a letter by mathematician Benjamin Peirce, who was a close friend and colleague, and by interpreting Bache's response. This chapter serves to illustrate the particular type of relationship fostered by Bache and his circle of friends and colleagues—a group of influential science administrators and university-based researchers that called itself the scientific "Lazzaroni." Against the backdrop of the preceding chapters, chapter ten will provide an opportunity to test the overall thesis of Bache's interest in national consolidation against his rationale for founding the National Academy of Sciences in 1863. In chapter eleven I conclude by arguing for a new paradigm for the study of American science in the nineteenth century.

Chapter 2

Family Background

The Franklin and Bache Families

Contemporary observers as well as historians have routinely associated Alexander Dallas Bache's name with that of his great-grandfather, Benjamin Franklin. In Europe, Franklin's name evoked an image of the quintessential American, of the humble but educated and intellectually resourceful citizen, an icon of the republic in the wilderness. In the United States, his name is intimately tied to the essence of American nationhood. Such associations, however, do little to reveal the substance of Alexander Dallas Bache's family background. They provide little information about the peculiar opportunities and restrictions for Bache's way into adulthood, and the relative success and failure of his ambitions.

As an adult, Bache would certainly identify with this Franklin but what aspect of his great-grandfather's many-sided life did Bache seek to emulate?[1] Was Bache a scientist in the sense that he had come early to a fascination with exploring nature? Did he appreciate Franklin's interest in research as a means to develop applications? In view of such questions, it seems appropriate to try to comprehend Bache "from the ground up," i.e. to consider his career against the backdrop of the choices available to him in the context of his parents' and grandparents' biographical decisions, their expectations, and their social, cultural, and political milieu. Hugh R. Slotten has observed that Bache's "commitment to science as an intellectual and social activity did not exist separately from his family background, his educational experiences, and his cultural ties."[2] Beyond this obvious

1 For the relation of Franklin's scientific and political thinking, see I. Bernard Cohen, *Science and the Founding Fathers* (New York: Norton, 1995), 135–95.

2 Hugh R. Slotten, *Patronage, Practice, and the Culture of American Science: Alexander Dallas Bache and the U.S. Coast Survey* (Cambridge: Cambridge Univ. Press, 1994), 21.

connection, a detailed consideration of Bache's background reveals just how his career trajectory and its apparent deviations (such as his educational work in the 1830s and early 1840s) matched and expanded his family's biographical and cultural pattern. The following consideration of his family background is not intended to provide an "explanation" of Bache's biography. Given his family's peculiar role in the nation's founding period, however, it is difficult to see how it could not have been of great significance for him.

Alexander Dallas Bache was born in Philadelphia on July 19, 1806, son of Richard Bache Jr. and Sophia Burrell Dallas Bache. On his father's side, Bache's great-grandfather was Benjamin Franklin. Because of the latter's clear significance for both the Bache family and for the United States, my discussion of Bache's background begins by recalling aspects of Franklin's life even if many aspects of it may have seemed remote to Bache's contemporaries and perhaps even Bache himself.

Franklin had made a fortune in Philadelphia by publishing *Poor Richard's Almanac* and by editing the *Pennsylvania Gazette*, the most successful newspaper in the American colonies, from 1723 until he retired from the newspaper business in 1748. Franklin left a long record of civic engagement, instigating the American Philosophical Society, Pennsylvania Hospital, and an academy that evolved into the University of Pennsylvania. Franklin served in the Pennsylvania assembly from 1751 to 1763, and was dispatched to England in 1757 as a negotiator for the proprietary colony. When he was again sent to London in 1764 in order to petition that Pennsylvania be made a royal province, he was representing other colonies as well. Prior to American independence, which he publically and strongly supported, Franklin came to represent all thirteen colonies, assuming a perspective that would naturally prefigure American nationhood. In 1776, as ambassador to the court of Louis XVI., he in fact represented the Continental Congress in Europe. Two years later, he was able to secure French support for American independence and to win an important ally in the fight against England. In Paris, Franklin assumed the role of the humble Quaker, inviting mythical and romantic projections, and thus became the perhaps "quintessential" American. In 1783, Franklin, John Jay, and John Adams signed the Treaty of Paris that ended the war for independence and secured to the emerging nation the Northwest Territory and the trans-Allegheny West. After an absence of eight years, Franklin returned to America in 1784 and continued to engage in politics.

After independence had been won, American politics turned to questions of how to organize the states and the confederacy. As a delegate to the Philadelphia Constitutional Convention in 1787, Franklin stood for democratic principles and against a proposed freehold suffrage. Franklin argued that the privileges of the electors should not be limited.[3] His nephew Benjamin Franklin Bache, and in fact the entire Republican opposition movement, would later connect to this political perspective. Franklin's name would be associated with an interpretation of American independence and of the U.S. Constitution that implied a broad application of democratic principles.[4]

Alexander Dallas Bache's grandmother Sarah ("Sally") Franklin (1743–1808) was Franklin's single legitimate daughter. Bache's grandfather bore the same name as his father: Richard Bache Sr. (1737–1798). Seven years older than Sarah, he been born in Settle, Yorkshire (England), and moved to New York in 1765 to join his brother there. He then relocated to Philadelphia and became a merchant and insurance purveyor, though never a successful one.[5] It seems unlikely that, two years after his arrival, he had been able to create the financial foundation for raising a family in America. He subsequently married Sally Bache in 1767, reportedly against the wishes of her father who feared that Richard was marrying for money. Franklin, who had amassed a fortune as a newspaper publisher and author, would indeed have to support his daughter and her family for the rest of his life. He helped in other ways as well. In 1776, Richard Bache became Postmaster-General of the United States, a post secured with the help of his predecessor Benjamin Franklin who had held this post under the British crown and then by the authority of the Continental Congress from 1775. The Postal Service, in the absence of modern bank transfers, served an important function by transporting money in addition to regular mail. Richard Bache lost this position in 1782 when his father-in-law, who was in France, became the victim of political attacks in America.

3 Sean Wilentz, *The Rise of American Democracy: Jefferson to Lincoln* (New York: Norton, 2005), 33 f.

4 See the changing perception of Benjamin Franklin and his association, by many conservatives, with "dangerous" democratic tendencies on the occasion of the Franklin's death. Ibid., 38 f.

5 Website of the Benjamin Franklin Tercentenary, http://www.benfranklin300.org/frankliniana/people.php?id=81.

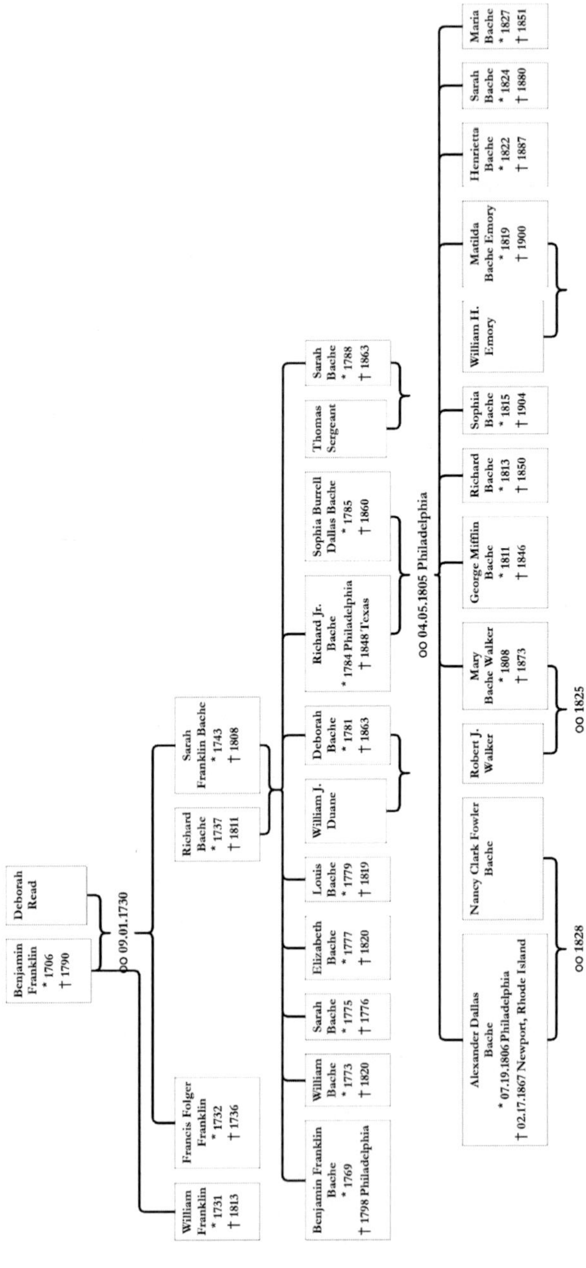

Fig. 2. Descendants of Benjamin Franklin

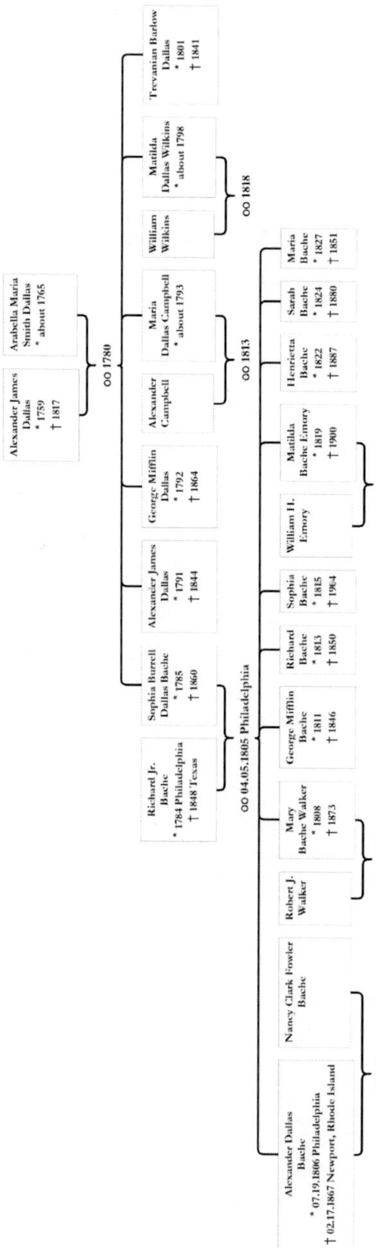

Fig. 3. Descendants of Alexander James Dallas

Richard and Sarah had eight children, seven of whom survived childhood. Their oldest son, Benjamin Franklin Bache was born in 1769, a year after they had been married. That they named their son after Sarah's father indicates that Sarah had no intention of breaking away from her father. Her son Benjamin Franklin Bache would indeed grow close to his grandfather. At the age of eight, he joined Benjamin Franklin in Europe and did not see his parents until he had turned sixteen. He was schooled in France and Switzerland before returning to Philadelphia with his grandfather in 1784. As will be discussed later, Richard Franklin Bache stepped into his grandfather's shoes and entered the newspaper publishing and printing business as well as politics in the 1790s. He helped shape the emerging Democratic-Republican political opposition to anglophile policies pursued by President George Washington and his Federalist supporters.

Richard's and Sarah's second child, their son William (1773–1820), and their fifth child, Louis (1779–1819), were not named after their father; it was Richard Bache (Jr., 1784–1848), their seventh child and Alexander Dallas Bache's father, who would be so named. Following the older Sarah's death in 1776, the parents passed on two opportunities to name another daughter after her mother. It was toward the end of the child-rearing years that Richard and Sarah Bache chose to name their children after themselves. This suggests that children were considered a service and tribute to the larger family system, and their names imply that within this family system, the mother's lineage remained dominant. When Richard Bache (Jr., Alexander Dallas Bache's father) was born in 1784, Sarah Franklin Bache was forty-one years old. While large families were common at the time, Richard and Sarah may have expected Richard Jr. to be their last child. That they chose to name their first son after Sarah's father, Benjamin Franklin, their first daughter after Sarah's mother, and their sixth daughter after Sarah's maternal grandmother (Franklin's wife Deborah Read), also points to the prominence of the maternal lineage. Richard Jr.'s name stands out because it represents the paternal side of the family. To the extent that Richard's name implied a dedication, Alexander Dallas Bache's father must have sensed that among his siblings, he represented the weaker and the publicly less significant strand of the family.

After his return to Philadelphia in 1785, Benjamin Franklin had moved in with his daughter's family. After several more years during which he was active in politics, philanthropy and scientific pursuits, he died in 1790. Sarah's family, who owed him much and had taken care of him, was now free to move on, and Richard and Sarah, who had inherited the bulk of Franklin's estate, decided to leave the city. Richard Jr. was ten years old when his family relocated to a farm on the Delaware River four years later. Few details are available about Richard's youth, his education, and early employment. What survives is a letter that seventeen-year-old Richard wrote to his mother, a teenager with few interests outside of his active social life.[6]

The Dallas Family

In 1806, Richard Bache Jr. married Sophia Burrell Dallas and followed his father's pattern of marrying into the second generation of an emerging American family dynasty. In order to comprehend the character of their marriage, a closer look at Sophia's background is in order.

Sophia Burrell Dallas' father, Alexander James Dallas, had been born in Jamaica in 1759. He was the son of Dr. Robert Dallas, a physicist who had emigrated there from Scotland. Alexander had two older brothers, Robert Charles and Stuart George, a younger brother, Charles Stuart, and two sisters, Henrietta Charlotte and Elizabeth.[7] Alexander James Dallas received his education in Scotland and England, where the family returned when he was about five years old. Alexander's mother had died before the family returned to England, and Robert Dallas had remarried.[8] When Robert Dallas died in 1769, Alexander was left an orphan. His step-mother then married one Captain Sutherland of the British Navy. Alexander's relationship to Mrs. Sutherland must have been strained, for while school

6 Richard Bache Jr. to Sarah Franklin Bache, January 18, 1801, Bache and Wistar Family Correspondence, Princeton University Library.

7 Raymond Walters Jr., *Alexander James Dallas, Lawyer, Politician, Financier, 1759–1817* (Philadelphia: Univ. of Pennsylvania Press, 1943), 7.

8 There is ambiguous information on whether it was Alexander James' mother whom Dr. Robert Dallas took back to England in 1765 or whether, after Alexander James' mother's death, Robert Dallas took his second wife (not Alexander James' mother) back to England. Compare John M. Belohlavek, *George Mifflin Dallas: Jacksonian Patrician* (University Park and London: Pennsylvania State Univ. Press, 1977), 10, and Walters Jr., *Alexander James Dallas, Lawyer, Politician, Financier, 1759–1817*, 7. Because of Alexander James Dallas' rather distant relationship to Sarah Cormack, whom Walters considers his mother, it seems that Alexander James Dallas, after his father's death in 1769, was an orphan. This would be in line with him spending his vacations with his aunt, not his step-mother, though this still implies that the relationship with the latter was probably tense.

was in recess, Alexander went to live with his aunt, not his step-mother. When he was eighteen and ready to pursue his plans to study law, he learned that he could not pursue these plans because, as a result of his step-mother's lifestyle, his father's estate had dwindled. Having lost his parents at an early age and with no funds for supporting his studies, Alexander began working in an uncle's business in London. But that business soon failed. With little financial room for maneuver, he decided to move back with his step-mother in Devonshire.[9]

In Devonshire, Alexander had earlier met Arabella Maria Smith (called Maria) who was now fifteen years old and a friend of his sister Henrietta Charlotte. Alexander had had ambitious educational plans which indicate that he sought to live up to his family's professional standards. Even though (or because) Alexander had no parents to support him and a step-mother who had done little or nothing to compensate for it, the young man was eager to have his own family. He proposed to Maria, who accepted, but the couple then faced opposition from Maria's aunts who had a say in this matter. Maria's family was willing to consent to marriage only on certain conditions, including the consent of Maria's father, a British army major in Jamaica. In addition, Alexander was to complete his studies and be admitted to the bar, which would take him about three years and Maria would have turned eighteen. Maria's aunts were obviously concerned that Alexander had no means to support a family.[10]

The couple agreed and Alexander left for Jamaica where he had intended to go with his stepmother, but, while in Barbados, he decided to return with his sister who asked him to accompany her back to England where *she* was going to get married. At the age of twenty, Alexander decided to break the bargain with Maria's aunts, to travel back to England, and to get married immediately. He wanted to pursue his intention even though he had not finished his education and had no certain means of income. Perhaps it impressed Maria that Alexander chose to return to England even though he could not know whether their families would support them. They were married, and the young couple then set out for Jamaica where Maria was to see her father for the first time since childhood. That both families happened to be connected to this island was perhaps considered by them to be a fateful coincidence.

9 Ibid., 6–9.
10 Ibid., 9–13.

Were Alexander and Maria rash and irresponsible? Were they neglecting a realistic assessment of their prospects? Alexander was unwilling to put considerations of income before those of love and marriage. But the couple did not bolt, but rather sought to win the approval of their families. Their decision to get married against the wishes of Maria's aunts was not a decision against authority in general. There is no indication that they felt they would later regret their decision or that they considered they had given up their reputation. Their decision was romantic and the success of their early ambition must have reasserted and emboldened them.

Alexander and Maria Dallas tried to settle among British imperial elites on the island of Jamaica but when the opportunity came up to move to America instead of returning to England, the couple chose to embark for New York City.[11] This decision was made in 1783, the very year the Treaty of Paris was signed and American independence was won. In Jamaica, Alexander was practicing law and had an income but Maria had difficulties adjusting to the climate. Alexander and Maria chose to tie their fate to that of the emerging and untried American confederation rather than return to England. The twenty-four-year-old and his young wife thus joined England's progressive nemesis. For the second time within just a few years, they had chosen to rely on their unproven instincts and their ability to succeed on new terrain. One could predict in 1783 that if the couple's ambition proved successful, it would embolden their characters and personal confidence.

Alexander and Maria moved straight to Philadelphia, the political hub and capital of the American colonies, where Dallas became a lawyer and politician. A few months after his arrival, he took the oath of allegiance to the state of Pennsylvania (the United States not yet being in existence) and he was admitted to the bar two years later. British-trained lawyers were in demand to help the independent states create a legal system in line with their British tradition. The common law had evolved heterogeneously in the various colonies but remained the basis for them all. The ready availability of English literature on the subject further assured the continued influence of British law in which Dallas had been trained in Jamaica.[12] The significance of precedence in common law required that records about previous cases would be available, but no such reports on decisions handed

11 Belohlavek, *George Mifflin Dallas*, 10–12.

12 Lawrence Meir Friedman, *A History of American Law*, 2nd ed. (New York: Touchstone, 1985), 144.

down by American courts existed. Between 1790 and 1807, Dallas published his first volume of "Reports of Cases Ruled and Adjudged in the Courts of Pennsylvania, Before and Since the Revolution." In his second volume, Dallas added cases from the U.S. Supreme Court, creating a tradition of reporting from that court.[13] In 1791, Pennsylvania Governor Thomas McKean appointed him Secretary of the Commonwealth, a position he retained for a decade. Dallas continued work on his four-volume report. He saw an opportunity to create and advance an independent judicial tradition and in doing so, he moved toward the center of his profession and of the political community he had joined.

Considering Dallas' ambitions, it is perhaps not surprising that in the political conflict that emerged after the ratification of the U.S. Constitution, the lawyer not only emerged on the Republican-Democratic side, but became a leader of that movement. It was here that the paths of the Bache and Dallas families intersected years before Alexander Dallas Bache's parents were married in 1806.

Against the backdrop of the French Revolution, underway since President George Washington's first year in office, there emerged an opposition movement culminating in what historians used to call the "first American party system." The administration's critics pointed to what they perceived as the federal government drifting away from the wartime alliance with France, and they disliked the strengthening of the central government through the assumption of state debts and the creation of a national bank. After 1793, "Republican" or "Democratic" societies provided a platform to discuss politics and to criticize President George Washington, Secretary of the Treasury Alexander Hamilton, and other Federalists. Within the federal government (designed by Constitutional framers on the assumption that it would include men of distinction and learning who stood above political factions), Thomas Jefferson and James Madison emerged as the leaders of this opposition movement, though they did not actively seek such a role in public. Neither they nor the Federalists welcomed the emergence of a political opposition. That such an opposition would be beneficial to the political community (as long as it accepted political institutions when defeated at the polls) was an idea that was only then evolving.

13 Ibid., 322 f.

Alexander James Dallas was among the most active proponents of the Republican-Democratic idea and among the founders of the Democratic Society in Philadelphia. In the 1790s, he cooperated in this movement with Benjamin Franklin's grandson Benjamin Franklin Bache (publisher of the *Aurora General Advertiser*).[14] As mentioned above, Bache had been educated in France and Switzerland and spoke French fluently. He had contacts and friends in Europe who could provide him with information about political developments there. Perhaps only a handful of Americans were as familiar with France as Bache was, and his newspaper (among the first to publish political commentary) became a natural focus for the Francophile political movement he helped bring about. While Bache was the movement's propagandist, Dallas became one of its key organizers. Their cooperation fizzled out after the president had signed Jay Treaty in 1794. While Dallas was ready to move on, Bache and others in the group continued to confront the administration over this matter. Dallas' relationship with the "Radicals," despite occasional cooperation under the Republican banner, grew distant.

Tertium Quid

That Dallas chose to take a less radical stance was perhaps connected to his experience during the Whiskey Rebellion in 1794. Western Pennsylvania was separated from the state's urban centers by mountains, and the distance had translated into cultural and political differences. When a federal excise tax was to be collected, and western Pennsylvanians refused to pay, Washington decided to make an example of it. For the first time since independence, the president used the federal government's powers to call out the state militia, an army of 13,000, roughly the size of the revolutionary army. When the troops had been raised, Dallas went West with them and there met the president in camp. A detailed interpretation of a letter from the field sheds light on Dallas' views on Pennsylvanian and American statehood. It provides an opportunity to further decipher this topic's significance for the Dallas family.

"Two men have been killed," Dallas wrote to his wife,

14 Wilentz, *The Rise of American Democracy*, 40–71. Wilentz gives an overview on literature on this topic. See also Jeffery Alan Smith, *Franklin and Bache: Envisioning the Enlightened Republic* (New York: Oxford University Press, 1990).

—one by the Jersey line, and one by the Pennsylvania line. I believe that both either provoked or deserved their fate; but the events are calculated to excite alarm, particularly when exaggerated, as they have been, by the enemies to the measures of the government.[15]

Dallas reports that some are surprised by the federal troops' use of force. The conflict tests the government's authority. Dallas is concerned that those who are unwilling to accept federal power will exploit the situation. By speaking of "enemies to the measures of the government" instead of "enemies of the government," Dallas leaves open the possibility that these insurgents are not opposed to government as such and that they can eventually be reintegrated. Dallas' comments suggest, however, that he was aware that federal authority as such was at stake, not a particular government policy.

The president, in a candid and manly manner, regretted that the deaths had happened, and observed that men who are engaged in the duty of supporting the law should be the last to violate it.

President Washington was not speaking to the citizens of western Pennsylvania but to federal troops—citizen soldiers—assembled from other parts of the country. That Dallas speaks of Washington's "candid and manly" manner suggests that it took courage to raise the issue of the appropriate use of force, and that it would have been easier to congratulate the troops on their decisive action. Perhaps there were indeed doubts about the appropriateness of shooting the two men. In this case, the problem was not the insurgency itself but keeping in line those who were engaged in suppressing it. In Dallas' view, some troops lacked a natural and intuitive understanding of an appropriate use of federal force. This could further diminish the federal government's credibility and standing.

This declaration was, fortunately for me, made in the presence of several who thought me lukewarm for inculcating a similar doctrine a day or two before the President's arrival. I enjoyed the triumph.

Dallas shares Washington's perspective and is glad that the president vindicates his insistence on obedience to the law. What was at stake was Dallas' leadership and charisma. Some considered him "lukewarm" and viewed him as an opportunist who invoked the law as an ideological prop. We may

15 Alexander James Dallas to his wife (n.d., ca. 1794), in George Mifflin Dallas, *Life and Writings of Alexander James Dallas* (Philadelphia: Lippincott, 1871), 34.

infer that they had opposed Dallas' criticism on the ground that the enforcement of federal authority, even in cases where it had no legal basis, was important for its own sake. If Dallas had been critical of such a view, how could he feel vindicated by Washington's support? The president had probably commented on these matters, not in front of the rank and file, but before a select group of officers. Dallas considers Washington's statement, and his perhaps unintentional retrospective endorsement of his view, to carry sufficient weight to be accepted by all of them. This suggests that a federal (or national) perspective that was not buffered by the charisma attached to the presidency, a *generalized* federal or national perspective, remained fragile at best. This deficiency is perhaps reflected in the fact that, instead of feeling satisfied or content that the law was upheld by the president, Dallas "enjoyed the triumph."

"The army took up their line of march yesterday," Dallas continued.

> The rear is at this moment leaving the town. I viewed the parade of their departure from a lofty hill. It was grand. The infantry, about six thousand; the cavalry, about two thousand; and the baggage wagons, about seven hundred. The expense and waste of such an army are inconceivable; but I think the government will be amply compensated by the effect which the prompt appearance of such a force upon such an occasion must produce throughout the continent and throughout Europe.

Dallas was impressed by the show of force of which the federal government was capable. In an almost Federal perspective, his evaluation of the army's size was not limited to the immediate need to quell an insurrection in Western Pennsylvania. He considered it a signal to Indians resisting American expansion and to Europeans doubtful of the American project that the federal government could muster an overwhelming force. If Dallas finds such signals useful, he implies that such doubts were perhaps well founded and could not be taken lightly. His own letter, after all, provides evidence for contemporary difficulties in assuming a generalized federal perspective.[16]

As an immigrant, Dallas had decided to join the emerging republic as it was just setting out, and he had then moved up into the country's political leadership. His letter indicates that he was keenly interested in building the American nation-state and to see it grow to eye-level with European countries. Dallas' biography attests to significant personal ambition. In 1794,

16 For an account of the Whiskey Rebellion, and of the situation witnessed by Dallas, see Thomas Paul Slaughter, *The Whiskey Rebellion: Frontier Epilogue to the American Revolution* (New York: Oxford Univ. Press, 1986), 205 f.

eleven years after he had arrived in America, Dallas was involved in helping develop a reliable political and national framework for it.

Considering Dallas' perspective in his letter to his wife, it is perhaps not surprising that his political views and allegiance shifted. In the country's first transfer of power from one political faction to another, Thomas Jefferson was elected president in 1800. Dallas was offered and accepted the position of District Attorney for Eastern Pennsylvania in 1801. He stayed in his home state but left its payroll and transferred to a federal post. This was not merely a matter of accepting a political reward for his efforts in Jefferson's election. (Pennsylvania was one of the mainstays of the Republican movement that he had helped get underway.) The new post came with the added incentive of putting Dallas in closer touch with his profession.

Dallas' professional loyalties, in fact, were the backdrop to his estrangement from the more radical sections of the Republican movement. Following Jefferson's election, it was suggested by some Republicans in Pennsylvania that lawyers be made redundant in order to facilitate a more "democratic" legal system, and to impeach certain Federalist judges. Dallas opposed such changes to Pennsylvania's judicial system. The compatibility of legal systems had facilitated his occupational transfer from the English colonies to the United States. Between 1790 and 1810, the United States population had almost doubled from fewer than four million inhabitants to 7.2 million—tremendous challenges for a country that was developing its political and cultural institutions. In the context of Dallas' parting of ways with the radicals, Henry Adams later observed that throughout America, "the bar was a sort of aristocracy, conservative to a degree that annoyed reformers of every class." Adams was referring to Dallas, Pennsylvania Governor Thomas McKean and other "educated leaders of the Republican party."[17] During the so-called Pennsylvania court fight, which began in 1802, Dallas refused to give up the idea of a strong and independent judiciary while radicals such as William J. Duane (now editor of the *Aurora*) favored a removal of Federalist judges. But while Dallas was closer to Federalism than Duane, he remained a committed anti-Federalist.[18] And even though Dallas had been rewarded for his role in the presidential election of 1800, he later argued on behalf of Samuel P. Chase in the United

17 Henry Adams, *History of the United States of America during the Administration of Thomas Jefferson* (New York: Literary Classics of the United States, 1986), 433.

18 Wilentz, *Rise of American Democracy*, 122 f.

States Senate, taking a pro-judiciary stance in opposition to Jefferson. Dallas remained a radical as long as issues of policy were involved, but he was much more hesitant about changing the political ground rules. From Duane's perspective, Dallas and his group were neither radical enough to be Democratic-Republican nor conservative enough to be Federalists, something without even a name: a "Tertium Quid."[19]

By 1805, Alexander Dallas Bache's grandfather could look back on an impressive political career in Pennsylvania and on successful work in his profession.[20] He lived with his family in a mansion on Fourth Street in Philadelphia (next door to Benjamin Rush), and a country house ("Devon") on the Schuylkill River that allowed the family to escape the hot city summers and regular outbreaks of yellow fever.

Sophia Dallas Bache

Born in 1785, Sophia Dallas Bache was Alexander James Dallas' and Arabella Maria Smith Dallas' first child. Sophia was born the same year that her father, after a mandatory post-immigration hiatus of two years, was admitted to the Pennsylvania bar. At that point, Alexander James and Maria Dallas had waited four or five years before they decided to have a family, and after Sophia was born, they waited another five years before they had their second child. Dallas was appointed secretary to the Pennsylvania governor in 1790 and thereby managed to establish himself, not only in his profession, but in politics as well. Sophia's brother Alexander James Dallas Jr. was born the next year. This timing suggests a connection between Dallas' career and financial fortune and the couple's careful decision to have a family, while it also sheds light on the character of their marriage.

19 Ibid., 124.

20 George Mifflin Dallas later described his father as "little, if at all, short of six feet in height, and erect without stiffness," explaining how he "patiently underwent every morning the careful curling of his silvered hair, stiffened with pomatum over his coat-collar, in a rose-knotted 'club.' With such a structure upon his shoulders, though lighted by a briskness and ease; quietude of bearing was alone inculcated, if not compelled. Besides, there were the drab-colored shorts, with small gold knee-buckles; the white-topped boots, leaving exposed an inch or so of the white silk stocking within; the white and ample vest, relieved by gairish crimson roll of velvet forming the edge of an interior flannel; and the unvarying white cravat. … He walked lightly, but with a natural dignity, which riveted the inquiring gaze of a stranger." Dallas, *Life and Writings of Alexander James Dallas*, 144.

That Maria was a mere fifteen years old when she married Alexander may have been premature, yet they remained cautious.

But Sophia's birth in 1785 preceded the couple's financial and economic confidence. It was not until 1790 that her parents felt secure enough to have another child. In this way, Sophia "represented" the family's uphill work before 1790 when her father's legal and political career was only beginning to get underway. When her brother Alexander was born in 1791, Sophia had been an only child for six years. None of her siblings would have her parents' undivided attention for as long. Her position among the Dallas-children suggests that Sophia was considered and considered herself to be a comparatively "grown-up" daughter, too close to her parents to be part of either the duo of Alexander and George Mifflin and too old to join Maria and Matilda. Trevanian was too young anyway. Sophia was set apart from her siblings by age, and by having been around when her father was still pushing his career forward, and perhaps this put her in a closer understanding, subjectively, with her parents and their struggle. Not only was she the oldest, but the oldest by six years and this implied that she aspired to high standards of responsibility in a responsible family.

To Sophia, a career in the modern sense was not an option. But considering her parents' ambition in establishing their family among the state's cultural and political elite, the stakes for choosing a husband were high. If her family's career trajectory was to be extended, it would have been adequate for her to marry an aspiring politician, but only if he could be seen climbing to the peak of American politics. Someone who aspired to professional success in medicine or law she would probably have considered too mundane. For Sophia, it would be important to heed the family's ambitious political role and its interests in cultural solidification. It was Richard Bache Jr. who qualified.

What did Richard have to offer? There are a number of perhaps less important connections. Sophia was the oldest daughter and Richard one of the youngest among seven surviving siblings. Such opposites may have attracted one another. There is also the political context of their families. As we have seen, Sophia's father had cooperated with Richard's brother in politics. What stands out, however, is Richard Bache Jr.'s background and how it could serve the aspirations of the Dallas family: To marry into the Bache family was in line with the family's expanding cultural and political role because it could thus connect to everything Benjamin Franklin's name stood for: national and cultural leadership based on a Democratic vision of

America. Given her family's political role, it was not an option for Sophia to marry into Philadelphia's Federalist elites that, under different circumstances, may have signaled the family's "arrival." By marrying into the Bache family, Sophia (and her father) could feel that her wedding confirmed all that her family had achieved. The Dallas family had been part of the city's and the state's elites, and there could now be little doubt that it was also part of an aspiring national leadership. She must have felt proud that it was Sally Bache, Benjamin Franklin's one legitimate daughter, who gave away her son Richard on April 5, 1805.

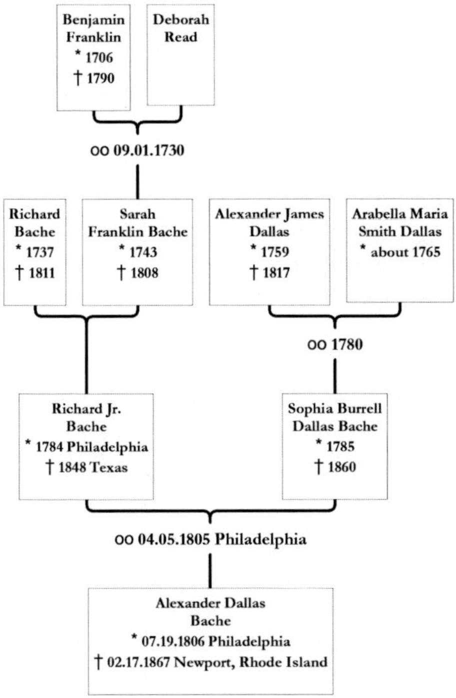

Fig. 4. Ancestors of Alexander Dallas Bache

The difference between the Bache and the Dallas families, however, was that the trajectories of their public significance were inverted. Richard Bache's oldest brother Benjamin Franklin Bache was the one who had continued Benjamin Franklin's newspaper tradition and political criticism.

He had died in 1798, leaving the *Aurora* to William J. Duane (who married one of Richard Bache's sisters and, as a radical Republican, became one of Alexander James Dallas' opponents within the Republican machine). Richard Sr. had little to show for a career. Whereas the Bache family looked back to Franklin, the Dallas family's success had been more recent. But Alexander James and Maria Dallas were immigrants and by marrying not just any American but a descendent of the all-American Benjamin Franklin, Sophia certified her family's ambition and reputation.

In order to grasp the relative weight of the two converging family lines, consider the first names that Richard and Sophia chose to give their children: Alexander Dallas Bache, the object of this study, was born in 1806 and named after his grandfather "Alexander," the center of the family dynasty whose name invoked expansionism and military prowess. He bore the added responsibility of living up to the successes of both the "Dallas" and the "Bache" family traditions. The couple's second child, Mary Blechenden Bache, born in 1808, was given her maternal grandmother's name. ("Blechenden" probably has an English reference, quite possibly pointing to Devonshire, where Alexander James Dallas and Arabella Maria Smith had met.) Richard's and Sophia's third child, George Mifflin Bache (born in 1811) was named after his maternal uncle, George Mifflin Dallas. Richard Bache Jr., born in 1813, would then bear his father's (as well as his paternal grandfather's) name. The references of Matilda's (1819–1900) and Henrietta's (1822–87) names are unclear.[21] Sarah Bache, the couple's eighth child (born in 1824) was the only other child (out of nine) whose first name represented the paternal lineage.

On balance, the children's first names indicate that the maternal lineage was of particular relevance to the family. Not only was Alexander Dallas Bache named after his maternal grandfather, his younger brother George Mifflin Bache was named after his uncle (Sophia's brother George Mifflin Dallas, 1792–1864), whose name, in turn, recalled that of Pennsylvania Governor Thomas Mifflin who had appointed Sophia's father secretary of the state in 1790. Mary was named after her paternal grandmother, Arabella Maria, and Richard's and Sophia's fourth child, their son Richard, was named after his father. This list of names corroborates previous observations: Richard had qualified for marrying Alexander James Dallas' oldest

21 Thomas Cushing, *History of the Counties of Gloucester, Salem, and Cumberland, New Jersey: With Biographical Sketches of Their Prominent Citizens* (Woodbury, N.J: Gloucester County Historical Society, 1974).

daughter through his family heritage, but this heritage was no longer a driving force. He was taken on by the Dallas family instead of Sophia merging into Richard's family tradition.

Richard Bache's Failure

Little is known about Alexander Dallas Bache's upbringing in Philadelphia. This investigation of the familial, professional, and political trajectory in which he found himself as he was growing up serves to delineate the backdrop against which he made his own decisions into adult life. It provides an opportunity for assessing the extent to which he carried on or broke away, and to understand the legitimacy that he could subjectively deduce for his own biographical route. Alexander Dallas Bache's name (the reference to his grandfather, to Alexander the Great, and to both the Dallas and Bache families) suggests that his parents had ambitious expectations. Before I turn to the career options that Alexander Dallas Bache, in keeping with these aspirations, must have entertained, I will return to a consideration of his father's career during the time his son was growing up in Philadelphia. Apart from what we know about the Dallas family, what did Alexander Dallas Bache's father add to his son's equation for choosing a career?

As suggested above, Richard seems to have received some education but no professional training. For the period between 1805 and 1812, we have no information on Richard's employment. Little is known about other family events or crises during that period. Other than the birth of his two oldest sons, Alexander Dallas and George Mifflin in 1806 and 1811 (five perhaps significant years apart), we know only of the death of his mother in 1808 and of his father in 1811. After war broke out in 1812, Richard followed in the footsteps of his two older brothers and served as a first lieutenant and aid to brigadier general Joseph Bloomfield in the 32nd Infantry, the Franklin Flying Artillery of the Philadelphia Volunteers.[22]

22 *Register, and Rules and Regulations of the Army for 1813, Communicated to the Senate, December 29, 1813, American State Papers, Military Affairs*, vol. 1, 13th Congress, 2nd Session. See *Handbook of Texas Online*, s.v. "BACHE, RICHARD" (by Lura N. Rouse), http://www.tshaonline.org/handbook/online/articles/fba02. The latter article seems to be erroneous in a number of details and is therefore used with caution. For example, Richard Bache is said to have left his family in 1836 but a number of other sources provide 1828 as the date of his departure.

Bloomfield had been Republican governor of New Jersey since 1801 and resigned this post in 1812 to become brigadier general. He was commandant at New York City, and after an excursion toward Canada in the fall of 1812 (during which Bloomfield fell ill and saw no action) he was assigned to the command of the Fourth Military District headquartered in Philadelphia.[23] In 1814, as British troops threatened the city and decisive action was needed, Bloomfield was relieved from his responsibilities there.[24]

His in-laws organized a job for Richard Bache. In 1812, Richard, as a member of the committee for Philadelphia, helped his father-in-law Alexander James Dallas organize James Madison's presidential campaign.[25] Again, with the help of his father-in-law Alexander James Dallas who was then secretary of the treasury in James Madison's cabinet, Richard, in 1815, was appointed postmaster of Philadelphia.[26] More lucrative, however, must have been his involvement in state politics: Upon his return from Russia, where he had been an aide to U.S. envoy Albert Gallatin, Richard's brother-in-law George Mifflin Dallas entered politics in Pennsylvania and took Bache under his wing. The younger Dallas brought with him his father's name and reputation, an education from Princeton, and his European experience. Considering his aristocratic taste and his family's connections among the professional elite of the state's foremost city, George Mifflin Dallas could perhaps have been a Federalist but he would have given up his family's credibility with Republicans. Together with Richard Bache, Thomas Sergeant (who was married to Richard's younger sister Sarah), and Trevenian Barlow Dallas (one of Dallas' brothers who provided the group with a Pittsburgh hub), the "Family Party" became an influential "New School Republican", pro-business element in Pennsylvania politics.[27]

23 Maxine N. Lurie and Marc Mappen, *Encyclopedia of New Jersey* (New Brunswick: Rutgers Univ. Press, 2004), 81; John K. Mahon, *The War Of 1812* (Gainesville: Univ. of Florida Press, 1972), 93.

24 Ibid., 316.

25 Walters Jr., *Alexander James Dallas*, 175.

26 Ibid., 235. John Binns seems to have suggested Bache for the post. M. Ruth Kelly, *The Olmsted Case: Privateers, Property, and Politics in Pennsylvania, 1778–1810*, Pennsylvania history and culture series (Selinsgrove, [PA]: Susquehanna Univ. Press, 2005), 144.

27 On the Family Party, see Walters Jr., *Alexander James Dallas*, 235; Belohlavek, *George Mifflin Dallas*, 18, 23–35. I will return to a discussion of Philadelphia and Pennsylvania politics in chap. 5.

Its influence was initially founded on its hold on the Republican caucus that nominated candidates for elected office, thereby controlling a significant portion of patronage. In 1817, the gubernatorial race in Pennsylvania provided the group with an opportunity to fight John Binns' position of influence as the outgoing governor's newspaper editor, confidante, and patronage counsel. There emerged a contest between Binns and the Family, both competing to boost the same gubernatorial candidate, William Findlay. It was Binns' paper that did the trick. Findlay was elected, but it was the Family, not Binns, that came away with the influence and patronage, turning the alienated Binns into a bitter enemy: Thomas Sergeant, Richard Bache's brother-in-law, was appointed secretary of the commonwealth; George Mifflin Dallas, another in-law, became deputy attorney general of Philadelphia; and Richard Bache founded the *Franklin Gazette*, a lucrative enterprise because, as the "government paper," it could rely on printing jobs from the state and from the post office.[28] George Mifflin Dallas was the Family Party's intellectual leader, and Richard Bache its propagandist (perhaps with allusions to his more famous grandfather and to his brother with whom he shared his name). Binns' later attacks on the governor only cemented the Family's influence in the state, which would last another ten years.[29]

The *Franklin Gazette*, under Bache, was edited by experienced newspaperman John Norvell (1789–1850). Unfortunately for Bache, Norvell moved on to edit the well-established *Aurora* in 1824 and he took the

28 Carlisle Republican, November 23, 1819; Ludwig Gall, *Meine Auswanderung nach den Vereinigten-Staaten in Nord-Amerika: Im Frühjahr* (Trier: F.A. Gall, 1822), 144–48. For an example of Richard Bache's style, see his *Oration, delivered at Spring Garden, July 5, 1813: to a very numerous and respectable company of Democratic Republicans, of the city and county of Philadelphia* (Philadelphia: Printed by John Binns, 1813).

29 The Family Party's role in the election of Findlay is detailed in Philip S. Klein, "John Binns and the Impeachment of Governor William Findlay," *Northumberland County Historical Society Proceedings and Addresses* XI (October 1, 1939): 51–66. The spoils are mentioned on p. 56. See also Philip Shriver Klein, *Pennsylvania Politics, 1817–1832: A Game without Rules* (Philadelphia: Historical Society of Pennsylvania, 1940), 128 ff. In a letter to James Madison, in which he announced the newspaper, Richard Bache wrote: "Could I succeed in putting an end to News-paper wars in Penna, as it is at present conducted, and abolish all personal + private abuse from the presses, I shall gain one of the objects for which I establish the paper, + relieve my native State from what may at present be termed, the horror of an election." Richard Bache to James Madison, January 8, 1818, microfilm reel 18, James Madison Papers, Library of Congress (http://hdl.loc.gov/loc.mss/mjm.18_0936_0938).

Family's lucrative support with him.[30] Bache had to confine himself to his job as postmaster of Philadelphia and he continued to depend on the support of the family network, especially his in-laws. It is perhaps not surprising and certainly in keeping with this general pattern, that Richard Bache took to the bottle. "Bache was much more of a liability than an asset to the family," concludes Philip Shriver Klein, referring to the larger Family network and its political interests.[31] And in 1828 the problems at the post office became too significant to be ignored and Richard lost this post as well.

The year prior to these events, perhaps out of desperation, Bache had physically attacked John Binns. Binns, who had become a Philadelphia alderman, had suggested to U.S. Secretary of State Henry Clay that he and his newspaper could deliver Pennsylvania to the administration in the 1828 election.[32] President John Quincy Adams' administration was in trouble after the alleged "corrupt bargain" in the 1824 presidential election and Adams now stood to lose against war hero Andrew Jackson. The administration was willing to meet Binns' demands and to provide him with lucrative government printing contracts, especially those from the post office.[33] As secretary of state, Clay was superior to John McLean, the general postmaster in Washington D.C., and Clay sought indeed to bolster Adams' political fortune by getting rid of McLean so as to get a handle on the post office patronage. But Clay was unsuccessful and McLean stayed in office until 1829.

Perhaps Bache's attack on Binns in 1827 was linked to federal pressure on the patronage-rich Philadelphia post office. If McLean had been removed, Bache may have had to go as well. Or Bache was trying to show his dedication to the party machine which, at the "eleventh hour," chose to bet on the most promising presidential candidate, Andrew Jackson. In any

30 The *Baltimore Patriot* reported on May 31, 1821 (p. 2) that Richard Bache had withdrawn from the *Franklin Gazette*, leaving the paper in the hands of his partner John Norvell. Stephen Girard refused a loan to Richard Bache in 1818, the same year the *Franklin Gazette* was first published. Merle Middleton Odgers, *Alexander Dallas Bache, Scientist and Educator, 1806–1867* (Philadelphia: Univ. of Pennsylvania Press, 1947), 6.

31 Klein, *Pennsylvania Politics*, 129.

32 Binns sued Bache for attacking him. Mary W. M. Hargreaves and James F. Hopkins, eds., *The Papers of Henry Clay* (Lexington, KY: Univ. Press of Kentucky, 1992), 6:1301. Bache was fined 100 dollars (*Norwich Courier*, April 4, 1828, 2).

33 Robert V. Remini, *The Election of Andrew Jackson* (Philadelphia: Lippincott, 1963), 45; Henry Clay to William Jones, January 23, 1827, in Hargreaves and Hopkins, *Papers of Henry Clay*, 6:110 f.

event, in 1828 Bache was accused of using government funds to fight the Adams administration.[34] The Family Party, with Richard Bache's brother-in-law George Mifflin Dallas at the rudder, could not afford to lose its influence. The Philadelphia postmaster was fired and replaced with Richard Bache's brother-in-law Thomas Sergeant. The machine thus retained control of their important patronage and propaganda post. Without his newspaper and post office job Richard Bache had difficulties providing for his wife and nine children.[35]

The general pattern seems to have been that Bache could initially rely on the support of Sophia's father, Alexander James Dallas. George Mifflin Dallas then built up the new Family Party in 1817, with Richard relying on its spoils. Richard's family background, his ties to the post office and to newspaper publishing, helped him for some time for he was selected less for his achievements and prospects than for his name and family loyalty. As a newspaper publisher, Bache depended on government printing and had no independent voice. He did not seek a more active and directing role in the family machine. George Mifflin Dallas had the advantage of a Princeton degree and the political training provided by his father's close political ally Albert Gallatin. George Mifflin Dallas had traveled and he was more immediately connected to the origins of the Republican movement that his father had helped to build. Richard Bache's parents had been much less eager to provide for his education. They had lived in Benjamin Franklin's public shadow. They had been sheltered by his wealth whereas George Mifflin Dallas' parents had been immigrants who were trying to help establish and join a national American culture.

34 An investigation was never conducted. Belohlavek, *George Mifflin Dallas*, 26. See also Secretary of State Henry Clay's letter to President John Quincy Adams, May 8, 1828, which seems to imply a personal use of embezzled funds: "I cannot omit respectfully to suggest the propriety of considering the expediency of removing, after the termination of Congress, the P.M. General [John McLean]. This is a suggestion not the result solely of the recent appointment, 'though if some allegations respecting the defalcations of the late post master [Richard Bache] be well founded, they afford sufficient ground for the measure. I believe that the sum of injury would fall far short of the sum of benefit from the removal." Robert Seager II, ed., *The Papers of Henry Clay* (Lexington, KY: Univ. Press of Kentucky, 1982), 262 f.

35 As early as November 23, 1819, the *Carlisle Republican* reported that Richard Bache had been "a defaulter to the amount of thirty thousand dollars, and that the Post Master General dare not trust him with the postage of letters received at his office." Bache had earlier tried to borrow money from Stephen Girard, but the loan was refused.

Chapter 3

A Career in Science?

West Point

In view of the family configuration outlined in the preceding chapter, we are now in a position to deduce some of the requirements that Alexander Dallas Bache will have had to take into consideration in deciding on a career after finishing private school in Philadelphia.[1] In fact, the decision was probably prepared by his family and then accepted by him, given that he was fifteen years old.

Bache's grandfather Alexander James Dallas had been one of the nation's leading politicians and lawyers, and after he was admitted to the bar in 1813, and became an assistant to Albert Gallatin, his uncle George Mifflin Dallas was likely to enter politics as well. Alexander Dallas Bache's uncle Benjamin Franklin Bache was also involved in politics by publishing the *Aurora*, from which the Dallas side of the family had since moved away politically. Alexander's father entered newspaper publishing in 1818, at about the time his oldest son must have thought about his own career. Perhaps the idea came up that Alexander Dallas Bache should join him. If it was considered, it must have struck Sophia (and perhaps her son as well) as being insufficiently ambitious and even risky, for Richard Bache was feeding on his in-law's political spoils. Business success depended on political windfalls. The Dallas tradition, however, stood for national consolidation. Beginning with the Whiskey Rebellion in 1794, Sophia's father, on the basis of his training as a lawyer and in loyalty to his profession, had assumed a more national and conservative outlook on political develop-

1 Alexander Dallas Bache attended Claremont Academy (or Seminary), a private school under the direction of John Sanderson who stressed classical education. See Merle Middleton Odgers, *Alexander Dallas Bache: Scientist and Educator, 1806–1867* (Philadelphia: Univ. of Philadelphia Press), 7 f.

ments. The newspaper business must, therefore, have seemed unattractive if it was even considered, and while law and politics were accessible avenues they would not have broadened the family's overall contribution to the historical success of the emerging United States, the one perspective, after all, in which both the Bache and the Dallas family traditions quite obviously converged.

Other lucrative or honorable careers were perhaps considered to be unattractive: The medical profession had produced important state-builders such as Benjamin Rush (Alexander James Dallas' neighbor in Philadelphia), but Rush's political involvement, his relevant contribution to nation-building, cannot be explained by his professional role alone. Most of his colleagues remained inactive politically. The same holds true for other professions such as art (including literature) and architecture, even though Alexander Dallas Bache's age cohort significantly reshaped these areas.[2] Considering these choices, one becomes aware that to "merely" enter a profession would have been a pale choice as it would have been difficult to live up to the Dallas and the Bache traditions, both of which pointed in the direction of state building.

The military met this criterion particularly well. Before the war for independence, a standing army had been a contentious issue and after 1783, the militia system, instituted by the confederated states, was preserved. Things had changed by the 1810s. The navy had played a limited role during the War for Independence but had since assumed an international and diplomatic role in protecting and setting claims for American business. During the War of 1812, furthermore, the navy seemed to make up for the army's defeat in Detroit and along the Canadian border. The Louisiana Purchase, a vast acquisition of territory controlled by the federal government, created a new role for the American army. Westward expansion required protection of settlers from Indians in "sparsely settled" areas. This was considered an internal affair as the United States claimed title to the land even before it had been settled, and because Indians were not considered to have the rights of nations. Unlike the army, the navy assumed an international role because, in addition to pirates along the African coast, it dealt with its British and French counterparts. It was the British Navy's impressment of sailors aboard U.S. vessels that had played a decisive role in the deterioration of diplomatic relations before 1812. From the war, the

2 Ralph Waldo Emerson, for example, was born in 1803, Edgar Allan Poe in 1809, architect Thomas Ustick Walter in 1804. (More on Walter in chap. 5 below.)

navy emerged as the prominent arm of the American military, for it seemed to have won the decisive or at least the prominent battles. So why did Alexander Dallas Bache *not* join the navy? We have no evidence to prove this, of course, but his family's dynastic perspective implies that it would probably not have appreciated Sophia's oldest son going into the navy if this area was already covered. In 1815, Alexander Dallas Bache's uncle Alexander James Dallas (Jr.), his famous grandfather's oldest son, was twenty-four years old. He was a commodore in the navy and had served in the recent war with distinction.[3]

So it was the army, perhaps a second and somewhat less glamorous choice that opened a new and relevant field of engagement for Bache's family. The end of the Napoleonic Wars in Europe had shifted attention away from the Atlantic and, because of Thomas Jefferson's diplomatic coup in 1803, toward the West. What was needed now was a standing army, an efficient alternative to the militia system (that would nevertheless remain the mainstay of the U.S. Army throughout the nineteenth century). Considering these particular circumstances, it is entirely within his family's trajectory that fourteen-year-old Alexander Dallas Bache concurred with his family's decision to enter the United States Military Academy at West Point in 1820. Unlike his father, Alexander Dallas Bache thus avoided the militia, which offered no career and was considered an addendum to civilian activity. Bache became a cadet at West Point, and this opened up a career in the army, though the school's conception was peculiar and it would be a mistake to view it solely as an officers' training institution.

The Military Academy at West Point had been founded during Thomas Jefferson's presidency in 1802, a signal that a Republican president, despite his former anti-Federalist rhetoric, would seek to create a strong and effective central government on the basis of an expansionist view of the United States.[4] Jefferson sought to establish institutions to train a national elite. In a letter in which he sketched his ideas for a University of Virginia, Jefferson expressed his ambition that the United States not only catch up but surpass European nations in science and in learning. The task was to

3 *Dictionary of American Naval Fighting Ships*, s.v. "Dallas," http://www.history.navy.mil/danfs/d1/dallas.htm (accessed October 30, 2010).

4 Peter Onuf points out that the "caricature of Jefferson as an antistatist libertarian does not hold, either at the federal or state level." Onuf, "Introduction", in: Robert M. S. McDonald, ed., *Thomas Jefferson's Military Academy: Founding West Point* (Charlottesville: Univ. of Virginia Press, 2004), 16. At West Point, New York, a small school had been in operation prior to 1802 though it had no legal status.

educate the citizenry; most of all, it was to train a national elite in order to develop the national culture from the top down.[5] Unlike the University of Virginia, the academy at West Point was a federal institution. It was under the supervision of the U.S. Army Corps of Engineers and thus under civilian control. While it was to create an officer corps for the army, West Point graduates were to provide the skills and expertise for the country's rapidly expanding infrastructure. Indeed, West Point would remain the sole conveyor of engineering knowledge until about 1840. Even as late as 1850, nearly all civil engineers would be trained here.[6] Civilians controlled the academy and this included appointment of instructors. West Point was, in other words, separated from the regular army and its purpose was to create an officer corps, not for a standing army, but for the militia. It was hoped to avoid the dangers of a standing army such as the English Army prior to American independence because such an army was considered a potential danger to the republic.[7] The discussions on the feasibility of a school for officers had thus turned, under Jefferson, toward technical and "scientific" training as a solution. Upon completion of their education at West Point, graduates were not expected to join the army but to return to civilian life and to take charge of militias in case of war.[8]

In this way, West Point graduates were sought to represent and diffuse the national idea among the country's citizenry. In keeping with the overall intentions in creating the academy at West Point, Thomas Jefferson had appointed "Republican" faculty and cadets who shared his political views.[9] After Congress authorized the president to create the Army Corps of Engineers and the Military Academy in 1802, Jonathan Williams (a grand-nephew of Benjamin Franklin who was familiar with fortifications through his European training) was appointed first chief engineer and the academy's first superintendent. When Alexander Dallas Bache decided to apply, Sylvanus Thayer, superintendent since 1817, was reshaping West Point.

---

5 Thomas Jefferson to Littleton Walter Tazewell, January 5, 1805, in Thomas Jefferson, *Writings* (New York: Literary Classics of the United States, 1984), 1149–53.

6 Richard Ernest Dupuy, *Sylvanus Thayer, Father of Technology in the United States* (West Point, NY: Association of Graduates, United States Military Academy, 1958), 10, quoting A. Riedler, *American Technological Schools*. H. R. Doc., 2d Session, 53d Congress. vol. 5, part I.

7 Stephen E. Ambrose, *Duty, Honor, Country: A History of West Point* (Baltimore: Johns Hopkins Univ. Press, 1966), 7–23.

8 Ibid., 6 f.

9 Allan Reed Millett and Peter Maslowski, *For the Common Defense: A Military History of the United States of America* (New York: Free Press, 1984), 10.

Thayer, who had been lieutenant in the corps of engineers since 1808, had travelled to Europe after the War of 1812 in order to investigate possibilities for improving U.S. military training. Impressed by Napoleon, he went to France and spent a year at the *École Polytechnique*.[10] Following his return, Thayer was appointed superintendent, took the French school as his model, and made civil engineering the foundation of the West Point curriculum. He also made French a mandatory part of the cadet's education.[11] To go to West Point in 1820 implied joining the military, but Alexander Dallas Bache would also be exposed to a Republican perspective on what would later be called "science," natural philosophy with a utilitarian bent in the context of an expansionist American agenda.

West Point bundled several strands and perspectives for American national development and for the military's role within it. Bache's own family had been in the midst of these developments. The "Revolution of 1800" had brought the Republican "opposition" into power, and Thomas Jefferson and his supporters took on a responsibility of solving problems on the national, federal level, an area that had seemed reserved to the Federalists whom they criticized. Alexander James Dallas had joined federal troops in 1794 to help undo the Whiskey Rebellion. Dallas had been afraid of Federalist designs for a federal army but his experience had led him to assume a more conservative Republican perspective closer to Federalism. The creation of the U.S. Army Corps of Engineers and of the U.S. Military Academy was part of this trajectory. Given the perspective of westward expansion, both engineers and an expanded and well-trained army would be needed. And given Alexander's first name, reminiscent of Alexander the Great and of a growing empire of a different time as well as Alexander Dallas Bache's family record of Republicanism, West Point would be an ideal match if the family's achievements were to expand into the military and, more specifically, into the army. What this shows is that in 1820, science was in the picture, but not at the motivational center, as Alexander Dallas Bache left his parents' home in Philadelphia to travel north to West Point on the Hudson River.

---

10 Dupuy, *Sylvanus Thayer*, 2 f. Thayer had been asked by the U.S. War Department to "'avail yourself of such opportunities as may be presented for acquiring by personal observation as well as thru enquiries, any information concerning roads, canals, bridges, the improvement of rivers and harbors, construction, labor-saving machinery, etc., which would be new to this country and of sufficient importance to render its acquisition desirable.'" Ibid., 18 f.

11 A. Hunter Dupree, *Science in the Federal Government*, 36 f.

No letters by Alexander Dallas Bache survive that would shed light on his youth and his thoughts on going to West Point in 1821 but the decision, likely prepared by his family rather than himself, seems plausible and even logical when taking into account his position within the larger family "system." Alexander (or "Dallas" as he would be called, aptly highlighting his mother's lineage and his grandfather's role[12]) would feel the responsibility of living up to the standards of the family. He attended neither the University of Pennsylvania in Philadelphia nor Harvard University, two other prominent educational options at the time. The former would have been close to home and would have added nothing new on a level compatible with the family's history; and the latter would have been too much of a Federalist choice, going against what his family, as immigrants (successful ones, but nevertheless consciously immigrants) believed in and had worked for politically. During the War of 1812, when Alexander James Dallas, then secretary of the treasury in James Madison's cabinet, reorganized the nation's finances, New England (and Harvard) had revived an anglophile Federalism in keeping with its merchant interests. The Dallas and Bache families were committed Republicans, of course, and the war with England had been brought about in the aftermath of Jefferson's election in 1800. To send their son to Harvard would perhaps have been considered a vindication of Federalism. Even more pertinent: From the perspective of the dominant Dallas-side of Alexander's family, Harvard and its emerging fraternities stood for learning and leisure but insufficient practical relevance.[13] There had been leisurely interests in the family, to be sure. Alexander Dallas Bache's grandfather Alexander James Dallas had befriended an actor and theater director in Jamaica who first recommended to him that he move to America. In Philadelphia, Dallas stuck to his interest, worked towards the legalization of theater, and wrote a play himself.[14] But these interests had always taken a backseat to roles of expanding legal and political responsibility. Alexander Dallas Bache was the oldest son of

12 See, for example, letters by his brother Richard Bache, written in 1847. Correspondence of A. D. Bache, Superintendent of the Coast and Geodetic Survey, 1843–1865, RG 23, roll 21, National Archives.

13 There are numerous accounts of Harvard in this period. Among them are Richard J. Storr, *Beginnings of Graduate Education in America* (Chicago: Univ. of Chicago Press, 1953), and Samuel Eliot Morison, *Three Centuries of Harvard, 1636–1936* (Cambridge: Harvard Univ. Press, 1936). For an excellent recent guide, see Roger Geiger, ed., *The American College in the Nineteenth Century* (Nashville: Vanderbilt Univ. Press, 2000).

14 Dallas, *Life and Writings of Alexander James Dallas.*

Alexander James Dallas' oldest daughter. How could one expect him to break away from this tradition? Certainly not at age fifteen when "Dallas" went to West Point.[15]

What does all this imply about Alexander Dallas Bache's motivation for a career in science? At this point, there is little evidence to support the idea that Bache was headed for a scientific career. Perhaps such evidence would be difficult to identify even in those cases where letters or diaries could be consulted. It may also be inadequate to expect that such interests could become relevant at such a young age, particularly if we take into consideration that there existed few contemporary avenues into science. There were posts in colleges and universities such as the University of Pennsylvania or Harvard but these were teaching posts, which left little time for independent investigation. In Philadelphia, there existed a tradition of investigation in the philosophy of nature and Alexander Dallas Bache's great-grandfather Benjamin Franklin, of course, had perhaps been its most famous exponent. Franklin had been a founding figure of many of the city's science-related institutions, such as the American Philosophical Society and the University of Pennsylvania. Even if available sources would reveal that Alexander Dallas Bache was interested in natural phenomena, no obvious educational choice would have resulted from it, even though West Point would certainly not have been a bad one.

Superintendent Thayer, whom one historian has described as "unbending, aloof, and even cold," had significantly reshaped the institution when Bache arrived there in 1821.[16] The curriculum reflected West Point's practical bent and sought to put it on a sound foundation. The four-year program included mathematics, "Natural and Experimental Philosophy," "Engineering Science of War etc.," French, and (with less emphasis) drawing, history, and "Moral Philosophy and Elements of National and Political Law."[17] It did not include ancient languages, botany and zoology,

---

15 In a letter by U.S. Secretary of War John C. Calhoun (March, 26, 1821), Bache was informed that he would be examined for admission to West Point in June of that year. Box 5, Record Unit 7053, Alexander Dallas Bache Papers, 1821–1869, Smithsonian Institution Archives (hereafter cited as "Bache Papers, SIA"). Calhoun knew whom he was writing to. Bache's father (and the 'Family Party') had supported Calhoun in Pennsylvania.

16 Ambrose, *Duty, Honor, Country*, 68.

17 "Course of Studies and Rules for the Classifications of the Cadets of The United States Military Academy," [1820], John Lawrence LeConte Papers, 1825–1883, American Philosophical Society, Philadelphia.

anatomy and medicine, political economy, rhetoric, and fine arts—subjects that were taught elsewhere.[18] By introducing subjects such as mathematics and French, Thayer had turned the academy into a college, a college that provided little room for a perusal of subjects or a leisurely engagement with interesting fields of inquiry. West Point had some qualities of a "total institution:" Thayer abandoned summer furloughs except for a student's third summer, in order to keep a closer watch over his pupils. During the summer, students were off to camp. Bache did not see his family from June 1821 (when he departed for West Point) to the summer of 1824 (three years later). Thayer took the stand that students had no rights. Discipline was tight. Students "recited" every day. A merit system evaluated the student's performance inside as well as outside the classroom. Cadets were graded daily and weekly to keep them on their toes. Examinations took place in January and June, with no second opportunity for those who failed.[19]

Again, no letters remain to explore Bache's contemporary response to his experience at West Point. We know that he excelled. Against keen competition, he finished at the top of his class and was praised as a model student.[20] During his junior year, he was an assistant to the instructor of mathematics and during his senior year, he taught chemistry. In July 1825, Bache was appointed brevet second lieutenant and second lieutenant engineers in the U.S. Army Corps of Engineers.[21]

That Bache would excel is perhaps not surprising or at least in keeping with his family trajectory and motivational logic. State building was the theme, and West Point provided an avenue into the military, an area in which the Bache and Dallas families had not yet developed roles that

---

18 Storr mentions these subjects for the University of Virginia. See Storr, *Beginnings of Graduate Education in America*, 13.

19 Ambrose, *Duty, Honor, Country*, 71–86.

20 Competition was keen that year. The Class of 1823 and the Class of 1824 each had one student selected for the Corp of Engineers, whereas four students (including Bache) were chosen for this branch in 1825. Class size had remained about the same: thirty-five graduated in 1823, thirty-one in 1824, and thirty-seven in 1825. See Francis B. Heitman, *Historical Register and Dictionary of the United States Army From Its Organization, September 29, 1782, to March 2, 1903* (Washington, D.C.: Government Printing Office, 1903), 144. See also a letter by Bache's former instructor, David Bates Douglass, n.d., Clements Library, Ann Arbor, MI, online at http://www.math.usma.edu/people/Rickey/dms/DeptHeads/Douglass-David-Bates-Clements-AA.htm (accessed May 19, 2008).

21 Secretary of War to ADB, July 9, 1825, box 6, folder 9, Bache Papers, SIA. Regarding Bache's military record, see also Heitman, *Historical Register*, 178.

would have allowed them to significantly advance the emerging nation. The decision to go to West Point indicated that this was indeed his perspective. Moreover, as a successful student at West Point, it was quite natural that Bache would turn to engineering: The merit system recently instituted by Superintendent Thayer provided for the most successful students to join the "aristocratic" corps of engineers instead of the "respectable" artillery or the "prosaic" infantry. The latter were reserved for cadets with lower achievements.[22] That Bache was to join the engineers, even if it coincided with his interests, was an automatic consequence of his success as a student. That Bache taught mathematics and chemistry perhaps hints at his interests. Among American colleges, West Point alone taught these subjects. Regardless of his preferences, however, Bache was on track to becoming an engineer.

In June 1825, at about the time when the final examinations took place at West Point, Secretary of War James Barbour wrote to Sophia that Alexander Dallas Bache

> ought to be to you as I am sure he will be a source of the greatest consolation. I know not whether it has been your lot to have your cup of life drugged in any degree with calamity. The draught must have been severe indeed if it is not sweetened by the blessing of your excellent son. I knew and loved your father, his great paternal ancestor I knew only by his works. I thought I permitted myself to ~~think~~ [—] that I saw the excellence of both branches about to be united in your son.[23]

Instead of writing to Richard Bache or to both parents, Barbour wrote to Sophia only. It is difficult not to consider this a confirmation of Sophia's prominent role, or of Richard's negligible one, in the eyes of a public official who probably knew the family well.[24] Barbour suggests that there may have been a "calamity" in Sophia's life, quite likely a veiled reference to her husband's difficulties—perhaps prompted by Richard Bache losing control of the *Franklin Gazette* that year. This reference would also be in

22 Ambrose, *Duty, Honor, Country*, 73.

23 James Barbour to Sophia Bache, June 10, 1825, box 3, folder 1, Bache Papers, SIA. Throughout this book, when quoting from *handwritten* sources I am using [brackets] for illegible words, a dash within brackets when a word is not legible at all, and ellipsis points in brackets […] mark my own omissions. Wherever possible, I will also convey text formatting (~~struck through~~, underlined, etc.). My own comments within quotations are in {braces}, and later additions to a text by the original author *in italics*. In quotes from *printed* sources (such as in chap. 4, 6, and 8), the original formatting is preserved.

24 Merle M. Odgers is also left with this impression, though he relies on hearsay. See Odgers, *Alexander Dallas Bache*, 10.

line with Barbour addressing Sophia alone instead of both parents. Barbour feels that Alexander Dallas Bache's success at West Point may help cheer up his mother. He refers to Sophia's father, Alexander James Dallas, Barbour's predecessor as secretary of war, and to Benjamin Franklin, her son's "great paternal ancestor." Barbour consoles Sophia by confirming that her son is the inheritor and protagonist of two great family traditions of national achievement.[25]

The U.S. Army Corps of Engineers

In 1828, Alexander Dallas Bache returned to Philadelphia. He had been an assistant engineer in the construction of Ford Adams in Newport, Rhode Island, since 1826, but an offer from the University of Philadelphia provided him with an opportunity to improve his financial situation and to get married. Joseph Henry later wrote that Bache's army salary allowed him to take care of his mother and dependent siblings but that it was insufficient for an own family. The need for a more substantial income, however, may have also had to do with his father's departure. There are conflicting versions about when Richard Bache left his family. Henry's observation suggests that this occurred sooner rather than later, i.e. in 1828 and not in 1836, or that his father, even while he was still with the family, could or would not support the family. As a lieutenant in the corps of engineers, Bache had earned less than he could expect to make at the University of Pennsylvania.[26] In Newport, a town grown wealthy from the maritime shipping business, Bache had met Nancy ("Ency") Clark Fowler, and the new post in Philadelphia allowed him to propose to her. They were married on Tuesday, September 9, 1828.[27]

Ency was the daughter of Christopher and Ann Clarke Fowler. Her father was one of Newport's prominent merchants. Newport had been founded during Colonial times. Its convenient location had facilitated an

25 This letter is mentioned in ibid., 14, and in Gould, "An Address in Commemoration of Alexander Dallas Bache." That Gould would later refer to this letter suggests that Bache's family and friends considered it important. Balfour's comments go along with observations in the preceding chapter and the family may have considered Balfour's letter an emblem of Bache's biographical trajectory and logic.

26 Joseph Henry, "Biographical Memoir of Alexander Dallas Bache," *Biographical Memoirs, National Academy of Sciences* 1 (1877): 183.

27 *Providence Patriot & Columbian Phoenix*, October 4, 1828, 2.

important harbor town that profited from the triangular trade in rum, molasses, and slaves. In 1820, Newport remained among the country's most active ports.[28] Christopher Fowler's Federalism was an extension of his successful commercial interests.[29] Nancy's mother, Ann Clarke Fowler (ca. 1768–1820), was a daughter of Peleg Clarke, a prominent Newport merchant and ship's captain who had made a fortune in the slave trade.[30]

Considering Bache's financial obligations in connection with his responsibility to take care of his mother and his siblings, it is remarkable that he chose to accept the post at the University of Pennsylvania, and that he did not look for employment in the emerging businesses of building canals or other infrastructure that were booming in the 1820s. He would have had a "role model," for as a senior at West Point, Bache had been the assistant of David Bates Douglass (1766–1849), a professor of civil engineering who was particularly interested in internal improvements. While teaching at West Point, Douglass took on consulting work and in 1825, he was offered the post of supervisor to the construction of the western section of the Erie Canal. This work was evolving in 1828, though Douglass did not resign from the faculty until 1831 when he chose to devote all of his time to the canal project.[31] Bache was perhaps aware of the strong demand for engineers in the thriving United States economy and that his valuable engineering degree could generate significant income.[32] Even though most of Bache's correspondence from this period is lost, a detailed consideration of remaining letters sheds light on why Bache chose not to go this way. As we have seen, Bache's family background set him up for

28 With an estimated tonnage of 10,950. Edward Peterson, *History of Rhode Island* (New York: J. S. Taylor, 1853), 258.

29 *An address, to the citizens of Rhode-Island, on the choice of electors of president, and vice-president of the United States* [Providence, R.I.?, s.n.], Nov. 1808.

30 *New Bedford Mercury*, October 20, 1820, [3].

31 See biographical notes on Douglass at http://www.clements.umich.edu/webguides/D/Douglass.html (accessed August 16, 2006, page no longer available). See also Douglass' letter of recommendation for Bache (n.d.) in the David Bates Douglass Papers, William L. Clements Library, University of Michigan, which was transcribed by S. W. Jackson in 2003 (http://www.math.usma.edu/people/Rickey/dms/DeptHeads/Douglass-David-Bates-Clements-AA.htm, accessed July 25, 2006). "[A]mong the many young men I have had charge of in both those departments [mathematics and engineering] I have seldom known a brighter example of talent to application united", Clements wrote.

32 Though perhaps not when starting out in this branch of work. Mark Aldrich, "Earnings of American Civil Engineers, 1820–1859," *Journal of Economic History* 31, no. 2 (June 1971): 407–19.

public service, and an extrapolation of the following letter throws additional light on the role of the corps of engineers in shaping Bache's career, and in steering him towards science.

Shortly before his departure for Philadelphia, Bache received a letter from his engineering colleague Alfred Mordecai. Just like Bache two years later, Mordecai had graduated at the top of his class and then became a second lieutenant in the corps of engineers in 1823.[33] "My dear Bache," Mordecai wrote from the Engineering Department in a letter that was marked "Private,"

> I kept your letter applying for a furlough in order to receive Col. [Grahish's] instructions about it before answering you. he [sic] has just arrived here + says that he is not empowered to grant a furlough of the extent demanded + would therefore defer a reply until he can [—] the subject to the Secretary of War who is expected in the course of a week.[34]

Bache had received the offer for a professorship from the University of Pennsylvania and he was arranging his move from Newport to Philadelphia with his wife. Bache had asked for an unusually long furlough that prompted Mordecai to defer the decision to a superior who, in turn, wanted to consult the secretary of war. Bache was playing it safe by trying to hold on to his military post while he had not yet transferred to his new position.

> I need hardly express to you my personal regret at the prospect of losing your talents in the corps, but probably, <u>all circumstances considered</u>, you act wisely, + I am much pleased that those talents are not to be applied to the development of black [letter] Law but will still be preserved to the cause of science.
>
> May you prosper,
>
> yrs truly

Mordecai speaks of "losing" Bache, a reference to Bache's decision to leave the corps of engineers. To what "<u>circumstances</u>" was Mordecai referring? He suggests that the corps of engineers was the most suitable or at least the initial place for Bache's interests but he concedes that if "<u>all</u>" circumstances were considered, the university may have something to offer.

33 Heitman, *Historical Register*, 724

34 Alfred Mordecai to ADB, October 1, 1828, RH 1897, box 12, Alexander Dallas Bache Correspondence, Rhees Collection, Huntington Library (hereafter cited as "Rhees Collection, HL").

Mordecai is either pointing to the limits of a professorship when compared to a post in the corps; or he is aware of Bache's increased income.

Mordecai juxtaposes "black letter law" and "science," implying that compared to the University of Pennsylvania, the corps of engineers was the more prominent venue for the pursuit of scientific interests. His remark reveals the esprit de corps of the army engineers, a self-conscious avant-garde dedicated to the development of the nation's infrastructure and of science. By referring to "black letter law" (i.e. undisputed principles of a particular legal field), Mordecai metaphorically characterizes universities as facilities for conveying ideas second-hand. The juxtaposition is that between standards that are not meant to be developed, on the one hand, and of active engagement and transformation, on the other. It is the corps of engineers (and West Point as its foundation), not the University of Pennsylvania, which Mordecai assumes to be dedicated to a transformational perspective.

To the extent that Bache shared Mordecai's perception, the Philadelphian's decision to return to his native city and to live up to his family obligations may indeed have come at a price. How could Bache henceforth partake in the development of this broad and national aspiration for the practical use of science? In Mordecai's view, after all, the University of Pennsylvania was a mere local institution that could hardly compete with the corps' significance.[35]

Two weeks later, Mordecai again wrote to Bache. The furlough had been granted and Mordecai was enclosing the official document. Mordecai then turned to Bache's private life:

When I last wrote, tho' aware of your approaching fate, I did not know how near your happiness was to [communication], or very most sincere congratulations would not have been withheld until now. Trust me, that altho' I am not prepared to act on the belief, I do believe your late step not only conducive, but necessary to

35 Mordecai was correct in assuming that Bache was expected to be a conveyor rather than a developer of ideas. In his letter offering Bache the post at the University of Pennsylvania, Horace Binney explained to Bache that it was "the Professor's duty to teach, in addition to the prominent subject of his chair [i.e. natural philosophy], astronomy + chemistry …. Such an elementary course as collegiate classes require," he added, "may doubtless be prepared by one of liberal education, who at the commencement of a professorship knows nothing of it." Horace Binney to ADB, August 29, 1828, RH 934, box 12, Rhees Collection, HL.

the enjoyment of some of the best feelings of humanity, to the possession of the most certain pledge, of rational happiness.[36]

Mordecai explains why he is late in congratulating Bache on his wedding. This sheds an interesting light on his previous letter because Mordecai cannot have meant that Bache's financial obligations were a consequence of his decision to get married. These obligations may have instead arisen from Bache's commitment to provide for his mother at a time when his father had left her.

While Mordecai is not ignoring the enrichments attached to being married, he speaks of "fate" and thus takes into view the limitations attached to matrimony. This corresponds to his comment that he was not ready to "act on the belief" in the enriching qualities of marriage.

I had a letter not long since from [Mahan] in regretting the loss of so 'fine a talent' (how French!) alluding to [Courtenay] he had little idea [that] we are in danger of sustaining another so soon—does he write to you? He seems to have become, like almost all the young men who have gone to France, much attached to their manners + customs, but bids me, in reply to my [cautions], to dismiss the idea of his being too much so, + of his making a worse citizen of the U.S. on acct of such attachment.

Dennis Hart Mahan (1802–1871), West Point class of 1824 and a second lieutenant in the corps of engineers had been studying in France since 1826. Mahan regrets the loss of a colleague, probably Edward Henry Courtenay who graduated from West Point in 1821. (Like Bache and Mahan, Courtenay had graduated at the top of his class.) At the time Mordecai wrote his letter, Courtenay had likely decided to leave the corps and to become a professor at West Point. (He would do so in February 1829.[37]) That Mahan regretted Courtenay's departure for West Point indicates that the corps and the academy had distinct loyalties, and that the latter was not merely a feeder system for the former.

France plays a significant role here, and against the background of her attraction, Mordecai raises the issue of national loyalty. In the context of Bache's and Courtenay's pending departure, he looks around and sees another colleague whose interests may turn him away, not only from the corps, but from the country the corps was designed to serve. If the corps allows limited fulfillment in either of these areas, what remained? There is

36 Alfred Mordecai to ADB, October 16, 1828, RH 1898, box 12, Rhees Collection, HL.
37 Heitman, *Historical Register*, 330.

the pursuit of science, but it, too, could be advanced in France. It comes down to a personal dedication to use one's "talents" and learning to advance the American nation. A sense of belonging, however, was only emerging, or in any case not yet secure, for otherwise Mahan could have dismissed or played with Mordecai's warning. Mahan seems to have assuaged his colleague by asserting that his feelings for France left his American loyalty intact. A United States citizen, in other words, adhered to his country abstractly, by choice rather than intuitive attachment.

National Purpose

In his eulogy of Alexander Dallas Bache many years later, Benjamin Apthorp Gould would point out that the friendships Bache "formed within the precincts of the Military Academy seem to have been of remarkable strength, and were most tenderly guarded by him throughout his subsequent life."[38] After his return to Philadelphia and after he had moved to Washington D.C. in the 1840s, Bache would indeed continue to speak warmly of West Point and use it as a model for other institutions. His success and strong identification with the academy suggest that its rationale aptly matched Alexander Dallas Bache's peculiar position in his family. He was the oldest son of Alexander James Dallas' oldest daughter whose marriage to Richard Bache verified the Dallas family's perspective of pushing into the nation's political and cultural center. An analysis of Bache's family and of his position within it, therefore, has left us with little evidence to support Slotten's rather general contention that Bache's "commitment to science … did not exist separately from his family background, his educational experiences, and his cultural ties."[39] All of these aspects were relevant but it now seems difficult to put them side by side. There was little in Bache's family background that pointed to a career in science. Benjamin Franklin was an important figure in Philadelphia and several generations later, the promise of being associated with his name could convince Sophia Dallas to get married to Richard Bache. There is no immediate evidence, however, that science, as a fascination with natural phenomena, would have prompted Bache's initial educational choices. The decision to go to West Point followed rather logically from the Dallas family's project of

38 Gould, "An Address in Commemoration of Alexander Dallas Bache," 4.

39 Slotten, *Patronage, Practice, and the Culture of American Science*, 21.

national integration and consolidation, and from their political allegiance and interests associated with these efforts in which the legal profession had played a significant role. It was only after Bache arrived at West Point, therefore, that Bache could come to appreciate the esprit de corps associated with the institution, with its commitment to progressive national advancement, and with the standard of science so prominently unfolded in France. Perhaps his mother imagined her oldest son growing into a Franklin of his own time to vindicate her willingness to put up with the "calamity." Even if she did, she would have had to decide which Franklin her son was to be: publisher, politician, or scientist?

There is nothing, in short, that pointed toward science before Alexander Dallas Bache arrived at West Point. His family had been engaged in the political movements that had been responsible for instigating it. Bache may have felt that West Point was a family affair. This would only explain that Bache felt a particular commitment to the army. A commitment to science followed from the academy's orientation towards French models and served as an additional vehicle for an evolving esprit de corps.

In 1828, there was perhaps no better place in the United States for an ambitious young man interested in science and technology than Philadelphia. When he returned to that city, Bache brought with him the advantage and responsibility of being part of its cultural backbone. He knew the city and its institutions, and those associated with them knew of his family's historic role. Benjamin Franklin had been involved in the founding of the University of Pennsylvania, he had been a founding member of the American Philosophical Society, and the Franklin Institute, founded in 1824, was named after him. Within a few years after his return to Philadelphia and still in his mid-twenties, Bache brought to these institutions a sense of *national purpose*, of viewing the trajectory of these institutions from a perspective that transcended their strong regional character. The following chapter will provide room to test and refine this hypothesis.

Chapter 4

Early Research and Institutional Development

Scientist or Administrator?

Considering Alexander Dallas Bache's national prominence in the 1840s and 50s, it is surprising how little work has been done on his early career. It was in the 1830s that Bache established a reputation that would prepare him for the post of superintendent of the U.S. Coast Survey and his influence in national scientific organizations. There is no review of Bache's scientific work in his pre-Washington period. This implies that we know little about the connections between the different areas of his engagement. While Bache's emerging leadership and prominence in Philadelphia institutions are generally acknowledged, Bache's professorship at the University of Pennsylvania, his leadership at the Franklin Institute, and his publications have been treated in separate accounts focusing on what may thus appear to be discrete interests.

This is not the place to provide a comprehensive overview of Bache's work in this period. Rather, I will focus on samples of Bache's scientific papers, consider their motivational implications, and connect them to his institutional work. To the historian of scientific institutions in the United States, Bache comes into full view in 1842 when he was chosen for the post of superintendent of the U.S. Coast Survey in the nation's capital. Bache's role seems much less clear during the preceding twelve years, even if this was the period during which he built his initial reputation. In this chapter, I will focus on how Bache (who was twenty-three when he returned to Philadelphia and thirty-six when he moved to Washington D.C.) sought to distinguish himself, what type of questions he pursued, how he did so, and how this relates to what we know about his family background and his career decision. As a backdrop for this analysis, I will sketch the economic, political, and institutional context for Bache's return to his home town, the setting for science in Philadelphia.

Bache at the University of Pennsylvania

When Bache was elected Professor of Natural Philosophy and Chemistry at the University of Pennsylvania in 1828, he joined the faculty at a decisive moment in the institution's history.[1] Bache's family had been involved in founding many of the city's prominent institutions including the university. During the Great Awakening of the 1750s, Philadelphians had erected a building for George Whitfield to preach in and to create a school for the poor. As religious fervor cooled and the plans for a school did not materialize, Benjamin Franklin and other leading Philadelphians bought the building from the old trustees and created the Philadelphia Academy, a secular institution that Franklin envisioned to provide a utilitarian education in English rather than teaching Greek and Latin. In 1753, it was chartered by Thomas and Richard Penn, the colony's proprietors, and re-chartered as the University of the State of Pennsylvania in 1779.

In the early nineteenth century, the school's secular character began to limit its appeal. Competitors such as Princeton attracted students from Philadelphia families that sought a denominational education for their sons. The university combined a school for boys with an academy for those of college-age, sometimes teaching both groups in the same class. Afraid to lose students and their small income, professors seemed reluctant to implement discipline. In response, the board of trustees insisted that the faculty strictly enforce rigid rules but this made matters worse. Just before Alexander Dallas Bache joined the university, the board, under the leadership of Nicholas Biddle (then president of the Second Bank of the United States in Philadelphia about which there will be more to say in the next chapter) chose to reorganize the faculty completely and to install a new provost. Bache was part of a new faculty of five, and this faculty was now given the freedom that their predecessors had unsuccessfully pleaded for.[2] The old university building (originally built by the city for President John

1 Nicholas Biddle to ADB, September 16, 1826, box 3, folder 2, Bache Papers, SIA. He would resign from the corps of engineers the following year. See the corps' acceptance of Bache's resignation, June 6, 1829, box 3, folder 2, Bache Papers, SIA.

2 Edward P. Cheyney, *History of the University of Pennsylvania, 1740–1940* (1940; repr., New York: Arno Press, 1977), 28–220.

Adams but never used as an early "White House") was torn down and two new buildings put in their place on Ninth and Market Streets, one for the large medical school and another for the college.[3] The corner stone was laid in March 1829, at about the same time the twenty-three-year-old Bache was joining the faculty.[4]

Bache taught three seminars on natural philosophy and chemistry every day but Sunday,[5] a total of eighteen hours a week.[6] His students were boys between sixteen and eighteen years of age. Up to thirty students attended each of his classes. At the opening of fall term in 1828, a little over sixty students attended, more than ninety students started the following year. Student numbers increased and the reforms were considered a success.[7] In his second year at the university, Bache became the faculty secretary, adding an administrative burden to that of his teaching duties. For some of his seminars, he may have been able to draw on his teaching experience at West Point, or simply work through the text books he provided, but it still must have taken time to establish a teaching routine.

Bache was the youngest member of the faculty of arts (distinct from the faculty of medicine). His colleagues had been born between 1773 (Robert Adrain, professor of mathematics) and 1797 (William H. De Lancey, professor of moral philosophy and provost). In 1829, Bache's colleagues in the "Collegiate Department" were at least seven years older than himself.[8] Perhaps Bache felt that his age and early success matched the university's successful rejuvenation. The reorganization's design was not his but he had been hired to help implement it.

---

3 Edward P. Cheyney, "The Connection of Alexander Dallas Bache with the University of Pennsylvania," American Philosophical Society, *Proceedings* 84, no. 2 (1941): 151–60, quotation 152 f.

4 Cheyney, *History of the University of Pennsylvania*, 228 f.

5 Thomas Coulson, "Alexander Dallas Bache, 1806–1867," *Journal of the Franklin Institute* 263, no. 6 (June 1957): 489. Bache taught at all levels save the freshman class. University of Pennsylvania, "Catalogue of the Officers and Students of the University of Pennsylvania," n.d., www.archives.upenn.edu/histy/features/1800s/1830s/catalogue.html.

6 Merle M. Odgers, *Alexander Dallas Bache: Scientist and Educator* (Philadelphia: Univ. of Pennsylvania Press, 1947), 23.

7 Cheyney, *History of the University of Pennsylvania*, 225. For a photograph of the buildings occupied from 1829, see University of Pennsylvania, "University Archives Digital Image Collection," http://imagesvr.library.upenn.edu/cgi/i/image/image-idx?type=detail&cc=pennarchive&entryid=X-20050407004&viewid=1 (accessed October 30, 2010).

8 Biographical data on the University of Pennsylvania's faculty is available at http://www.archives.upenn.edu.

Edward Potts Cheyney has observed that Bache had few difficulties keeping his students' attention and that there was much less disobedience in his classes than in those taught by his colleagues. All of Bache's colleagues in the Collegiate Department struggled with student misconduct, but "there is not a complaint emanating from Professor Bache's room." Because Bache's successor was not spared such problems, Cheyney concludes that it was Bache's teaching style rather than his subject matter that caught the student's attention and sparked their interest.[9] Bache integrated into his sophomore and senior seminars subjects on which he was working outside of the classroom, such as magnetism and steam-boiler experiments.[10] Cheyney suggests that "a good observer is apt to be a good teacher, and *vice versa*."[11] At the very least, Bache seems to have been an engaging teacher. This implies that he had a convincing interest in a subject matter he was able to convey particularly well.[12]

In January 1829, Bache was elected a member of the Academy of Natural Sciences of Philadelphia, an organization specializing in natural history and a depository for artifacts.[13] In April of that year, he became a member of the American Philosophical Society (APS).[14] Even more important, however, was his new membership in the Franklin Institute for the Promotion of the Mechanic Arts. While Bache taught natural philosophy at the university, the APS and the Franklin Institute provided organizational platforms for research that the nineteenth-century American university was not yet designed to facilitate. While the Franklin Institute became essential for Bache and for the national development of science and technology during the 1830s that he helped shape there, Bache became a decisive figure in the APS as well. The university was considered a teaching institution and there was little room for research and organizational development. These interests had to be facilitated elsewhere. The APS and the Franklin Institute related to different phases of the city's and the country's

9 Cheyney, "The Connection of Alexander Dallas Bache with the University of Pennsylvania," 155.

10 See the catalogue for 1830, "Penn 1830: Catalogue," http://www.archives.upenn.edu/histy/features/1800s/1830s/catalogue.html (accessed October 1, 2010).

11 Cheyney, *History of the University of Pennsylvania*, 159.

12 Bache brought out the first American edition of David Brewster's *Treatise on Optics* that he then used as a textbook. David Brewster, *A Treatise on Optics*, 1st ed. (Philadelphia: Carey, Lea, & Blanchard, 1833), with an introduction by Alexander Dallas Bache.

13 S. George Morton to ADB, January 28, 1829, box 6, folder 9, Bache Papers, SIA.

14 J. R. Kane to ADB, April 17, 1829, box 6, folder, 9, Bache Papers, SIA.

history, and Bache's choice to focus his efforts on helping develop the Franklin Institute rather than the APS provides important clues about the kind of work Bache stood for.

The Urban Setting

Unlike most American cities, Philadelphia was able to look back on a strong tradition of scientific and technological inquiry, and the American Philosophical Society had been part of it. In the late eighteenth and early nineteenth centuries, the APS as well as David Rittenhouse's observatory, William Bartram's botanic garden, and Charles Willson Peale's Philadelphia Museum gave the city "a good claim to being the cultural center of the new republic."[15] Even if Philadelphia had limited national influence, Boston, the other science-minded American city, had always provided a regional rather than a national center. New York City had not developed a similar scientific culture. While the fledgling national government was still located in Philadelphia, politicians supported the city's scientific organizations. Thomas Jefferson, as Secretary of State (1789–94), cooperated with the University of Pennsylvania in evaluating patent applications. Jefferson was vice president of the United States (1797–1801) while he was also president of the American Philosophical Society, and he served as an unofficial liaison between the two. The federal government left for Washington in 1810, however, and subsequently weakened its support of science in Philadelphia. Henceforth, cooperation with the national government, such as the American Philosophical Society's preparation of Meriwether Lewis for his exploration of the trans-Mississippi west, remained exceptions.[16] In the 1820s, the city's scientific role seemed to fade.

Despite occasional cooperation with the federal government, the APS had always had a regional focus. The organization had been founded in the wake of a growing national self-awareness in 1743 and was rejuvenated in 1769 when Benjamin Franklin became its president. By 1829, its activities had flagged. The *Transactions* had not been published for several years. Bache chose to accept membership in the APS after his return but he chose not to focus his institutional efforts on this organization. As will be

15 A. Hunter Dupree, *Science in the Federal Government: A History of Policies and Activities to 1940* (Baltimore: Johns Hopkins Univ. Press, 1986), 8.

16 Ibid., 8–28.

seen, Bache had significant resources for institutional development and the fact that he developed interests that could not be facilitated by his professorship at the University of Pennsylvania alone is perhaps little surprising. Had he wanted to preserve and strengthen his family's legacy, he may well have concluded that he should throw his support behind the APS in which his great-grandfather had played such a memorable role. But Bache focused his attention on a much younger organization.

The Franklin Institute was created in response to a problem quite different from the one the APS had been intended to address. Whereas the latter was Philadelphia's attempt and claim to provide an emerging nation with a scientific organization on par with its European equivalents, the Franklin Institute's founding was a response to the increased significance of technology in the context of industrialization.

Prior to 1800, Philadelphia's wealth had derived from its role as one of the major American ports for trade with Europe. By the 1820s and 30s, several factors eroded the city's former advantage. Philadelphia was no longer the country's largest city and it lost important shipping business to its competitors on the Atlantic seaboard. It now competed with Baltimore for trade from the Susquehanna Valley. The Erie Canal provided New York with access to vast inlaying areas and this funneled the European trade to this northern competitor.[17] Philadelphia's harbor, one hundred miles from the coast, was subject to blockage by ice in wintertime while the harbors of Boston and New York City could be reached all year. Many ship owners directed their vessels to the latter cities where they avoided the risk of being marooned. By the 1830s, Philadelphia merchants had come to rely on New York's shipping and banking services, markets in which Philadelphia had previously been the leader.[18]

Between 1810 and 1840, the city managed to retain its position as one of the wealthiest in the country, however, by shifting its focus to manufacturing.[19] Steam engines and locomotives played a decisive role in this development. Philadelphia had the advantage of a plentiful and cheap supply

17 James Weston Livingood, *The Philadelphia-Baltimore Trade Rivalry, 1780–1860* (Harrisburg: Pennsylvania Historical and Museum Commission, 1947).

18 Diane Lindstrom, *Economic Development in the Philadelphia Region, 1810–1850* (New York: Columbia Univ. Press, 1978), 23–44; John Majewski, *A House Dividing: Economic Development in Pennsylvania and Virginia before the Civil War* (New York: Cambridge Univ. Press, 2000).

19 Philadelphia surpassed New England as the leader in engineering in the 1840s, and was in turn passed by New York City in the 1880s.

of coal and iron from its hinterland. The lack of water-power created a need for steam engines in the Midwest, and the city became a leader in producing them. Philadelphia's preeminence in manufacturing, according to Andrew Dawson, was not due to the production of textile machinery but to that of steam engines that could be used to power large tools to cut metal, for locomotives, and on steam powered ships.[20] In both New York and Philadelphia, capital from commerce was available for such investment-heavy undertakings. "The social, political, and economic elite," David R. Myer writes, "… allied their capital and influence with talented engineers and mechanics, some of whom also came from the elite, to propel steam engine manufacturing from feeble efforts in the 1790s to the great industrial machinery works of the New York and of Philadelphia in 1820." Because of the large size of investments necessary in large-scale manufacturing, sophisticated engineering was required to help minimize the risk for investing significant sums in foundries and machine works.[21]

The Franklin Institute had been founded in the context of this strengthening interest in large-scale manufacturing. Upon his arrival in Philadelphia in the 1820s, Samuel Vaughan Merrick (1801–1870) had received, as a gift from his uncle, a bankrupt company that produced fire-engines. Merrick was trying to learn about the production of large-scale machines and iron founding but the local mechanics' organizations turned down his application for membership. He decided to organize a school himself and cooperated with William H. Keating in founding the Franklin Institute in 1824. Keating, a professor of mineralogy and chemistry had traveled to Europe in 1822 after completing his studies at the University of Pennsylvania and was then appointed to a chair at his alma mater.[22] During the following years, the Institute became indispensible for the development of Philadelphia's technology-driven industries.

When Bache returned to Philadelphia in 1829, the Franklin Institute, with its focus on the development of technology, had assumed a role dis-

---

20 Andrew Dawson, *Lives of the Philadelphia Engineers: Capital, Class, and Revolution, 1830–1890* (Aldershot, Hants, England; Burlington, VT: Ashgate Publishing, 2004), 13 ff.

21 David R. Meyer, *Networked Machinists: High-Technology Industries in Antebellum America* (Baltimore: Johns Hopkins Univ. Press, 2006), 40–44, quotation 44. This development was accompanied by new social issues as an increasing number of Philadelphians became part of a labor force that worked in the emerging factories. Sean Wilentz, *The Rise of American Democracy: Jefferson to Lincoln* (New York: Norton, 2005), 210 f., 282–87.

22 Bruce Sinclair, *Philadelphia's Philosopher Mechanics: A History of the Franklin Institute* (Baltimore: Johns Hopkins Univ. Press, 1974), 4 ff., 29.

tinct from that of the American Philosophical Society: The latter had been founded in a time of growing cultural self-awareness moving towards political independence. It represented the ambitious claim of an American contribution to universalistic scientific progress. Following the War of 1812, the nation's political independence had been reasserted and cultural attention could shift towards the country's internal development. The Franklin Institute's focus was distinct from the APS in that it was designed to convey technical knowledge and expertise. In this sense, it fell in line with Bache's career decision to look inward rather than to the international arena and to attend the Military Academy at West Point instead of joining the navy.

Bruce Sinclair has argued that Bache, as one of the members shaping the Institute's science policy during the 1830s, was interested in "employment, research facilities and funding, and a greater public estimation for men of science." Sinclair writes that "tied to all those needs was the strong desire to make the intellectual standing of science in America equal to the standing it had in the Old World."[23] But Bache was not so much concerned with developing a financial basis for careers in science, even if this would of course be an important issue for the profession's nineteenth-century development; rather, his main concern was to establish, support, and develop rational standards and ideals for science, technology, and culture, and to make them indispensable and authoritative components of life in the United States. In the 1830s, such interests could only be pursued on the city and state levels even if similar developments could be anticipated for the national arena.

The Franklin Institute's Raison d'Être

The Franklin Institute of the State of Pennsylvania, for the Promotion of the Mechanic Arts had been founded to advance the diffusion of expert knowledge. It was modeled after similar institutions in England.[24] "Its purpose," Bruce Sinclair writes,

23 Ibid., 152.

24 Julius Adams Stratton and Loretta H. Mannix, *Mind and Hand: The Birth of MIT* (Boston: MIT Press, 2005), 1–31. The most comprehensive study of the Franklin Institute is Sinclair, *Philadelphia's Philosopher Mechanics*. Other works include Henry Butler Allen, "The Franklin Institute of the State of Pennsylvania," *Transactions of the American Philosophical Society* 43, no. 1, New Series (1953): 275–79, and papers referred to in footnotes below.

was to promote the useful arts by diffusing a knowledge of mechanical science at little cost to the membership. That object would be carried out by a program of popular lectures, the formation of a cabinet of models and a library, and by the awarding of prizes for useful improvements in the arts.[25]

Both education and exhibitions clearly pointed to a public role of "self-improvement" by means of education and instruction.[26] Sinclair has suggested that when "the Institute was founded in 1824, there was no clear plan for the conduct of scientific research." He adds that the "organization was begun as a mechanics' institute with popular education as its primary goal."[27] But the board of managers, in their first report, referred to "'the establishment of an Experimental Workshop and Laboratory'" that was to support the Institute's educational mission. The workshop was to help in teaching and in evaluating machinery from all parts of the state.[28] And in an early memorial, the Institute's leadership mentioned a museum and a laboratory as possible areas of engagement. They suggested that the organization's role was not going to be limited to conveying and evaluating ideas but to help develop them.[29] Because of financial restraints, and probably also because of a lack of members who could live up to these ambitious ideas, these plans gained little traction.

Membership in the Institute was open to anyone willing to pay the small membership fee. Lecturers ("professors") were appointed to provide evening education with an emphasis on practical utility but not excluding "natural philosophy." The Institute published a *Journal* that, during these early years, printed articles that had been taken from other publications. In this sense, too, the *Journal* served the diffusion of knowledge rather than its development. Many considered the Institute's educational program and its exhibitions to be its prominent activities with 20,000 visitors in 1826 and 40,000 in 1831.[30] Exhibitions provided exposure to the Institute and its

25 Sinclair, *Philadelphia's Philosopher Mechanics*, 32.

26 Bruce Sinclair, *Early Research at the Franklin Institute: The Investigation into the Causes of Steam-Boiler Explosions, 1830–1837* (Philadelphia: The Franklin Institute of the State of Pennsylvania, 1966), 3 f.

27 Ibid., 3.

28 Sinclair, *Philadelphia's Philosopher Mechanics*, 138 f.

29 Sinclair refers to a memorial to the legislature which contained the idea of pursuing research: "The Memorial of the Officers and Board of Managers of the Franklin Institute of Pennsylvania for the Promotion of the Mechanic Arts, 26th February, 1824," in the Franklin Institute Archives. See Sinclair, *Philadelphia's Philosopher Mechanics*, 33 f.

30 "Quarterly Meeting," *Journal of the Franklin Institute* 6, no. 5 (November 1831).

mission. In 1832, the Board of Managers wrote in their annual report that the "experience of each succeeding year adds to our conviction that next to education, and second only to that in importance, are the displays of domestic manufactures."[31]

Fig. 5. A 2003 photograph of the Franklin Institute's building in Philadelphia, which was designed by architect John Haviland.

The Institute was able to muster very strong support in Philadelphia as large numbers attended its lectures and joined as members.[32] The lecturers were to provide regular classes, a marked departure from what other mechanics' institutes were offering. A year after its inception, the organiza-

31 "Report of the Board of Managers," *Journal of the Franklin Institute* 9, no. 2 (February 1832).

32 Sinclair, *Philadelphia's Philosopher Mechanics*, 36 f.

tion's early enthusiasm had brought about a well-received and well-attended exhibition on American manufactures and plans for a building to be erected as the Franklin Institute's new home.[33] The Institute thus combined a strong interest in self-improvement with the ambition to become a visible representative of, and clearing house for, technological innovation in the city and the state.

Bache joined the Franklin Institute at a decisive moment, much like when he had accepted a post at the University of Pennsylvania at a moment of transformation in that institution's history. The Institute's affairs had largely been controlled by James Ronaldson, Matthew Carey, and Peter A. Browne. All three represented variations of a strong interest in developing the Institute in ways that would benefit the state's manufactures and its industry. Just as Bache was returning to the city in 1829, the dynamics within the Institute's leadership shifted. During the preceding years, Samuel Vaughan Merrick, the institution's initiator who came from a successful family of entrepreneurs and educators, had been relegated to the sidelines but was increasingly able to assert his influence.[34] He had strongly opposed the establishment of a scientific school that would exclude the teaching of Greek and Latin because he viewed the Franklin Institute's role in the context of other institutions, such as the University of Pennsylvania, where these languages remained a requirement. After Bache had joined the Institute in 1829, he teamed up with Merrick to introduce a number of changes.

While the Institute had shown ambitions to function as a statewide review board of technological innovation, it had lacked the financial basis to test large-scale apparatus. In March 1829, just as Bache was taking up teaching duties at the University of Pennsylvania, Merrick proposed that the Institute undertake experiments on waterwheels. Generating power from water was important to American industry but there existed little systematic understanding of what kind of wheel would be most efficient. Merrick ingeniously tackled the financial issue by proposing that those who stood to benefit from the research pay for it. An appeal was sent out to mill owners and others interested in effectively using water power. The

33 Ibid., 40 f., 46–49.

34 On Merrick's background, see Edward Digby Baltzell, *Philadelphia Gentlemen: The Making of a National Upper Class* (Glencoe, IL: Free Press, 1958), 101 f. See also John K. Brown, *The Baldwin Locomotive Works, 1831–1915: A Study in American Industrial Practice* (Baltimore: Johns Hopkins Univ. Press, 1995), 5.

subscriptions of funds allowed the Institute to engage in its first large-scale research project.[35] More than 1,500 dollars were subscribed, and supporters included organizations such as the New England Society for the Promotion of Manufactures and Mechanic Arts which meant that even the powerful Boston textile interests underwrote the idea. The first report on the committee's work was published in March 1831 and was generally considered a huge success. Bache had joined Merrick and a handful of colleagues in working on the waterwheel project just after he was elected a member, and he was a co-author of the committee's report.[36] The notes for the waterwheel reports are among his papers at the Smithsonian Institution Archives.[37]

The project was a significant juncture for the Franklin Institute for its leaders now felt emboldened to expand its activities. In 1830, Bache proposed that the Institute conduct experiments on the causes of steam boiler explosions. This project is of particular significance. Even though one historian has argued that the experiments provided much data and little else, Bache and his friends would later refer to it as a significant achievement.[38] Joseph Henry later suggested that the steam boiler experiments put Bache on the shortlist of American science.[39] What, then, was their contemporary significance?

The Institute's financial plan for this investigation differed only slightly from the earlier one used for the waterwheel project. In June 1830, the Board of Managers appointed a seventeen-member committee that chose not to rely on financial contributions from industry and to work with eyewitness reports of steam boiler explosions instead of conducting experiments. Apparently without being aware of such activities, the U.S. secretary of the treasury, following a request by the House of Representatives, also commenced to collect data on steam boiler explosions. When the secretary learned of the Franklin Institute's activities, however, he offered to financially support that project. This proposal provided the Franklin Institute

35 Sinclair, *Philadelphia's Philosopher Mechanics*, 140–47.

36 Ibid., 150.

37 Waterwheel notes, box 5, folder 38 ("Water Wheel Report"), Bache Papers, SIA.

38 Dupree, *Science in the Federal Government*, 50.

39 Henry included Alexander Dallas Bache's steam boiler experiments, James P. Espy on meteorology, Henry D. Rogers on the geology of New Jersey, and Bache's intensity apparatus. Both Espy and Rogers were active members of the Franklin Institute. See Joseph Henry, *The Papers of Joseph Henry* (Washington: Smithsonian Institution, 1972–2008), 3:68.

committee with an opportunity to return to the idea of conducting experiments. The final decision on whether such experiments could indeed be made hinged on the quality of the experimental designs that the Institute was to submit to the secretary. Among the committee's members were scientists such as James P. Espy and Robert Hare, as well as experienced mechanics and manufacturers such as Mathias Baldwin. And yet it was Alexander Dallas Bache who, in 1830, took charge of developing the experiments.[40]

Following his involvement in the Institute's waterwheel experiments, Bache emerged as the guide and manager of the steam boiler project. His research design was accepted and the Institute embarked in what was the first federally sponsored research project. After little more than a year, Bache took on the chairmanship of the general committee (to oversee several subcommittees).[41] He coordinated the project, represented it to the secretary of the treasury, and wrote the final report in which he brought together the conclusions of the subcommittees assigned to specific experiments.

Between 1829 and 1836, the period under discussion here, Bache published thirty-four papers. Between 1836 (when he would travel to Europe) and 1843 (when he moved to Washington D.C.), he published only fifteen.[42] A number of Bache's papers during the early period related to his work for the Franklin Institute's committee on steam boiler explosions.[43] Other publications were connected to Bache's work in the classroom, such as his paper on demonstrating the absorption of heat by different types of material.[44] A number of publications dealt with a variety of physical and meteorological phenomena and some conveyed recent

40 For the steam boiler committee's work, see Sinclair, *Philadelphia's Philosopher Mechanics*, 173–94.

41 Sinclair, *Early Research at the Franklin Institute*, 17.

42 Cheyney, "The Connection of Alexander Dallas Bache with the University of Pennsylvania," 158.

43 These include: Alexander Dallas Bache, "Alarm to be applied to the interior flues of steam boilers," *Journal of the Franklin Institute* 14, no. 4 (October 1832): 217–23; Alexander Dallas Bache, "Safety apparatus for steam boats, being a combination of the fusible metal disk with the common safety valve," *Journal of the Franklin Institute* 11, no. 4 (April 1831): 217–21; Alexander Dallas Bache, "Experiments on the efficacy of Perkins' steam boilers, or circulators," *Journal of the Franklin Institute* 19, no. 6 (June 1835): 379–86.

44 Alexander Dallas Bache, "Experimental illustrations of the radiating and absorbing powers of surfaces for heat, of the effects of transparent screens, of the conducting power of solids, &c.," *Journal of the Franklin Institute* 19, no. 5 (May 1835): 303–09.

European work.[45] Bache republished, for example, William Whewell's *Treatise on Optics* with a new appendix, a book he probably used for teaching. At least one of his papers (on "Experimental illustrations of the radiating and absorbing powers of surfaces of heat"[46]) had a decidedly didactic bent and was perhaps also deduced from his classroom work.

During his most active period in terms of papers published, therefore, Bache's work on steam boiler explosions remains his most significant contribution. This report won him national and even international acclaim, and it had a significant impact on the Franklin Institute's developmental trajectory. Historians have observed that Bache fared less well in developing or modifying groundbreaking theory.[47] A detailed consideration of samples of Bache's publications from this period will confirm the larger argument that his impetus for the pursuit of science derived less from a fascination with nature than from his biographical responsibility to help develop American nationhood. A detailed consideration of three samples from this period will clarify this point: Bache's report on steam-boiler explosions, his interest in weights and measures, and his involvement in a debate about the causes of meteor showers.

The Report on Steam Boiler Explosions

Bache's 1836 "General Report on the Explosions of Steam-Boilers" constituted part II of the Committee's Report. The way in which Bache introduces the subject and presents the report illustrates the Institute's role in the context of an emerging American culture.

"The Committee," Bache wrote,

> undertook the task imposed upon them by the Franklin Institute, with a deep sense of the responsibility which it involved. On the one hand, a series of disasters by which human life was sacrificed, called loudly for an investigation of the causes

45 Alexander Dallas Bache, "Note relating to the hardening of lime under water, by the action of carbonate of potassa, &c., and to the hardening of carbonate of lime in the air, by potassa and soda," *Journal of the Franklin Institute* 19, no. 1 (January 1835): 6–8; Alexander Dallas Bache, "Proposed forms of diagrams for exhibiting to the eye the results of a register of the direction of the wind," *Journal of the Franklin Institute* 22, no. 1 (July 1836): 22–27.

46 Alexander Dallas Bache, "Experimental illustrations of the radiating and absorbing powers of surfaces for heat."

47 Robert V. Bruce, *The Launching of Modern American Science, 1846–1876* (New York: Knopf, 1987), 17.

which produced them; on the other, an untimely or ill-directed interference with a branch of industry, which has been a source of unparalleled advantage to our country, was truly to be deprecated.[48]

What caused the committee to feel this sense of responsibility? Bache does not focus on the consequence of the report for the reputation of the Franklin Institute and instead highlights how the committee's work affects the relationship between public interests and those of private industry. He is taking into account the regulation of the production and operation of steam boilers, or the publication of dangerous practices, either of which would put pressure on the industry. Such "interference," Bache presupposes, would not be bad per se, but it could be "ill-directed" or "untimely." He invokes a consensus that American industry, at this stage of its development, should remain unregulated.

The opening paragraph shows that the committee, under Bache's chairmanship, consciously tackled an issue of national dimension and thus conveyed the impression that it was confronting the most advanced technological developments bearing significant economic implications. In this sense, it boldly claimed national leadership for technological evaluation and review.

Bache continued:

Emanating from an Institute 'for the Promotion of the Mechanic Arts,' the Committee felt advantageously situated. They could not justly be suspected of a desire to trammel the progress of any art; and yet the public confidence, which had always been accorded to the institution, would naturally attach to a committee selected by it.

In his opening paragraph, Bache had first mentioned the public's and then the industry's interests. In the following sequence he addresses the industry's and then the public's concerns, thus bracketing the former with the latter. This corresponds to the public's larger and less particular role in society.

Bache asserts that the committee stood above such interests. This credibility he deduces not from the federal government's support of the

48 The Franklin Institute, "General Report on the Explosions of Steam-Boilers (1836)," in *Early Research at the Franklin Institute: The Investigation into the Causes of Steam-Boiler Explosions, 1830–1837*, ed. Bruce Sinclair (Philadelphia: Franklin Institute of the State of Pennsylvania, 1966).

project on steam boilers, but from the Franklin Institute's particular mission and record.

The Committee further believed that the apprehensions of the public, aroused by the frequent recurrence of accidents, could only be satisfactorily allayed by an investigation, which would show that such accidents were not unavoidably incident to the useful agent which they distrusted, but resulted from a want of due regulation of its power, or from circumstances incidental to its use which could be foreseen, and therefore guarded against.

Bache develops his point further. He suggests that the public may perceive accidents to be "unavoidably incident" to steam boilers. Taken literally, this implies that the public considered the technology to be dangerous regardless of how it was designed or handled, and Bache implies that this would jeopardize the technology's legitimacy and perhaps even lead to its abandonment. Bache's committee thus served two purposes: It sought to find ways to improve the safety of steam boilers by understanding the causes of explosions and it sought to use review and regulation as tools for legitimizing technological development in America. Even though the state is the obvious candidate for assuming such responsibilities, and despite the federal administration's support of the committee's work, Bache has not mentioned the federal government.

With these views the Committee commenced, actively, the collection of information upon the subject intrusted {sic} to them. The replies to their circular were canvassed in their meetings, and finally laid before the public. It occurred most opportunely for the ultimate success, though not for the rapid completion of their labours, that an opportunity was afforded them for experiment, by which to test many of the suggested causes of, and preventives to, the explosions of steam-boilers.

Bache reports that the committee had initially not planned to conduct experiments but that an occasion to do so had then presented itself. Without experiments, the committee's initial role seemed to be restricted to a deduction of possible causes from accounts of the circumstances of boiler explosions. Bache merely speaks of "an opportunity" for experiments here and still does not mention their sponsor.

These experiments, originally proposed by our public-spirited fellow-citizen, S. D. Ingham, Esq., then Secretary of the Treasury of the United States, have been brought to completion and presented to the public under the auspices of the present Secretary.

Bache now mentions that the committee's work was supported by the federal administration. By referring to the secretary of the treasury as a citizen first, however, and by referring to his role in the federal government second, Bache suggests that Samuel D. Ingham's support of the committee's work was somehow disconnected from his official duties, and that Ingham's interest in advancing research on the causes of steam boiler explosions was a dimension of his enlightened citizenship. Bache could have enhanced the committee's public standing by stressing that the federal government considered it worthy of sponsorship. This would have helped him respond to his concern about reviewing and legitimating technology. But Bache does not seem to consider the government's sponsorship a relevant sign of effective regulation.

In some sense, of course, Bache's presentation seems sensible because politicians, as politicians, cannot evaluate the quality of scientific work or of technological design. But unless they have a marked interest in such matters and the experience to match, neither can private individuals. Bache refers to Ingham in a way that highlights the secretary's personal acquaintance with scientific and technological matters. Departing from such an informal commonality, Bache seeks to establish the Franklin Institute's authority on its own right, without reference to a higher political authority and its endorsement. The committee chose to stand and to guarantee the merit and quality of its work for itself. This suggests that the committee chairman and the Institute he represented were assuming a strong sense of cultural authority and self-reliance.

> The Committee trust that they have, by the experiments just referred to, shown not only what are some of the causes of explosion, but, which is quite as important, what are certainly not causes. In this way they hope to have turned away the attention of ingenious men from false hypotheses which cannot furnish the remedies they are in quest of, and to have pointed out some directions in which their labours may be profitably bestowed.

Bache's reference to "ingenious men" points to inventors interested in the application of principles rather than their deduction. He considers the committee to assume the vicarious role of identifying and evaluating such principles. In that sense, the committee represents the principle of rationality without immediate attachment to a particular sphere of application. This view, though "elitist," is not condescending or dismissive, but concerned with the proper development of American technology. Bache considers the committee to be legitimized by the quality of its work and through the intellectual shortcuts it provides for anyone interested in its object of investigation.

Bache uses the past tense in "hope to have turned away." He assumes that some of those who are interested in steam boilers rely on the committee's previous publications in the *Journal of the Franklin Institute* rather than on the committee's report. Therefore, while a rather closely knit group of inventors and mechanics followed the committee's work from up close, the report was intended for a broader audience.

> A desire to complete the reports upon their experiments, has induced a delay in the present report, which, thus far, however, the Committee are satisfied will be found to have been judicious. This conclusion they rest upon the many references, which will appear in the following pages, to those experiments, which have given an authority to recommendations and suggestions, that could not have been claimed for them unless thus strongly supported by facts.

At the beginning of the paragraph, Bache offers an explanation for a delay in publishing the results of experiments. He deduces the quality of these experiments from the use that others had already made of them. This confirms that the Franklin Institute's *Journal* had already taken notice of the committee's reports and that Bache distinguishes between an informal and a larger audience. His presentation of scientific and technological authority was geared, not toward the circle of inventors and mechanics familiar with the Franklin Institute's ongoing work, but towards the general public. As chairman of the committee, and as a representative of the Franklin Institute, he not only presents investigative results to the community but provides a model for a specific type of authority and expertise in America.

Bache deduces the experiments' soundness from the number of references by inventors and other colleagues. He appears to be deducing quality from popularity, but the intention here is different: Bache underlines the benefits of expertise for those who accept the committee's work and takes for granted that the reliability of results naturally creates such "authority."

In a paragraph I am not quoting here, Bache next explains that the committee was unable to include in their report the results of tests on the strength of material and that it would supply them later. He then moves on to discuss the public perception of steam boiler accidents:

> In this report the Committee have endeavoured, by examining the different accounts of explosions on record, and the writings on collateral subjects, to ascertain what causes have been operative in producing these disasters. The difficulty of

procuring satisfactory testimony in regard to them, has been often pointed out. Most frequently those from whose mismanagement or want of vigilance they have immediately resulted, have been victims to them, and when they have survived, the precise state of things before the occurrence was imperfectly known to them; and, however honest, their minds have received a bias towards the non-existence of certain circumstances judged likely to have produced the results.

Bache recounts the difficulties of attempting to deduce from the description of witnesses the causes for the explosion of steam boilers. In those cases in which the operators of a boiler (those aware of the more technical aspects of its operation) survive, they were either unaware of what was about to happen and could not relate the particulars of the moment before an accident, or they may have indeed noticed a problem but would then shield themselves by misrepresenting the situation to the investigators. Bache does not seem to be interested in blaming engineers, though he assumes that disasters are frequently caused by their negligence. His reference to "however honest" indicates that Bache's interest is limited to learning about the causes of boiler explosions, and that he views the engineers' reports as possible sources for relevant information. Bache's perspective is free of moral judgment and focused on the explanatory problem that his committee set out to solve.

It hence follows that in regard to many explosions, either none of the circumstances which immediately preceded them, and bearing upon them, are known, or by inaccurate statements of them, an appearance of mystery is thrown around the whole matter, calculated to baffle research, and to alarm the community, who are exposed to a recurrence of the same dangers.

Bache's conclusion logically follows from his observation even though he adds a new dimension by suggesting that the engineer's account of the circumstances just before the accident was "calculated" to mislead. It makes sense to assume that an engineer will be interested in clouding the causes of the accident if he feels that he might bear responsibility for it. But how does it serve the engineer's interest to "alarm the community?" If indeed, as Bache writes, the engineer *intends* to alarm the community, he benefits from the courage that the boiler's danger reflects on him. The engineer turns risk into reward. Again, Bache does not evaluate this behavior. He pragmatically considers its relevance for studying boiler explosions.

Bache continues as follows:

> Thus it appears that of the numerous explosions on record, few are made to subserve the cause of humanity, by a knowledge of their proximate causes. The details of the number of killed and wounded, and of the more or less entire destruction of the boilers and of the boats, are given in the daily prints, and public curiosity is satisfied.

A common response to disasters, Bache suggests, is not to investigate their true causes. Regulation is a matter of state and national politics, and because the investigation took place in the United States and because it was sponsored by the federal government, it would not have been surprising had Bache restricted himself to speaking about the American situation. But Bache conceives such investigations to be a matter that the public owes, not to Pennsylvanians or Americans alone, but to all of humanity.

An implication of Bache's view is that if few explosions are put to use as sources of data for understanding the operation of boilers, a larger number of explosions would provide additional occasions for gathering such evidence. In line with the perspective he assumed above, he considers the fatalities of an explosion to be particular and incidental. Bache investigates explosions as tokens of general principles rather than as a service to the deceased or as a response to sensational interests.

In the following paragraph, which I am not quoting in full, Bache asserts that in selecting their evidence, the committee having "themselves no theory, or theories, to support, […] have of course not been biased, by such views, in the elections made." He suggests that an investigation is flawed if it serves to bolster a preordained theory and that the committee serves no interest other than trying to identify true causes. In the next paragraph, Bache argues that the committee has made a relevant contribution to the understanding of boiler explosions:

> This mode of proceeding is, obviously, not calculated, by one effort, to exhaust a subject. But the Committee believe that they are able to make a decided step forward in the knowledge at present existing, in a connected form, on this subject. That to the causes pointed out by a Committee of the British House of Commons, in 1817, namely, improper construction or material of a boiler and undue but gradual increase of pressure, they will be able to add others as important, and as fully proved as the former. Nor will any cause for alarm result from this extension, since it will be found that it is only ignorance of these circumstances which constitutes their danger, and that they may be prevented from occurring and remedied when they occur.

Bache suggests that the committee's work expanded on findings by similar British investigations. The committee's work was not restricted to providing technical advice but it also served the purpose of helping advance knowledge about the causes of explosions. The Franklin Institute thus claimed to make relevant contributions to an international discourse on this matter.

Bache anticipates some readers' expectation that there was no protection against some causes of steam boiler explosions. His austere tone suggests that the committee was not prone to presenting anything in a more favorable light than warranted by circumstances, and it prepares the ground for stressing the relevance of the education of engineers.

> It will be full time after the well-ascertained causes of explosion have been duly guarded against, to look for others more occult in their nature, if indeed there are such.

Bache's tone is laconic. He doubts that explosions cannot be explained rationally and implies that an adequate procedure for trying to explain phenomena such as explosions consists in identifying and dealing with deducible issues, and not to concede at the outset that the underlying problems defy a rational approach. But Bache's insistence suggests that he considers such doubts to remain current among the American public. The committee's work and its report, beyond its immediate goal of explaining steam boiler explosions, serves the wider purpose of helping establish in the United States a culture of rationality: the expectation that problems may be broken down to their basic elements, and solved with the help of underlying principles.

In the next two paragraphs, which I am not quoting, Bache stresses that the committee's work was limited to recommending measures aimed at improving public safety. He then asserts that the committee considers it entirely feasible that their recommendations be put into law. "In submitting this project," Bache writes,

> the Committee obviously do not entertain a doubt of the competency of Congress to legislate on the matters embraced in it. The several discussions in that body on the subject, the recommendation of the President of the United States, and especially the very detailed provisions of the bill recently proposed in the Senate, fully sustain them in this opinion. They consider the question now to be, not whether any regulations may be made, but how those to be made may be rendered most efficient and complete. For this completeness the very respectable Committee who reported the bill referred to, in the Senate of the United States, have expressed themselves anxious; and the labours of this Committee, so far from

> being an interference, will, no doubt, as far as they may be approved, be looked upon as forwarding the views thus expressed.

The very fact that Bache emphasizes Congress' legislative competence indicates that it could not be taken for granted. Bache's use of the adverb "obviously" serves to make his analysis seem unchallenged. Bache considers the discussions and suggestions in Congress and by the president as supporting the committee's view, not the other way around.[49] He suggests that political discussion had shifted away from the question of the legitimacy of regulations to the consideration of what such regulation could consist of (leading to "detailed provisions"). The committee assumed leadership in a trend Bache considers well underway. Instead of imposing its view of the federal government's role, the committee merely supported an ongoing movement. By contrast, the opposition to an amplified governmental role is characterized as backward-looking. Bache was anxious to have his country evolve in a direction in which technological and scientific insights would be turned into policy. He anticipated a more rational cultural development, not for the expansion of government for its own sake, or the financial support of scientists, but for the development of rationality on solid grounds. The choice of words throughout this introduction nevertheless suggests that Bache and the committee considered themselves to be proponents rather than followers of such a perspective.

Before Bache moves on to discuss the experiments, he closes his introduction with the following paragraph:

> The good effects which have attended the adoption of partial preventives in England, and the excellent effects from the more complete ones in France, should urge us, as Americans, to do our part in preventing further destruction of life and property by these disastrous explosions. And while we apply means for this purpose, experience and reason both teach us that they will produce no undue or severe restraints upon mechanical skill or commercial enterprise, but rather that they will aid both, by increased confidence on the part of the public.

Bache could have argued that it was in America's best interest to enhance the security of steam boilers but he relies on a comparison instead. He implies that France, England, and the United States shared a responsibility to provide such safety. His perspective is that of humanity organized in dis-

49 For a detailed account of discussions on these matters, including those in Congress, as well as remarks by President Andrew Jackson in 1833, which led to the regulation of steamboats in 1838 and 1852, see Louis C. Hunter, *Steamboats on the Western Rivers: An Economic and Technological History* (New York: Dover, 1994), 520–46.

crete nation-states, each with a distinct responsibility to implement universal standards of rationality valid across national boundaries. Bache suggests that in comparison, the United States was behind other countries, and his criticism implies that America should lead, not follow, in realizing such standards. Bache and the Franklin Institute's committee were missionaries of a perspective that embraced a transnational and universalistic technological and scientific culture that was to be implemented through national political efforts.

Had the attitude towards regulation indeed been as positive and firm as Bache suggested above, his pledge to follow the English and French models would hardly have been necessary. Bache feels that the United States is behind, and he addresses the public like an intellectual rather than an expert.

In the second sentence of this paragraph, Bache concludes his introduction by returning to his initial topic. On the surface, his argument is directed at those opposing regulation, a group that includes American manufacturers. Given the commission's work for the Franklin Institute, an organization with strong ties to manufacturers, this emphasis suggests that the particular direction assumed by Bache and his colleagues was not shared by them all. But Bache goes beyond emphasizing the benefits of regulation. As suggested above, the issue of public "confidence" is central to his introduction.

In conclusion, Bache's introduction to the report on steam boiler explosions provides a glimpse of the clarity with which this thirty-year-old member of the Franklin Institute sought to establish for his colleagues and himself a self-reliant authority that was independent of political endorsement and support. In a situation in which we would today perhaps not have been surprised had Bache boosted his committee's result by highlighting the governmental support it received, he did no such thing. Even though it was later pointed out again and again that the report's significance in part stemmed from the federal government's support of the project, Bache apparently sought to have the evaluation of factors leading to explosions of boilers verified and legitimized by qualified experts such as himself and his colleagues. What motives explain Bache advancing such claims? There is nothing in his introduction that would point to personal enrichment or advancement. Bache received no income from his work on behalf of the committee, and neither did he receive remuneration for his other activities in the Franklin Institute. He declined to accept payment for articles submitted to the *Journal of the Franklin Institute* even though such payment was customary.[50] Instead, the introduction to Bache's *Report* contained the motivational premise that the United States was in need of catching up to standards of rationality adopted elsewhere. Bache viewed the American situation from the perspective of universalistic developments, and he considered each country to be responsible for implementing and developing them.

Looking back on the previous chapters of this investigation, it is of course evident that this outlook followed naturally from Bache's family background: National development becomes possible against standards of a rational discourse that transcends a particular nation, and it requires a dedicated effort by a particular nation-state to connect to it. Bache had emphasized that the committee's work on the causes of steam boiler explosions expanded on ideas developed by an earlier British committee on the same subject. To adopt rational standards, therefore, was to engage in a rational discourse that transcended national borders. The expansion of knowledge was a side effect. To catch up with France and England, Bache urged Americans to dedicate and ready themselves for participation in such a discourse. The Franklin Institute's research, by providing insights into what was perceived to be a highly complex if not mysterious problem, not only sought to provide recommendations for a practical problem, but to help establish the idea that America was participating in a universalistic rational discourse.

The forty-eight-page report for which Bache had written the introduction contains limited original investigation on physical principles and instead summarizes the committee's testing of certain hypotheses that it took from contemporary research and from common public beliefs. The report consists of five sections of which the first three contain the bulk of the argument. The first part deals with "Explosions from undue pressure within a boiler, the pressure being gradually increased." Bache thus put at the front of the report a section in which a popular theory on the causes of the explosion of steam boilers was disproved: By way of experiments, some of them violent, the committee could show that boilers burst not only when pressure inside of them was increased rapidly, but that they would also burst if pressure was raised gradually. The point had been made before, by the select committee of the British House of Commons in 1817,

---

50 Sinclair, *Philadelphia's Philosopher Mechanics*, 215.

but Bache and the committee apparently felt that the idea had not sunk in.[51] Because of the significance of pressure inside the boiler, the committee highlighted the role of measuring the pressure inside a boiler and of safety valves and their protection from interference from the crew. According to another popular theory, insufficient water levels caused boilers to explode. A number of sub-theories existed to explain why this would result from low water levels.[52] The committee tackled this problem next, and focused on possible reasons why metal, "unduly heated" in areas not covered by cooling water, would cause explosions. The committee explored two possibilities: Weakening of the material when heated; and metal serving as a "reservoir of heat to furnish highly elastic steam," i.e. the injection of water into the boiler and the subsequent build-up of pressure too large to be handled by the safety valve.[53] Another popular theory, the decomposition of water into hydrogen that was then considered to be the true cause of explosions, was discounted. The committee was able to show that such a decomposition of water did not take place. Gases might indeed lead to explosions, but they were of other origins.[54] In these first sections of the report, Bache occasionally highlights not only the results of the committee's work, but stresses its train of reasoning. This further illustrates his efforts to provide an example for proper research. Once he has made the point that elastic steam may develop with explosive force if water was flushed onto unduly heated metal, for example, he writes that with "such a powerful agent present, … it might have been supposed that no other cause for explosion would have been looked for, than the action of this steam." He continued that the "case is, however, otherwise, and the Committee must turn aside from their direct course to examine briefly" the theory that explosions were caused by hydrogen gas.[55] Bache was explicit not only about what the committee did, but also about what it could have done. It seems important to him to point out that a given phenomenon may have more than one cause, and that all possible causes need to be taken into consideration even if one proven cause appears to be sufficient for an explanation. The committee's report was directed at the public, and

51 Bache himself points to the study by the select committee. See Franklin Institute, "General Report on the Explosions of Steam-Boilers," 6.

52 On prevailing theories on steam boiler explosions, see Hunter, *Steamboats on the Western Rivers*, 282–304, particularly 292–95.

53 Franklin Institute, "General Report on the Explosions of Steam-Boilers," 13.

54 Ibid., 15 f.

55 Ibid., 15.

one would think that a report on the committee's conclusions would have been sufficient. But Bache's committee presented its report as a model of a thorough, rational, and circumspect consideration of a problem. This goes along with the observation that Bache, in his introduction, sought to establish independent authority for the type of investigation undertaken by the committee. Had he been able to rely on such authority, he may have found it unnecessary to add such methodological commentary.

Bache moved on to tackle questions deduced from the initial set of problems. What caused metal to heat in such a way as to become a danger to the boiler's safety? An insufficient supply of water has already been mentioned. Bache also worked through other possibilities, including deposits by water (that insulated the metal from cooling water), and the careening of a steam boat (that lowered the water level on one side of a boiler, or in a set of connected boilers). In another part of the report, Bache discussed the construction of boilers including questions related to the strength of materials. In each instance, concrete proposals were deduced for ensuring a safe operation of boilers. Human ignorance, carelessness, and outright mischief (such as steam boat racing and blocking the safety valve) were treated as well. "In the present state of general education in our country," Bache suggested, "it would obviously be impracticable to insist that firemen, or even steam-engineers, should be versed in the scientific principles which regulate the use of steam."[56] The country provided opportunities for higher education but it lacked broad public education. In this sense, too, England and France were ahead of the United States. For the case of steam-boiler safety, this implied that if engineers (i.e. those operating a boiler) could not be relied upon to understand a boiler, appropriate procedures had to be enforced through regulation. "The public have … a right to expect from the higher authorities, beginning with the chief engineers, and rising to the captains of steamboats and masters of shops," Bache wrote, "that they should exert all the moral influence which vigilance can produce." And it could expect "from the law, that it should constrain all these, by appropriate penalties, to the discharge of their responsible duties."[57]

In keeping with this perspective, the report concluded with a (federal) bill that the committee had drafted for Congress, a "Bill for the regulation of the boilers and engines of vessels propelled in the whole or in part by

56 Ibid., 39.
57 Ibid., 39 f.

steam."[58] Its provisions could be traced to the results of the committee's investigations, such as that stipulating that there be a penalty for a bursting of a boiler "caused by deposite" (implying that the boiler had not been sufficiently cleaned) or providing for two safety valves, one inaccessible to the steam engineer. The bill contained proposals for a licensing and an inspection system that Congress eventually implemented in 1852.[59]

Weights and Measures

Standards of length and of weight provide unity, coherence, and reliability, and they do so only if they have been defined by the state. Because of its political implications, this topic was of great concern to Bache.[60] Very few letters by Bache survive from the period under discussion here. However, among them is a letter he wrote to his uncle George Mifflin Dallas in 1833 that will be the second sample for discussing Bache's scientific work.

Bache's uncle, a U.S. Senator from Philadelphia, had just suffered a political defeat when he received the following letter.[61] Dallas had earlier sent his nephew a report on weights and measures by Ferdinand Rudolph Hassler, the superintendent of the U.S. Coast Survey since 1808 and of

58 Ibid., 43–48.

59 A law passed in 1838 "failed and failed badly," according to Lewis C. Hunter. It was emasculated by taking out a hydraulic test for boilers (so as to effectively test their strength) and the licensing and examination of engineers was watered down so as to make them ineffective. Inspectors were appointed by federal district judges and accountable, in their judgment, standards, and collection of fees from the owners of vessels, only to themselves. The new steamboat law of 1852 implemented an effective licensing system and other ideas suggested by the Franklin Institute committee under Bache's direction. Inspectors were organized in boards, and they were selected on a merit basis. They could not only issue but also revoke licenses. For a review of efforts to regulate the operation of steamboats, see Hunter, *Steamboats on the Western Rivers*, chap. 13, "The Movement for Steamboat Regulation," 520–46. See also John G. Burke, "Bursting Boilers and the Federal Power," *Technology and Culture* 7, no. 1 (Winter 1966): 1–23; Richard N. Langlois, David J. Denault, and Samson M. Kimenyi, "Bursting Boilers and the Federal Power Redux: The Evolution of Safety on the Western Rivers," *University of Connecticut, Department of Economics, Working Paper Series* (May 1994), http://www.econ.uconn.edu/working/1994-01.pdf.

60 Bache would become superintendent of Weights and Measures in 1844. Hugh R. Slotten, *Patronage, Practice, and the Culture of American Science: Alexander Dallas Bache and the U.S. Coast Survey* (Cambridge: Cambridge Univ. Press, 1994), 79.

61 See chap. 5 for details concerning this episode.

Weights and Measures since 1830.[62] In 1833, the issue of uniform weights and measures remained unresolved even though the Constitution had granted Congress the power to deal with it. The different colonies had used weights and measures that corresponded to those in England. The standards used in the American colonies, however, diverged from one another because they had been brought from England at different times. By 1833, several congressional initiatives had gone nowhere. Apart from the adoption of the troy pound for coinage in 1828, Congress had not taken any action. In 1830, the Senate again considered the issue and ordered that a comparison be made between the weights and measures in use at the principal U.S. customs houses. The Treasury Department complied with this request. Ferdinand Rudolph Hassler, who assumed the duties of superintendent of weights and measures the same year, in March 1831 transmitted his report to the Senate. Issues of policy had been on the backburner for years. It had been the immediate problem of providing standards for use in all areas of trade and agriculture that had prompted the Senate to act. Hassler confirmed that there existed large discrepancies among the weights and measures used in the various ports.[63]

For Hassler, the English yard was the real standard of length in the United States. In his view, if copies available in the United States could be compared to the original in England, and if deficiencies could be found, they had to be replaced. Hassler deduced the standard weight of an avoirdupois pound from the troy pound in the Philadelphia mint (1 avoirdupois pound equals 7,000/5,700 pounds troy), thus accepting the English standard. The units of capacity adopted by Hassler were the wine gallon of 231 cubic inches and the Winchester bushel of 2,150.42 cubic inches. Both of these capacity standards had been abandoned in England but Hassler stuck to them because they were in common use in the United States. Hassler's recommendations would later be adopted by Congress. While the congressional resolution "does not specifically adopt the standards described above," one historian has observed, "its practical effect was to

62 Florian Cajori, *The Chequered Career of Ferdinand Rudolph Hassler* (1929; repr., New York: Arno Press, 1980), 153 ff. For Hassler's work as superintendent of weights and measures, see "United States Standards of Weights and Measures: Their Creation and Creators," Arthur H. Frazier, *Smithsonian Studies in History and Technology* 40 (1978), http://www.sil.si.edu/smithsoniancontributions/HistoryTechnology/sc_RecordSingle.cfm?filename=SSHT-0040; Lewis V. Judson, *Weights and Measures Standards of the United States: A Brief History* (Washington: U.S. Department of Commerce, 1963), 2–6.

63 On the history of weights and measures in the emerging U.S., see ibid., 2 ff.

make them the standards for the United States, inasmuch as the weights and measures distributed to the States in accordance with the act were in almost every case adopted by the State legislatures soon after their receipt."[64] All states received copies of these standards. When Bache wrote to his uncle in 1833, these developments could perhaps be foreseen. Hassler's report, while responding to the immediate need of providing uniform weights and measures, could not do so without adopting units, and even though Congress had perhaps not intended to act on this question in principal, it put into effect a train of events that would lead to such a decision. A few months after Bache had written the following letter, the Franklin Institute addressed the matter as well. The Pennsylvania legislature had asked the Institute to advise it on a bill relating to weights and measures in the state. Bache became part of a nineteen-member committee to discuss the proposals in the light of developments in England, France, and in the American states. Bache took on the responsibility of reporting on the latter.[65]

Bache's letter begins as follows:

Philad. Jany. 27th, 1833

My dear Uncle,

I thank you for the report of Mr. Haßler on weights & measures which I have studied with attention.[66]

Bache's handwriting is unusually clear and legible. This provides Bache's letter with an official note.

As a member of the U.S. Senate, Dallas could more easily access Hassler's report than Bache could in Philadelphia. Bache either requested the report because he was interested in it or his uncle had sent it for Bache's evaluation. The latter is more likely because Bache stresses that he read it

---

64 Ibid., 8.

65 "Quarterly Report of the Board of Managers," *Journal of the Franklin Institute* 12, no. 3 (September 1833): 159; Bache's report on the U.S. is in Alexander Dallas Bache, "Report of the Committee on Weights and Measures. Appendix to the Report of the Committee of the Franklin Institute on Weights and Measures. Abstract of the reports on Weights and Measures which have been submitted to the Congress of the United States, or to the Legislature of Pennsylvania," *Journal of the Franklin Institute* 8, no. 4 (April 1834): 232–47.

66 ADB to George Mifflin Dallas, January 27, 1833, George Mifflin Dallas Papers, (Phi)1460, Historical Society of Pennsylvania. Reprinted with permission by the Historical Society of Pennsylvania (HSP). Concerning formatting conventions used in transcribing this letter, see chap. 3, 58 n. 23 above.

"with attention." If his own interest had prompted him to ask his uncle for the report, he would have not reported back in this way. If he would then have proceeded to write more about the report, he could not have taken for granted that his uncle was interested in what he was writing, and a likely response would then have been to make a case for why it is interesting or of importance. But Bache writes back as though he had been asked to read it. Bache does not write about family matters but immediately addresses such official concerns. That he can do so without even hinting at more personal subjects suggests that he can assume that his uncle also views the two areas to be intertwined. Or Bache was writing his letter in a way that would make it possible for his uncle to share it with someone else. Both possibilities would harmonize with Bache's neat handwriting.

The experimental part of his labors seems to have been conducted with a minuteness + precision worthy of his reputation.

At twenty-seven, Bache is highly confident about his role as an emerging scientist and as an evaluator of senior colleagues in the federal administration. He considers Hassler's report not merely in terms of a scientific paper that anyone may criticize, but as indicative of the author's research abilities.

As superintendent of Weights and Measures, Hassler was merely doing his job by writing his report. Bache, however, does not refer to Hassler's role as superintendent and federal official, but as a colleague among scientific peers. Whatever he has to say about Hassler, it would, in a sense, be stronger if Bache invoked Hassler's responsibility as superintendent. This was not just any scientific paper but a report preparing federal political action on weights and measures, and if Hassler failed to discuss his subject properly, Bache could infer that he was perhaps not the right man for the job. Bache, however, speaks of Hassler as a colleague. He focuses on the "experimental part" of Hassler's report and he restricts his evaluation to the latter's "minuteness + precision." Is Bache inferring that Hassler is known to be exact in his calculations, but that his work is of limited quality in other respects?

In relation to the conclusions to which he has come I have felt both regret + mortification: it seems to me that if adopted they will go further to depreciate the science of our country than an ill conducted experimental inquiry could possibly have done.

Bache indeed takes issue with Hassler's report. He criticizes the report's conclusions rather than its calculations. That Bache felt "regret" shows that he anticipated more ambitious suggestions and was sorry that Hassler, thirty-six years his senior, has not lived up to this potential; "mortification" is even stronger and points to feelings, not of scientific appropriateness, of a lapse of reason and judgment, but to feelings of honor and dignity. These were not matters of strategic calculation and practical judgment, but of convictions and of pride.

Bache speaks of the possibility of an "ill conducted experimental inquiry" which suggests that there are inquiries that are not conducted experimentally. If such non-experimental inquiries go wrong, they are not perceived to harm science in the United States and Bache takes for granted that there would be critics to point to the mistakes. Bache also assumes that anyone engaged in experimental inquiries has a responsibility to get it right so as to protect American science, which implies that such an approach remained vulnerable in the United States. Bache considered himself to belong to an avant-garde, a "first generation" consciously aware of charting into the future a path that others would later follow.

What was the source of Bache's feeling of wounded pride? While it is possible that Bache felt mortified as a scientist, it is more likely that he felt mortified as a citizen dedicated to advancing science in his country. Bache had had particular hopes with respect to the development of science in his country; Hassler's report suggests to him that those in charge of shaping American science policy had much lower expectations and much humbler ambitions than he did himself.

> You will probably agree with me that a national system of weights + measures should rest upon ~~a national~~ independent scientific principles + not be wholly borrowed from abroad.

Bache's intention and ambition clearly point in the direction of an independent, national scientific culture. This is the premise for his judgment as he takes the view that American independence needs to be realized in all areas of life, including those affected and shaped by science. That this process was anything but complete in the eyes of Bache is an obvious implication here. It is interesting that he had started out to write about "national" principles, and then corrected himself to speak of "independent scientific principles" instead. He not only avoided repetition of the adjective but highlights his preoccupation with establishing a uniquely American cultural and scientific life.

But what are "*independent* scientific principles" (my emphasis)? To the extent that principles are scientific they will be universal, but Bache insists that there are principles that are both universal and different from European ones. Why does he not confine himself to arguing that weights and measures should be designed so as to be logical, coherent, and, if possible, established by tradition? The curious reader is left with the impression that American independence and cultural particularity was legitimized by scientific universality.

The last part of Bache's sentence ("+ not be wholly borrowed from abroad") seems superfluous, for the point that the principles ought to be self-sufficient had already been made. The use of the verb "to borrow" highlights Bache's intention to be self-sustaining, without using something that really belongs to others and has not been developed independently. He concedes that total abstinence in this respect is impossible ("wholly").

> The weights now in use as well as the measures we derived from the mother country originally, and the attempt of the French nation to change the current weights and its failure are a warning to others not to depart too far from the system in common use.

Bache refers to American standards in use in 1833: the yard, the bushel, the wine gallon, and the (avoirdupois) pound. (As mentioned above, the basis for the latter was the troy pound, a standard retained by the mint in Philadelphia and originally adopted as a standard for coinage.) These weights and measures were no longer in use in the "mother country" which had introduced, in 1824, the imperial gallon (10 pounds of water) and the bushel (eight gallons). France had tried to move to a metric system, but had reversed course, and the old and new systems were now used side by side there, with obvious problems in everyday life.[67] Even though the French system was perhaps more universalistic and certainly more logical than the English one, Bache argues here that the French failure to install that system should be a warning. He implicitly suggests that the United States stick closer to tradition and common practice. What, then, did Bache mean when he spoke of "scientific principles" above? Apparently he did not mean the rather logical metric system. Perhaps he implied that the United States should gain independence in these matters by adopting weights and measures that could be established without reference to Euro-

67 Judson, *Weights and Measures*, 3, 5.

pean standards (such as the troy pound), but through experiments? It is indeed such a solution that Bache had in mind:

> The object of a system is rather to produce uniformity in all parts of a country than to introduce novelty. In a practical point of view if a standard of length now adopted + multiplied so that it might be distributed all over the country; yard measures for example made exactly alike + acceßible to every one; if further, weights uniformly in their character should be likewise so distributed, + measures of capacity, the problem would be solved so far as new, [at large] in a busineß point of view, are concerned. In a scientific point of view the case is different, some unit of measure is to be provided by experiment, which can be recovered, if lost, by a similar experiment; the weights + measures of [—] capacity are to bear a determinate relation to the measure of length so that they, if lost can be recovered or if deterioration in the standard should take place, it may be ~~verified~~ discovered + corrected.

Bache distinguishes between a business and a scientific perspective, and he is concerned with the latter: establishing proper principles and basic coordinates. He implies that a "scientific point of view" is not practical, though it seems relevant to be able to deduce proper standards from experiment. Bache's use of the term "practical" entails the immediate application of principles only, not their underlying logic and recovery. It is the latter dimension that Bache stands for. Bache's reservations with respect to Hassler's suggestions become clear against the background of contemporary developments.

The metric system, introduced in France during the Revolution, had been designed as a decimal system. The older French system provided no such convenience for calculation, nor did the English system of weights and measures, even if it was standardized. At the time the French Academy decided on the metric system, the seconds pendulum had been used for establishing units, and at first, it seemed not unlikely that the Academy would continue to rely on it, but it chose to go with a length based unit instead: a part of the distance between the North Pole and the equator on a line close to Dunkirk and Barcelona. (Perhaps the Academy chose to base the meter on a measurement of the earth's meridian because it approximated the length of a seconds pendulum.[68]) Thus, in 1833, there existed two ways of deducing a length unit (and others from it).

68 Giulio D'Agostini, "The Earth based units of length and the birth of the metric system," http://www.roma1.infn.it/~dagos/history/sm/node4.html.

But in his report to which Bache responded in his letter to his uncle, Hassler had chosen neither. The Superintendent of Weights and Measures opted for English standards as standards for the United States—standards derived from experiments with the seconds pendulum in England, not in America. Readings differed according to the place where the pendulum oscillated because of variations in the earth's field of gravity. It also depended on other variables, including temperature. Hassler's preference implied that American standards could not be deduced independently. The metric system could be criticized for putting a somewhat random measurement (of the earth's meridian, not the equator, or other natural entities) at the center of the system. Because the meridian chosen by the French Academy ran through France, the United States and other countries could not deduce their standards independently, on their own territory. But the metric system, as Bache infers in his letter, had run into difficulties. Other systems offered the opportunity of deducing a standard measure from a seconds pendulum, thus offering an advantage in terms of national and cultural independence. Bache preferred the English to the French one, in other words, *but he wanted that system to be rooted in experiment* (i.e. the seconds pendulum).[69] This is what he implied when he wrote that it should be possible to "recover" a measure of length. In this way, the rationality of the French system would be replaced by an adherence to tradition, but this tradition would be put onto a new experimental (and thus both national and universalistic) footing: national, because the site of the experiment would need to be fixed by law; and universalistic, because anyone could perform the experiment to deduce the standard measure.

Bache proceeds to describe the English system:

> The English took the pendulum vibrating seconds to [~~then~~] furnish their measure of length. The length of the seconds pendulum in the Latitude of London + vibrating in a [— —] in a vacuum was ~~passed~~ determined by experiment + calculation + compared with the existing standard measure, so that thereafter the ratio of the standard yard to the pendulum was fixed by law. This yard being divided into thirty six parts, gave the legal inch, the ascertained weight of a cubic inch of water at 62° Fah. enabled them to establish the weight of the standard pound with reference to the weight of this cubic inch of distilled water. The gallon + bushel were in like manner declared to contain so many cubic inches, + might therefore be determined by the weight of water which they would contain. Observe then that

69 Paolo Agnoli and Giulio D'Agostini, "Why does the meter beat the second?," *physics/0412078* (December 14, 2004), http://arxiv.org/abs/physics/0412078.

> as little disturbance as possible was effected in existing weights + measures, while the whole system was placed upon a fixed basis.

All this is in line with our interpretation that Bache was looking for a combination of traditional measures that could be recreated from experiment.

In the following paragraph, Bache explains with what changes the English system was adopted in New York State, and he then contrasts these efforts with Hassler's proposals—to adopt, for the standard of length, "the truss yard in the possession of the [State?] Department [—] constructed by Troughton and conforming nearly to the British Parliamentary Standard." And for the standard of weight, "a pound weight (troy) made for the U.S. mint, in 1824, by Capt [Rater] of the British Navy, being a copy of the standard pound of Great Britain." Finally, for capacity, "the bushel containing 77.627413 lbs. Avoirdupois + the gallon 8.33888222 lbs again being the weights of water corresponding to their legal dimensions in inches." Bache concludes that this "seems to me degrading in a national point of view" and he does so for the reasons we have anticipated:

> First. That our standard of measure is not an absolute determination in our own capital.
>
> Second. That not satisfied with relying upon the British standard for ~~weig~~ measure and making an independent deduction for weight, we take their weight as well as their measure.
>
> I will not urge the objection which might recur to the standards of capacity in a scientific point of view and which Mr. Haßler was fully competent to have met and [— — —] instead of [—].

We can now see why the University of Pennsylvania professor described himself as being "mortified." Bache has an intuitive sense for the coordinates of cultural independence and self-reliance. The application of this perspective leads him to criticize Hassler for a missed opportunity in shaping the nation's cultural coordinates. What is interesting here is not only that Bache suggests that standards of measure be determined in Washington D.C. but that he then goes on to criticize Hassler's unnecessary reliance on English measures and weights. Bache sarcastically points out that Hassler seemed eager to maximize American dependence on English units instead of mobilizing scientific resources in the United States. This idea is further implied in the last sentence. Bache suggests that Hassler's problem was not that he did not understand the scientific issues; what Bache is implying is that Hassler lacks sufficient dedication to put the United States on its own cultural feet. What distinguishes Bache and Hassler, in this view, is that the latter (an émigré from Switzerland) is insufficiently American, whereas the former considers himself to be part of an intellectual avant-garde mobilizing universalistic laws of science to establish America, by way of properly informed legislation, as an autonomous and independent national entity.

Bache continues his letter as follows:

> Will you my dear Uncle inflict this upon yourself so far as to say what impression is made upon your mind by the statements. Would you oblige me by speaking with Mr. Adams in relation to the subject? to which he has given much consideration + time. Can you tell me whether any legislation is proposed on the subject by Congreß? A bill has been reported to our Senate by the commission on the laws of the commonwealth.

Had his uncle sent Hassler's report to his nephew for comments, Bache would hardly have asked his uncle to provide him with feedback. It is therefore more likely that it was Bache who had asked his uncle for the report and that he was looking for an opportunity to comment on it. This complements our impression of his interest in the development of a coherent and nationally independent application of scientific principles. Bache's interest in weights and measures falls in line with the role assumed by the Franklin Institute committee on steam boiler explosions and, in a slightly different way, in his perception of the army as a component of, or a model for, a national research organization. In retrospect, Bache, in the way he opened his letter, turned his uncle's favor of sending Hassler's report to Philadelphia into a token of his relative's interest in his nephew's opinion. It is in line with this interpretation that Bache asks his uncle to contact John Quincy Adams (1767–1848) in this matter. Adams, sixth president of the United States (1825–1829) had a marked interest in cultural and scientific matters. In 1821, as Secretary of State, he had submitted his own proposals for an American system of weights and measures in which he suggested the adoption of either the English Imperial system or the French metric system.[70] In 1830, Adams returned to Washington from Massachusetts when he was elected to the House of Representative. In referring his uncle to Adams, Bache sought to connect to a potential ally in federal politics. What are his motives? On the basis of this letter, there is evidence for his interest in a policy on weights and measures that corre-

70 United States, Department of State [John Quincy Adams], *Report of the Secretary of State, upon Weights and Measures, Prepared in Obedience to a Resolution of the Senate of the Third March, 1817* (Washington: Printed by Gales and Seaton, 1821).

sponds to his deep convictions of cultural sovereignty and scientific independence. Accordingly, any bill relating to weights and measures is of interest to him, and apparently, Bache cannot rely on the press to report such developments and refers to his family connections instead.

Bache's last sentence here throws an interesting light on his own perception of national politics. By speaking of "our Senate" he cannot have in mind the U.S. Senate because he had just implied that his uncle was informed about goings-on in the U.S. Congress. Bache is referring to the Senate of the state of Pennsylvania. He could have referred to it as "the Senate here" or "the state senate," but he instead highlights its proximity to himself and his uncle ("our"). Even though Bache's family, including his uncle, had assumed prominent roles in federal politics, this remark indicates that Pennsylvania remained the natural locus of political identification. As we have seen, Bache was emerging as a protagonist of American national consolidation. That even Bache would remain attached to Pennsylvania in this way provides a glimpse of how new and distant the United States appeared to the country's citizens.

Mr. Tyler returned exceedingly grateful for your kindneß to him, which I feel almost as much as he does himself. The aspect of his case was entirely altered according to latest news.

It would be useleß to say how they do at your house to you. Julia + Eliz. were here this afternoon. All well with us + your Casey better of her rash.

Very truly yours
A. D. Bache

Bache seems to refer to a client whom his uncle, a lawyer, was able to help. There are no details available here except that Bache knew Tyler. Bache then closed his letter by suggesting that his uncle was in close touch with his wife and children, thus acknowledging the fact that Dallas spent much of his time in Washington and away from his family. The closeness of his uncle's family to Bache and his wife Ency is suggested by the visit of two of Dallas' daughters, Julia and Elizabeth.[71] Bache hardly mentions Ency and himself at all; they appear rather to be part of the larger family.

71 Julia later published her father's letters. That Elizabeth was Dallas' daughter could not be ascertained and is merely suggested by Bache mentioning her alongside Julia. His uncle had eight children. Julia is mentioned by John M. Belohlavek, *George Mifflin Dallas: Jacksonian Patrician* (University Park: Pennsylvania State Univ. Press), 186.

As mentioned above, the Franklin Institute took up the issue of weights and measures a few months after Bache had written to his uncle.[72] The Pennsylvania House of Representatives referred a bill to the Institute for evaluation, and Bache became a member of a committee to deal with it.[73] Its report, published in the *Journal of the Franklin Institute* in April 1834, was split into three sections: Reports on systems of weights and measures in France and England were complemented by a report on the state of affairs in the United States, written by Bache.[74] In his contribution, Bache rehearsed the work that had been done by prominent politicians such as Thomas Jefferson and John Quincy Adams and he surveyed the solutions adopted by the various American states. He highlighted the quality of the report made by John Quincy Adams in 1821. "The course pursued by Mr. Adams in his general investigation," Bache wrote, "is bold and ingenious, pursued more in the manner, and with the views, of a legislator, a statesman, and a lawyer, than of {a} mere man of science."[75] He commented on the scope of Adams' investigation, his review of the problem "from the Hebrews and Egyptians, to the Greeks and Romans and, with modifications, through the complex, varying, and entangled system of the older English to the present standards." Bache was impressed by a consideration of policy issues against a broad historical canvas. His underlying ambition, we can infer from his insistence on the continued relevance of Adams' paper, was for the United States to aspire to similarly lasting historic significance. In his overall assessment, Bache continued to favor a pragmatic solution that combined available models with American mores. In reference to a report by Professor James Renwick, which had guided legislation on weights and measures in New York State, Bache again pointed to the idea of preserving existing standards and "to insure the recovery of the standard if lost, or its verification if required" by fixing the ratio of the

72 Sinclair, *Philadelphia's Philosopher Mechanics*, 192 f.

73 "Quarterly Report of the Board of Managers, 1833," 159. Other members of the committee: Samuel V. Merrick, William H. Keating, Mathias W. Baldwin, Sears C. Walker, Rufus Tyler.

74 Sears Walker, "Appendix to the report of the committee of the Franklin Institute on weights and measures," *Journal of the Franklin Institute* 17, no. 2 (February 1834): 94–109; T. McEwen, "Report giving an account of the system of weights and measures of France," *Journal of the Franklin Institute* 17, no. 3 (March 1834): 160–71; Alexander Dallas Bache, "Abstract of the reports on Weights and Measures which have been submitted to the Congress of the United States, or to the Legislature of Pennsylvania."

75 Ibid., 235.

standard of linear measure to a pendulum "with a brass rod vibrating seconds, in vacuo, at the temperature of melting ice, at the level of the sea, and in some given latitude and place." He also proposed that the standard of weight be derived from that of the standard for length.[76] While Bache did not explicitly comment on Renwick's solution, his support of the concept seems clear enough as he ends his own report with it.

Perhaps realistically, Bache pointed out that national action was not yet possible, even though he considered such action most adequate, and that, for the time being, the states needed to move forward independently. This was not in line, or in spirit, with what Bache had written to his uncle, but this is the assessment he provided in his report.[77] His more ambitious national perspective was shared by others on the committee. Bache had aligned with John Quincy Adams to "consult with foreign nations, for the future and ultimate establishment of universal and permanent uniformity."[78] In its final report, the committee of which Bache was a member referred to the same principle when it suggested that national solutions should precede solutions by individual states. It found this point very important, in fact, and gave three out of nine pages of text to it. "So impressed are the Committee with this view," Bache and his colleagues argued, "that … the most imperfect system of weights and measures which has ever been framed, would if applied in all the states of our union, be preferable to the most perfect system which should be adopted by any one commonwealth singly."[79] Just like John Quincy Adams in his report, the committee found it imperative that standards be fixed that would be universal both geographically and temporally. Unlike Adams, they did not assume the national perspective from which international cooperation appeared possible. They assumed a bottom-up view of the problem, urging national instead of either international or local state action. The committee recommended urging the State of Pennsylvania to become a proponent of coordinating action in this matter among the states, and to seek a federal implementation of standards. Most states anticipated national legal action anyway, the committee argued, hence the "exceeding importance of uni-

76 Ibid., 241.

77 Ibid.

78 Ibid., 238.

79 "Report in Relation to Weights and Measures in the Commonwealth of Pennsylvania," *Journal of the Franklin Institute* 14, no. 1 (July 1834): 13.

formity" should be pursued with rigor. Only in case such an attempt failed should the state act by itself.[80]

Instead of merely discussing the bill it had received, the committee reported to the Pennsylvania House of Representatives a revised version. As the state could apparently not foresee federal action and declined to ask for it, the committee's alternate plan was put into law in 1834. The Franklin Institute then became an agent of the state government by procuring standards pro bono.[81]

The Debate on Meteor Showers

Both the 1836 report on steam boiler explosions and the 1833 letter on weights and measures have shown that Bache considered science to be intrinsic to national development, that he took himself to be a leader in creating a national basis for cultural development, and that he viewed regional scientific institutions as microcosms of a universal scientific discourse. Under his chairmanship, both the Franklin Institute's boiler experiments and its report on weights and measures concluded with proposals for federal political action. A discussion of Bache's involvement in the 1834 debate on meteor showers provides further evidence for his organizational rationale.

The discussion in which Bache chose to participate took place in the wake of significant meteor showers in 1833. In the early morning of November 13, thousands of meteors could be seen against the clear moonless sky in North America.[82] The number of meteors was much higher than on a common cloudless night, and it was widely noticed and commented on in newspapers all over the country. Meteors had been seen before, of course, but not in such large numbers during anyone's lifetime. Today, it is known that meteor showers are caused by meteoroid particles burning up in the earth's atmosphere. Meteor streams are remnants of comets and every time such particles meet with an obstacle (such as the earth) they are scattered further. Over a period of time, a comet will thus be diffused along its path. It seems likely that all meteors were once part of such a stream. The intensity of a meteor shower will therefore relate to the "age" of the re-

80 Ibid., 12–14.

81 Sinclair, *Philadelphia's Philosopher Mechanics*, 193.

82 While meteor showers could be seen most clearly in the early morning of November 13, they were perhaps also visible very late at night on November 12, 1833.

specttive meteor stream or comet, and to their size, when their orbit intersects with that of the earth. Meteor showers tend to occur periodically, and this is the case with the Leonids, a relatively young stream, that caused the great meteor shower in the early morning of November 13, 1833. The origin of meteors or meteor showers, however, was not known then even if individual advances in understanding their origin had been made. They were largely perceived to be a meteorological phenomenon, and not to even have an astronomical explanation.[83]

The impressive sight of more than one thousand meteors per minute, clearly visible across North America, prompted Denison Olmsted, professor of mathematics and natural philosophy at Yale College, to collect accounts of the meteor showers and to publish them with his own conclusions in two 1834 issues of the *American Journal of Science and Arts*.[84] His article is today considered a major advance in the study of meteor showers for Olmsted introduced the idea that the origins of meteors may be astronomical, and that they recur periodically.

In the first, and most substantial, section of his article, Olmsted relayed reports of meteor shower sightings that were sent to him by observers from around the country. Some had been published in newspapers. Olmstead wrote that he presented excerpts from a geographically diverse sample, "from east to west and from north to south."[85] His idea was to collect and classify these reports so as to draw conclusions on the nature and origin of the meteors.[86] In his essay, he quoted from different accounts of meteor sightings.[87] In his own "Synopsis of the Facts," Olmsted summa-

83 See Littmann, *The Heavens on Fire*, chap. 3, 35–52. See also Joe Rao, "The Leonids: The Lion King of Meteor Showers," *WGN, Journal of the International Meteor Organization* 23 (August 1, 1995): 120–35.

84 Denison Olmsted, "Observations on the Meteors of November 13th, 1833," *American Journal of Science and Arts* 25, no. 2 (January 2, 1834): 363–411, continued under the same title, 26, no. 1 (April 1834), 132–74. Olmsted dated the storms November 13, i.e. to after midnight.

85 Ibid., 152.

86 Olmsted does so in his own initial account he published the day the meteor showers occurred, November 13, 1833, and which he reprints, with slight alterations, in ibid., 135.

87 Among them were Alexander C. Twining, civil engineer at West Point, Rev. Dr. Humphreys, President of St. John's College, W. E. Aikin, M.D., professor of chemistry and natural philosophy in Mount St. Mary's College whose account had been published in a Maryland paper, one F. G. Smith as quoted in a Lynchburg paper, Ashbel Smith, M.D., and James N. Palmer, "Practical Surveyor &c." Olmsted also quotes anonymous

rized these accounts and noted their statements on weather, time and duration, meteor varieties, etc. In this first paper he abstained from presenting his own conclusions. He noted that on the evening before the meteor showers took place, a sudden change of weather had occurred. From observations in Boston, he deduced that the total number of meteors must have been over two hundred thousand. He points out that according to most observers, the meteors fell silently.[88] He also noted that most reports located the meteor shower's radiant, the point in the sky from which the meteors seemed to emerge, in the constellation Leo.[89] Alexander C. Twining of West Point had collected accounts of the meteor showers from vessels having recently entered New York harbor. With reference to these observations, Olmsted suggested that "the phenomenon at a given stage, as at the maximum, for example, appeared to places differing in longitude 10, 20, 30, or 40 degrees, *at the same hour of the night*." He pointed out that this would have "an important bearing on the question, whether the origin of the meteors was terrestrial or astronomical."[90]

In his second paper, published in April 1834, Olmsted used these observations to develop a new theory of the causes of meteor showers. This theory had significant implications as Olmsted suggested that the November meteor showers were caused by a comet circling the sun, a comet whose path intersected with the earth's so that the two bodies meet at regular intervals. He rejected the idea that meteor showers were part of the weather and offered an astronomical explanation. Olmsted used the recorded positions of prominent meteors simultaneously observed in different locations, to triangulate their approximate height at 2263.5 miles. He sought to deduce the characteristics of meteors as precisely as he could, inferring that they consisted of light and transparent material that was highly combustible when encountering at high speed the density of the earth's atmosphere.

He did not stop there, however, and proceeded to explain why the meteor's radiant, despite the earth's rotation, remained in the same location of Leo for several hours, and why similar occurrences had been observed in that position several times in history. He deduced that the comet respon-

writers in the *Columbian Centinel*, in the Frederick Citizen, the *Ohio State Journal*, the *Georgia Courier*, and the *Salt River Journal* (ibid., 135–50).

88 Olmsted, "Observations on the Meteors of November 13th, 1833," 157–61.

89 Ibid., 163.

90 Ibid., 172.

sible for the meteor showers traveled around the sun, and, because the meteor shower's radiant seemed to remain in the same position in the sky and drift westward with it, that "*the body revolves around the sun in an elliptical orbit, but little inclined to the plane of the ecliptic, and having its aphelion near to the orbit of the earth*."[91] The comet, in other words, traveled around the sun on an elliptic path, and at its furthest point away from it (aphelion), its path was almost parallel to that of the earth at the very moment the comet was closest to it. To an observer on earth, the shower's radiant thus seemed to remain stationary for some time rather than moving across the sky. Finally, Olmsted concluded that because the November meteor showers had been observed in previous years, the period of the comet's elliptical rotation around the sun could be expressed in whole years or aliquot parts of it, and that the comet's and the earth's path intersected at regular intervals. Olmsted argued that the comet's period was six months, for otherwise, according to Kepler's observations, "the line of the apsides would not be long enough to reach the earth."[92] Hence he predicted that similar meteor showers could be observed in November 1834 as well, one year after the great meteor showers he and so many others had witnessed and which had prompted him to develop his theory.

Many aspects of Olmsted's theory have since been revised. Olmsted conceived of the comet encountered by the earth to be relatively compact, and he thus assumed that it had to travel in the direction of the earth, for otherwise the meteor showers would have been brief. He did not imagine a comet to consist of nebulous matter spread out over millions of kilometers.[93]

It was Alexander Dallas Bache's colleague and friend James P. Espy who was most immediately concerned with Olmsted's theory. Espy's evolving research was affected because Olmsted plucked the phenomenon of "shooting stars" from the emerging field of meteorology and put it squarely into astronomy.[94] A year after the large meteor showers of 1833, James P. Espy published a scathing critique of Olmsted's model in the January issue of the *Journal of the Franklin Institute*. Espy's language was stiff indeed. He repeatedly asserted that "I cannot give my assent to the cometic

---

91 Ibid., 165 f.

92 Ibid., 166.

93 Littmann, *The Heavens on Fire*, 25.

94 James Roger Fleming, *Meteorology in America, 1800–1870* (Baltimore: Johns Hopkins Univ. Press, 1990), 56.

theory."[95] In the first part of his paper, Espy concentrated on weather-related arguments. Olmsted had suggested that the great number of meteors affected, during their descent, not only the temperature, but the direction of the wind. "The first effect was the *westerly wind*, which suddenly succeeded the meteoric shower."[96] He thus proposed that the weather was affected by the meteor shower, not the other way around. Espy countered that observers had agreed that the westerly wind had not succeeded but preceded the meteor showers and that it could therefore not have been caused by it. He also argued that if the comet's path was interior to the earth's (relative to the sun), the meteor stream could not have been seen at night, for it would have been confined to that side of the earth that faced the sun. (It would later turn out that the comet's orbit was indeed exterior to the earth, but this did not disprove and merely adjusted Olmsted's theory.) According to Espy, such flaws were "fatal."

Bache's role in this debate was limited and closely associated with Espy's position. The two Franklin Institute comrades shared many interests, among them their research on steam boiler explosions for which Bache, after his teaching duties at the university, spent three nights each week. Espy's tart reply to Olmsted's theory in January 1835 had been preceded by two essays from Bache's hand. In 1834, the public with excitement had looked forward to the predicted recurrence of meteor showers in the morning of November 13, twelve months after the initial event that had prompted Olmsted's work. Even though Olmsted had suggested that his theory did not depend on the recurrence of meteor showers, it remained a significant test. Bache prepared the ground for Espy's refutation of Olmsted's ideas by stating that no significant meteor showers had occurred. In a brief paper of three and a half pages, he reported his observations.[97] He had read Olmsted's account, Bache explained.

On Saturday the 5th inst. my notice was drawn to a paragraph which I supposed to be from the pen of our mutual friend Professor Olmsted, calling attention to the Zodiacal light then visible for some hours before sunrise, and suggesting a query in

95 James P. Espy, "Remarks on Professor Olmsted's Theory of the Meteoric Phenomenon of November 12th, 1833, denominating Shooting Stars, with some Queries towards forming a just Theory," *Journal of the Franklin Institute* 15, no. 1 (January 1835): 9–19. The article was continued in another issue of the *Journal*.

96 Olmsted, "Observations on the Meteors of November 13th, 1833," 160.

97 Alexander Dallas Bache, "Meteoric observations made on and about the 13th of November, 1834," *American Journal of Science and Arts* 27, no. 2 (January 1835): 335–38; republished in *Journal of the Franklin Institute* (September 1835), 149–53

regard to its connexion with 'Falling stars' and to a change in its appearance on or about the 13th of November. This induced me at once to commence a series of observations, which were continued until the 19th inst. and on the morning of the 13th, the number of observations and their duration was increased.

Having witnessed the remarkable meteoric phenomenon of the 13th of November, 1833, and having been engaged during the summer, in conjunction with my friend Mr. Espy, in observing meteors, I felt competent, as far as experience could render me so, to the task which I had undertaken.

Bache, at the bottom of his paper, dates his letter November 21, 1834, but the above paragraph likely refers to July 5, 1834: After Olmsted's paper had been published in July, this was the only fifth of a month that year that was also a Saturday.[98] Bache reports that he and Espy observed meteors all summer. They read Olmsted's paper in July, became aware of its implications for Espy's work, and commenced their observations until after the meteor shower's November date. They did not limit their observations to November because a strong meteor showing in advance of that date would have weakened Olmsted's theory. They were also taking up Olmsted's suggestion, of course, to study the occurrence of zodiacal light in connection with meteor showers. The Yale professor had mentioned zodiacal light in connection with his discussion of why the cometary body causing meteoric showers returned at intervals and why it was seen at a particular and seemingly stationary place in the sky. Because of the latter, Olmsted had proposed that the comet circled the sun and not the earth, and that its elliptical orbit, while it was close to the earth, was nearly parallel to its orbit, so that it could seem to remain in the same position of the sky for two hours. He had gone on to argue that the comet's periodical time was close to six months, i.e. that during this period, it traveled once around the sun and the earth in an elliptical path.[99] Could such a comet, apart from the meteors it shed, be seen in the sky? Olmsted proposed that this may indeed be the case, and referred to zodiacal light: Like a comet for the path that he described, zodiacal light would "disappear by or before the first of

98 Olmsted's second part of his paper on "Observations on the Meteors of November 13th" was published in the *American Journal of Arts and Science* in April 1834; however, as Mark Littmann points out, the April issue was part of the journal's July volume. Bache, therefore, received the *Journal* containing Olmsted's essay in July or later. Littmann, *The Heavens on Fire*, 31. For a calender of 1834, see Microsoft Outlook.

99 Olmsted turned out to be wrong in assuming that the comet, when close to the earth, was at its aphelion, that it traveled around the sun, and that it traveled in the same direction as the earth.

May." He wrote that "after the month of May, if seen at all, it will appear on the western side of the sun and rise before him, until the month of August, when it may possibly reappear for a little while in the evening sky."[100] In mid-northern latitudes, zodiacal light "is best seen in the evening in February and March and in the morning in September and October."[101] Olmsted also pointed out that his argument on comets was independent of the occurrence of this phenomenon. That Bache and Espy initiated their observations during the summer suggests that they were taking seriously Olmsted's idea that zodiacal light may be seen before the predicted November meteor showers.

In his paper, Bache moves on to report the essence of his observations, the absence of meteor sightings in November:

> The conclusions to which my observations have led, and in which I feel entire confidence, are, that *at the city of Philadelphia there occurred on the 13th of November, 1834, no remarkable display of meteors of the kind witnessed in 1833, and that there was probably no similar occurrence on those mornings which were clear, just before and after the 13th inst.*[102]

Bache thus challenges Olmsted's theory, for he denies that any significant meteor activity could be observed during the night of November 12/13.

In the following sections of his paper not quoted here, Bache lists the results of his observations between November 9 and 13 in which he focused on a possible recurrence of the meteor showers. Bache, for every observation entry, records the direction of the wind and the temperature, which suggests that he considered the occurrence of meteors not to be independent of the weather. His observations usually began at 3 A.M., except for November 13, when he was at his post at midnight. He reports in his paper one meteor in the nights prior to November 13 (after 3 A.M.), and for the night of the 13th, some meteors after 2.40 A.M. For late in the night, after 5.15 A.M., Bache reports five "faint meteors in half a minute, and then very rare.—Three after those five in about fifteen minutes." This made him think that "an unusual meteoric display might be about to commence or had commenced," but that the radiant of subsequent meteors did not match the initial meteors' radiant, and that they could therefore not be part of the same meteor shower, if indeed Olmsted's theory was to be used. He concludes that "the only remarkable occurrence of meteors is

---

100 Olmsted, "Observations on the Meteors of November 13th, 1833," 172.

101 *Encyclopedia Britannica, 2005 Ultimate Reference Suite DVD*, s.v. "Zodiacal light."

102 The italics are Bache's.

that noted between 5h 15' A. M. and 5h 30'. But this was neither *in degree, nor in kind like a portion of the meteoric phenomenon noticed in November*, 1833."[103] And: "*These meteors were similar both in degree and kind to ordinary meteors.*" Bache had in view that Olmsted's theory was restricted to explaining the meteor showers of November 13, 1833, not meteors in general, for he differentiates between "ordinary" meteors and others that have a common radiant source outside the atmosphere.

Bache then compares his observations in November with those during the summer, "made by Mr. Espy and myself." While the two Philadelphians had restricted their field of observation to one fifth part of the sky in the summer and opened this field to include at least one third in November, the number was not much higher then. Today, astronomers would consider "eight meteors in fifteen minutes," as reported by Bache, a strong showing, whereas Bache measured meteoric frequency against the previous year's outstanding display.

What we can see, then, is that Bache's essay is what he announced it would be—a field report. His paper on meteors is not a deductive, carefully laid-out paper intended to make a contribution to the progress of theory. Bache's scope was limited to testing Olmsted's theory against new data, and he merely reports his observations and draws no conclusions from them. His comments indicate that he was engaging in a field in which he was not at home; it was Espy who had a stake in the discussion and who would attack Olmsted's theory in the same volume of the *Journal of Science and Arts* in which Bache's paper was printed.

Does this imply, however, that Bache was merely preparing the ground for his friend? To some extent this was certainly the case, for Espy's criticism of Olmsted's theory was confirmed by Bache. But Bache was not Espy; his focus had been on gathering accurate data so as to evaluate Olmsted's theory, and this implies that he was taking Olmsted seriously. And even if the general thrust of Olmsted's theory proved correct in the long run, Bache's insistence that nothing could be seen that night forced its proponents to explain such variations.

Through association with his friend Espy, Bache found himself on the losing end of an argument. He had engaged, not in theory-building, but in observation, and a second paper by Bache on this subject reveals that the

103 Alexander Dallas Bache, "Meteoric observations made on and about the 13th of November, 1834," 337. The emphasis is Bache's.

problem of reliable observation, concerted efforts of rational comprehension, attracted him.

Bache had closed his first paper by suggesting that it "will be interesting to have information on this subject from different quarters of our country as having a direct bearing upon the explanation of the meteoric phenomenon of last year." In a relatively short paper published in the September 1835 issue of the *Journal of the Franklin Institute*, Bache followed up on this idea. He had received replies from military posts where he had inquired about observations during the night of November 12, 1834:

> Having found that the inference drawn from my observations on the morning of the 13th of November, 1834,* at Philadelphia was directly opposite to that to which Professor Olmsted had been led, from the observations at New Haven, I felt naturally desirous to determine what might have been the extent of country over which the unusual display of meteors seen at New Haven had taken place, this extent having a direct bearing upon the question of the nature of the phenomenon.[104]

This question was similar to the one Olmsted had addressed in his first paper on the November 1833 meteor shower. His paper had contained extensive descriptions by eyewitnesses from around the country and from vessels at sea. With respect to the November 1834 meteor showers, Olmsted had published his own impressions in the *American Journal of Science and Arts* in January 1835.[105] The question Bache set out to answer in his second paper again did not concern the logic and deductive quality of Olmsted's theory. He confined himself to checking the theory against data. To be sure, Olmsted, too, was interested in understanding how many and what kinds of meteors could be seen in 1834, even if the observation of a smaller number of meteors in 1834 would do less harm to his theory than Bache might have supposed. The 1834 meteor display Olmsted had pre-

---

104 The article was originally published as: Alexander Dallas Bache, "Replies to a circular in relation of an unusual meteoric display on the 13th Nov. 1834, addressed by the Secretary of War to the Military posts of the United States, with other facts relating to the same question," *American Journal of Science* 28 (1835): 305–09. I am drawing on the reprint from the *Journal of the Franklin Institute* (September 1835), 149–53. The asterisk here refers to bibliographic information for Bache's paper in the *American Journal of Science and Arts.*

105 Denison Olmsted, "Facts respecting the Meteoric Phenomena of November 13, 1834," *American Journal of Science and Arts* 29, no. 1 (January 1835): 168–70; republished in: *Journal of the Franklin Institute* 16, no. 6 (December 1835), 367–69, where it was followed by Bache's "Meteoric Observations" (see above).

dicted did take place but was less pronounced than the previous year. There was less public attention and Olmsted could rely on fewer reports.

Bache continued:

> At my request, communicated through the kindness of the Chief Engineer, the Secretary of War, Gov. {Lewis} Cass, issued a circular to the commandants of the different military posts of the United States, requesting to be informed whether any unusual meteoric display had been witnessed at their respective posts, on the morning of the 13th of November, 1834.

One problem with Bache's idea, of course, was that the military personnel asked to report on meteoric activity had not been prepared to do so. Soldiers may not remember the meteors they saw while on duty during a particular night, unless they were aware that prominent meteor showers were expected. But for our purposes, it is significant that Bache would conceive of the national army as an instrument for scientific observation at all. Olmsted had relied, in his first paper on the meteor showers, on reports by private individuals, with an emphasis on informed observers who could identify the meteor's radiant. The army was not likely to be able to provide information as specific, but Bache conceived of it as a coherent network of nighttime observers in all areas of the continent. In doing so, Bache utilized an idea that had evolved in the context of meteorological research. There had long existed a tradition of gathering observational data on meteorology in Europe. In the early nineteenth century, college professors, the Army Medical Department, and the General Land Office had been involved in gathering information on long-term climatic phenomena in the United States.[106] In the context of evolving debates about the nature and causes of forceful phenomena such as storms, the Franklin Institute had established a committee on meteorological observations in 1831.[107] Bache and Espy were members of this committee, and in his controversy with Olmsted, Bache further develops some of its methodological ideas. By suggesting that the army provide data on meteor activity, Bache turned a branch of American government into an ad hoc research organization complete with a centralized command structure and administration.

---

106 Fleming, *Meteorology in America*, 1–22.

107 Ibid., 56 f. In September 1834, the Franklin Institute and the American Philosophical Society combined their efforts and created a joint committee on meteorological observations. Joseph Lovell, the army surgeon general, had offered to provide observations from military stations.

In "as brief a manner as possible," Bache proceeds to relay reports from military posts.[108] Bache provided a list of responses from army posts along the Atlantic coast, in the south, in the western territories, etc. In most cases, "no unusual meteoric appearances were noted." Major Churchill at Fort Johnston, Smithville, North Carolina, wrote that "no one was particularly engaged in watching for a recurrence of the meteors of 1833."[109] Other responses also indicate that the dramatic 1833 meteor showers provided a standard. This suggests that the soldiers who had been turned into meteor observers were unaware that meteor activity was thought to be imminent. The responses to Bache's inquiry merely showed that meteor activity in 1834 was less pronounced than it had been the year before, if there was any activity at all. Only Captain Clitz at Fort Mackinac, Straits of Michilimackinac, Michigan Territory, reported that "an intelligent young man, who was posted at the north angle of the fort, saw a shower of meteors in the north, between 12 and 1 o'clock, the duration of which, as near as he can collect, was about one hour."[110] Because this report remained singular, Bache concludes "against the occurrence of any extensive and remarkable display of meteors, so far as ordinary observation could have detected such a display."[111]

In this last section of his paper Bache added observations sent to him by colleagues from across the United States. In the title to his paper, Bache mentions reports from military posts only, and these observations he obviously viewed as the paper's substance. Additional observers, Bache argued, confirmed military reports that "no unusual meteoric display occurred." Those meteors that had been seen, were not "visible over an extensive region of country, like the phenomenon of November, 1833, but were local."[112] In line with Espy's interpretation of the meteor showers of 1833, Bache then subscribed, wrongly, to a weather-related or a non-astronomical explanation. He pointed out that the weather had been much more

108 Bache's brevity markedly contrasts with Olmsted's first essay in which accounts of the meteor shower were spread out over twenty pages. Olmsted, "Observations on the Meteors of November 13th, 1833," 134–56, with additional references to letters from observers elsewhere, including his second paper, published in July 1834.

109 Alexander Dallas Bache, "Replies to a circular in relation of an unusual meteoric display on the 13th Nov. 1834," 150.

110 Ibid., 151.

111 Ibid., 152.

112 Ibid., 152 f.

uniform across the country in 1833 than it was a year later. This seemed to explain the heterogeneity of observations in 1834.

Olmsted responded to Bache in a paper that was published in October 1835. He reported that on November 13, 1834, meteors had been seen in large numbers. "My respected friend, Professor Bache," Olmsted added, "has collected and published [...] a long list of testimonies of those who *did not* see the foregoing meteoric exhibition." He pointed to the details of several meteor sightings reported by vessels returning to American harbors. "Sailors," he suggested with reason, "are better observers of *celestial* phenomena than soldiers. Stars must fall thick and bright, to surprise the vigilance of a sentinel in time of peace."[113] He noted that the American Philosophical Society had recorded that "'no unusual meteoric display was seen at Philadelphia on the 13th November, 1834.'" With "the impartiality expected from learned bodies," Olmsted insisted, "they will also record the fact, that such display *was* seen at various other places, in both hemispheres."[114] This criticism was directed at Bache and at his friend Espy who were most likely responsible for the American Philosophical Society's judgment. Olmsted implied that if that organization restricted itself to recording observations made in Philadelphia, its horizon was limited indeed.

Olmsted's theory provided the basis for subsequent research on meteors and Bache's observations were eventually explained and integrated into the evolving theoretical model. But the significance of Bache's involvement in this debate is not limited to his support of Espy's position. Bache took part out of general interest in a phenomenon that was widely discussed in public. During the summer of 1834, in addition to his other responsibilities, he engaged with his friend in nightly observations of the sky. The November 13, meteor showers, furthermore, could be observed on the North American only. In line with Bache's interest in developing American research, he perhaps considered the debate on meteor showers an opportunity for a scientific discussion without European interference. Similar to another important contemporary debate in which Espy was involved, the debate on meteor showers took place between opposing camps in distinct regions: Espy and his supporters (such as Bache) in

113 Denison Olmsted, "Facts respecting the Meteoric Phenomena of November 13th, 1834," *Journal of the Franklin Institute* 16, no. 6 (December 1835): 367–69, quotation 369 [first published in the *American Journal of Science and Arts*]. The italics are Olmsted's.

114 Ibid.

Philadelphia vs. Olmsted and his supporters (such as Benjamin Peirce and Charles Henry Davis) in New England.[115] Bache would later move beyond such regional attachments, but the particular bent he gave to the presentation of his meteor observations connects to our earlier observations on Bache's work and background. In his second paper, Bache had tried to resolve the problem of how observations of large natural phenomena such as meteor showers could be controlled and thus made reliable over extended geographic areas. His perspective coincided with the national project of "cultivating" the continent, an idea that was engrained in the nation's founding, in his family's public engagement, and in West Point's mission. In keeping with his background, he had contacted the army, a corps of observers dedicated to a common ethos and representing it in all parts of the nation and of its territories. Bache's involvement in the 1834 debate culminated in his idea of turning the army, a prominent agent of the federal government, into a state-supported corps of scientific observation. Bache conceptualized citizens in even remote areas of the American continent as capable of to adhering to the rigors of observation and to contributing reliable scientific data.

Research Interests and Institutional Development: Common Denominators

Bache's scientific work fell short of the more theoretical and consequential work by Joseph Henry or Denison Olmsted. But this observation should not be used to diminish Bache's contribution to science because he was committed to resolving a different problem. Bache was easily able to stay on par with intellectual developments of his time. What our observations have repeatedly suggested, however, is that he was concerned with helping his emerging country assume a position to participate in a universalistic scientific discourse, and to connect it to the leading standards of rationality. As we have seen, this idea extended from both the Bache side of his family

115 This may help explain the violence of the debate on the nature and causes of storms (1834–1843) between Espy and his group including Bache, on the one side, and William C. Redfield and his group, on the other. For a while, Bache's endorsement of Espy was perhaps solidified by the support Espy received from French scientists (vs. support for Redfield's theories by scientists in England). Bache would distance himself from Espy in 1839 after his colleague sought support for his theories in public rather than among colleagues. Fleming, *Meteorology in America*, 23–54.

(Benjamin Franklin's legacy in politics and science), and from the Dallas side (Alexander James Dallas' early efforts at political nation-building), and Bache's efforts before 1836 are a surprisingly straightforward extension of this trajectory. A look at his institutional efforts in this period will help solidify this point.

When Bache joined the Franklin Institute, he had insufficient standing to bring about significant change. In 1831, he had supported Samuel V. Merrick's proposal that the Institute's exhibitions be moved to a biannual schedule. By effectively reducing the exhibition platform, it would have thus been possible to select from a larger sample of tools and machines and to increase the exhibition's overall quality. Merrick argued that the Institute's exhibitions had become routine, and Bache supported a policy that would provide the Institute with less exposure and better quality.[116] The proposal failed, however, which suggests that the Institute's leadership was not willing to let go the public attention attached to the event.

Three years later, Merrick and Bache put forward the same proposition, and this time it passed. What had changed was that Merrick was now chairman of the Committee of Exhibitions that Bache had joined as well.[117] That they were able to convince their colleagues, however, also reflected Bache's recent success in having his research design for the steam boiler investigation accepted by the secretary of the treasury and to secure the Institute's first research grant. In the absence of a national academy, the Franklin Institute increasingly sought to assume such a role. Bache was the decisive figure for making such ambitions possible.

Soon after he had become an active member of the Franklin Institute, Bache had joined the Committee of Instruction that was in charge of the Institute's lecture series and educational program. The Institute's monthly meetings had for some years included a period during which any question could be raised. This was conceived to provide an opportunity for self-education but it lacked focus. It is telling that such discussion was considered to fall under the purview of this Committee. As its chairman, Bache proposed in 1832 that the discussion section be replaced by separate conversation meetings during which any member could report on interesting apparatus or ideas rather than raising conversational questions ad hoc.[118]

116 Ibid., 101 f.
117 Ibid., 101–03.
118 Ibid., 104, concerning Bache's chairmanship.

This would then provide the focus for the discussion.[119] By separating it from the Institute's organizational meetings, Bache turned the conversation meeting into an important platform for an exchange of technological and scientific innovation. He turned these meetings into a natural gathering point for those interested in discussions of technology and science rather than administration.

This format was a success. Everyone could now feel obliged to report interesting findings, and everyone was there for ideas rather than business. The free-wheeling consideration of ideas that were connected to the presenter's immediate technical or scientific interests guaranteed that the discussion, whatever its merits, would be relevant to at least one person in the room who stood up for it.[120] "The managers congratulate the members of the Institute," Bache wrote as Chairman of the Board of Managers in May 1833 in a somewhat patronizing way, "on the success which has attended the commencement of the experiment of substituting conversation meetings for the formal monthly meetings heretofore held."[121] With very few exceptions, Bache, during the upcoming years and prior to his departure for an extended trip to Europe in 1836, presented an idea or an invention at every meeting.[122] Oftentimes these had been pointed out to him, or inventors had asked Bache to present on their behalf because they were curious about the committee's feedback. In September 1833 Bache wrote that the conversation meetings were a success because they "induced many to contribute to the information of their fellow members, who otherwise would hardly have come forward."[123] And a year later, he reported that the "absence of form" had been important in enhancing their appeal and attendance.[124]

---

119 Franklin Institute, "Annual Meeting of the Franklin Institute," *Journal of the Franklin Institute* 11, no. 2 (February 1833): 85–86, and Franklin Institute, "Annual Report of the Board of Managers," *Journal of the Franklin Institute* 11, no. 2 (February 1833): 89 f.

120 In this context, it seems inadequate to consider Philadelphia's Franklin Institute a professional organization. It lacked the embracing character of the later American Association for the Advancement of Science (AAAS). For an example of such a view, see Ruth Schwartz Cowan, *A Social History of American Technology* (New York: Oxford Univ. Press, 1997), 142.

121 "Thirty-seventh Quarterly Report of the Board of Managers of the Franklin Institute," *Journal of the Franklin Institute* 15, no. 5 (May 1833): 303.

122 See the *Journal of the Franklin Institute* for this period.

123 "Quarterly Report of the Board of Managers, 1833." The report was signed by Bache.

124 He added: "their organization requires only to be a little better understood to enable each one present to attend in succession to every one of the objects which may be, in

In 1834, Bache became chairman of the Franklin Institute's Board of Managers and set out to introduce the perhaps most significant change by replacing the Committee on Inventions with a Committee on Science and the Arts. The former had been responsible for reviewing inventions, to advise their proponents on possible improvements, and to report significant ideas to the Institute's members and to the public; the latter provided the same service, but did so on a slightly different basis. Whereas membership in the Committee on Inventions had been effectively restricted to members of the Board of Managers (because the Committee was a subcommittee to the Board), membership in the Committee on Science and the Arts was open to anyone deeming himself ready to undertake the task of reviewing innovative machinery and ideas, and pledging himself to do so by signing up for it. What the new committee needed, Bache wrote, was the "'scientific labors of those whom education, business in life, and habits render peculiarly qualified for the task.'"[125] He argued that

> Hitherto … the labours of the Institute have been directed principally to the instruction of its members while the community has not in turn been benefitted by the members to the extent which the amount of talent and information possessed by them would warrant the public in expecting. This defect may be attributed rather to want of opportunity than to want of inclination. By our present organization all labours calculated to increase the usefulness of the institution, are devolved upon its managers, and a few others who have shown zeal in its behalf, while the mass of its members retire from labour, probably because they have not been made to feel how useful their exertions may be to the public and to themselves.

The new committee, in other words, was to allow members to participate in the work of evaluating ideas and innovative machines. "It is believed," Bache wrote,

> that there are many of our younger fellow members, who, having been during years past in attendance upon the lectures and schools of the Institute, are now

the course of discussion, submitted for examination." Franklin Institute, "Annual Report of the Board of Managers," *Journal of the Franklin Institute* 13, no. 4 (April 1834): 228. Both Bache and William Hamilton (Actuary) signed the report, but it was Bache, the chairman, who wrote it.

125 Quoted in A. Michael McMahon and Stephanie A. Morris, *Technology in Industrial America: The Committee on Science and the Arts of the Franklin Institute, 1824–1900* (Wilmington, DE: Scholarly Resources, 1977), xxi.

ready to repay with interest from their acquired stock of knowledge, the benefits which they may have received.[126]

Bache sought to provide a platform for those imbued with an investigative sense and he relied on their judgment and on the critical discourse in the committee's meetings to preserve their quality.

The new committee provided a book into which members could enter their names so as to "pledge themselves to perform such duties … as may devolve upon them, and to sustain by their labours the scientific character of the Institute." It was to vouch for the quality of the Institute's scientific work and counsel inventors where necessary. When the Philadelphia city councils conferred on the Franklin Institute the right to choose the recipient of the Scott Medal, this responsibility was also taken on by the Committee on Science and the Arts.[127] Not only was the new commission independent from the Board of Managers, it was to represent, oversee, enforce, and develop all scientific aspects of the Institute's work. The responsibility for this function devolved upon members who were neither selected nor expected to meet a particular formalistic criterion. Even if such qualifications could not yet be verified by asking for a degree (graduate education in the modern sense did not yet exist), this shows that Bache considered scientific standards to be a matter of personal conviction and dedication. His idea for a Committee on Science and the Arts reflects his concern with stimulating a sense of responsibility within a research discourse and to help grow a culture of intellectual investigation.

Bache's reliance on the humility of the Institute's members proved to be well-founded. As the committee's first chairman, he urged members to sign up for the committee as "many, it is believed, have not yet come forward, from a sentiment of modesty, which, however estimable in itself, is deemed to be wholly inapplicable in this case, and to be mischievous in its tendency."[128] In September 1834, forty members had enrolled, and the Committee's monthly meetings were attended by about thirty.[129]

Not everyone favored the introduction of experimental research and scientific evaluation. When the Institute commenced research on the ef-

126 Franklin Institute, "Annual Report, 1834," 229 f..

127 Franklin Institute, "Annual Report of the Committee on Arts and Sciences," *Journal of the Franklin Institute* 15, no. 3 (March 1835): 177.

128 "Forty-first Quarterly Report of the Board of Managers of the Franklin Institute," *Journal of the Franklin Institute* 13, no. 6 (June 1832): 372.

129 "Quarterly Meeting," *Journal of the Franklin Institute* 14, no. 3 (n.d.): 166 f.

fectiveness of different types of waterwheels for the production of power, Isaac B. Garrigues who was then Chairman of the Board of Managers, felt compelled to point out that the Institute's tight financial situation had not been caused by costs related to these experiments. "The expenses of the important experiments on water power," Garrigues wrote, perhaps in response to doubts as to their usefulness, "were defrayed by individual contributions, not by the Institute. The later researches under the direction of the committee on [steam boiler] explosions, were commenced at the request of the Secretary of the Treasury of the United States, and are making {sic?} at the expense of the Department."[130] Bache's interests were hardly shared by everyone. Indeed, the following year Garrigues wrote that it was an "inference … founded in a misapprehension" to blame the Committee on Inventions for not reporting about their work. As "from their nature," he pointed out, "they are least understood, without the circle of those who have occasion to feel their [labours'] benefits." "The committee are to serve as counsellors {sic} to inventors, not always as reporters."[131] Garrigues struck a comparable tone when he reported on the work of the committee investigating waterwheels. "The preparation of the numerous tables already given to the public," he wrote, "has proved that the committee are not averse to labour," thus implying that they could be.[132] Garrigues countered criticism of delays in the work on steam boiler explosions by pointing out that the "committee are understood to be present twice in each week, at their room in the Hall of the Institute, and have repeatedly extended to the members of the Institute an invitation to inspect their proceedings."[133]

The creation of the Committee on Science and the Arts is usually considered to be a turning point in the Institute's early history and a major early accomplishment by Alexander Dallas Bache. The committee further developed the role that the Institute had assumed in its large research projects, a less spectacular but highly efficient way of providing its members with a platform for an earnest exchange of ideas, which even in cases were technological matters were concerned, took place for no other purpose

130 "Thirty-fourth Quarterly Report of the Board of Managers of the Franklin Institute," *Journal of the Franklin Institute* 10, no. 6 (December 1832): 373.

131 Franklin Institute, "Annual Report, 1833," 88.

132 Ibid., 89 f.

133 Ibid., 90.

than to test a given design.[134] The Committee on Science and the Arts has even been viewed as an early version of a national academy of sciences.[135] But the idea is usually restricted to highlighting its role in oversight, not in enhancing a scientific discourse. While Bache's contemporary comments on the committee have not survived, he would probably have appreciated such a comparison. His Franklin Institute colleague Merrick is known to have "urged the Institute to assume responsibility since the nation had no national scientific bodies like the British and French."[136] The Committee on Science and the Arts, however, would not pursue projects similar to the earlier investigation on steam boiler explosions.[137] Bache's activities, instead of bringing researchers together, remained focused on developing institutional contexts for training them.

While Bache was teaching natural philosophy there, the institution's medical school had become the nation's leading center for medical instruction. During the 1790s, the physician's response to repeated outbreaks of yellow fever in Philadelphia drew attention to the city and its medical institutions. The nation's foremost physicians, such as Benjamin Rush and Philip Syng Physick, had taught or were teaching in the medical school when Bache arrived. Its student numbers were much higher than in Bache's department. There were never fewer than 350 students after 1810 and sometimes as many as 485 before Bache joined the Collegiate Department that handled a fraction of these student numbers.[138] Professors of medicine had a much higher income than Bache and his departmental colleagues.[139] Their work in training medical doctors, of course, was of immediate relevance when compared to the classics or moral philosophy, but Bache must have been able to relate to the practical relevance of medicine because he had used his West Point training in the Army Corps of Engineers. His background prepared him to envision a role for physics,

134 A. Michael McMahon, "'Bright Science' and the Mechanic Arts: The Franklin Institute and Science in Industrial America, 1824–1976," *Pennsylvania History* 47, no. 4 (October 1980): 351–68.

135 Thomas Coulson, "The first hundred years of research at the Franklin Institute," *Journal of the Franklin Institute* 256, no. 1 (July 1953): 9. McMahon considers the Institute "a national research institution." McMahon, "'Bright Science'," 355 f.

136 Ibid., referring to Minutes, Board of Managers, 13 May and 10 June 1830, Franklin Institute Archives, Philadelphia.

137 McMahon considers this a "failure." Ibid., 357.

138 Cheyney, *History of the University of Pennsylvania*, 208.

139 Ibid., 210.

chemistry, and geography (the subjects he taught) similar in practical relevance to that of medicine in the department next door. There is evidence that during his period of prominent activity at the Franklin Institute, Bache frequently spoke out against and vetoed proposals for the Institute to engage in natural history or geology. We must assume that Bache would have been unable to effectively argue against such expansions on the basis of his personal research preferences alone. That he sought to limit the Institute's focus to the physical sciences goes hand in hand with his interest in relating research to politics, to placing at its center a scientific perspective that related to the nation's practical affairs, to Philadelphia's emerging industrial interests and to the development of the nation's infrastructure. As we have seen, his efforts were intended to boost a rational discourse relating to America's specific practical needs.

The significance of Bache's role for the Franklin Institute may perhaps be gauged by the institutional lull that followed his departure. Samuel Vaughan Merrick, who had been the Institute's other leading figure and one of its instigators, after 1836 fully concentrated on business. Bache, who had most vividly represented the scientific idea, left the University of Pennsylvania and the Franklin Institute in 1836, and the latter now entered a quiet period that would last into the 1860s.[140]

In an 1869 obituary for Merrick, an article published by the American Philosophical Society in its *Proceedings* granted that he, too, was a philosopher, implying that Merrick held that title despite his connection with the Franklin Institute. The author juxtaposed "pure science" with "action," and Merrick was considered to have been among the "men of action rising into the sphere of thought," clearly putting thought above action.[141] It is Bache's cooperation with Merrick and his work at the Franklin Institute rather than for the American Philosophical Society that stands out, and his interest in developing a rational discourse for solving practical problems provides a key to understanding Bache's institutional focus in the early 1830s. The Franklin Institute provided him with an organizational vehicle for realizing ideas implied in his scientific work. The aim was to make the Institute, with the Committee on Science and the Arts at its core, a self-sustaining and self-propelling agent of intellectual review. Bache sought to help expand and institutionalize the nucleus of an American "community

140 McMahon, "'Bright Science'," 358.

141 Daniel R. Goodwin, "Obituary Notice of Samuel Vaughan Merrick, Esq.," American Philosophical Society, *Proceedings* 11, no. 81 (January 1869): 584–97.

of investigators" in an institution that anticipated and sought to strengthen the role of the hesitant nation by helping secure the rationality and sustainability of its development.

Chapter 5

Girard College and Central High School, 1836–1842

Girard College as a Political Symbol

In 1836, eight years after he had returned to Philadelphia, Alexander Dallas Bache chose to leave his professorship at the University of Pennsylvania and assume the presidency of Girard College in Philadelphia. That this move appears to have little in common with his previous efforts has been reflected by the bifurcation of historical writing on Bache. While historians of science have focused on Bache's national role after 1842, historians of education have been interested in his work prior to his departure to Washington D.C. For our purposes of reconstructing the particular outlook that Bache brought to his work, and the background to his professional leadership, it seems relevant to inquire into the common denominators of these distinct biographical periods, if any such common ground may be identified. What prompted Bache to leave his professorship for the presidency of a college? Even if we concede that Bache had developed an interest in institution building, why would he agree to help found and administer a college rather than continuing to devote himself to institutions such as the Franklin Institute through which he had such a significant impact?

Hugh R. Slotten, a historian of science who has also written on Bache's educational interests, has argued that these interests were part of a contemporary "Whiggish culture." Drawing on Daniel Walker Howe's concept, Slotten suggests that despite his family's connections to the Democratic party, Bache adhered to the complementary goals of economic innovation and industrial development, on the one hand, and a conservative "social order and moral absolutism," on the other.[1] Science and education, according to Slotten, aligned with both of these aspects. He suggests that in his design of a curriculum at Central High School in 1839, Bache sought to convey both an innovative spirit and attach to it a moral-political element. Bache then transferred this dual approach to his Coast Survey work. Science, Slotten argues, "provided ideological support for both the conservative and innovative elements of the Whiggish cultural vision" as it helped provide a common view transcending political and social divisions.[2] Slotten depicts Bache's introduction of science to be prompted, at least in part, by its perceived role in "moral education that would mold the individual republican citizen."[3] To the extent that Bache was interested in nation-building, in other words, Slotten sees him forging citizens after moral standards that served economic class interests. Somewhat in line with Slotten, sociologist David F. Labaree has argued that Philadelphia's Central High School, which Bache directed between 1839 and 1842, "helped to promote the republican community … through the provision of a common set of academic experiences."[4] Even though citizenship training or political instruction was absent from Central's curriculum, Labaree contends that the school, while providing academic training, adhered to a moralistic and class-based agenda. Some of the more recent literature on Bache's role in the history of the American high school echoes these ideas. William J. Reese maintains that reformers such as Bache ignored "how much they differed from republican theorists of the early national period" to fashion "an ideology congenial with their particular social interests."[5] He points out that the high school was called "the 'crown' of the emerging public school system" and argues that this shows that its supporters favored "hierarchy, centralized power, and cultured authority" even though republicans of the revolutionary period had opposed such

1 Hugh R. Slotten, "Science, Education, and Antebellum Reform: The Case of Alexander Dallas Bache," *History of Education Quarterly* 31, no. 3 (Autumn 1991): 329. See also Daniel Walker Howe, *The Political Culture of the American Whigs* (Chicago: Univ. of Chicago Press, 1979).

2 Slotten, "Science, Education, and Antebellum Reform," 333, 335 f.

3 Ibid., 339.

4 David F. Labaree, *Making of an American High School* (New Haven: Yale Univ. Press, 1992), 29.

5 William J. Reese, *The Origins of the American High School* (New Haven: Yale Univ. Press, 1999), 41.

centralization.[6] This is at odds with our analysis of Bache's motivation for institution-building prior to 1836, and our reassessment of these ideas will have to focus on Bache's career decisions and the plans he developed and pursued, between 1836 and 1842, as president of Girard College, as superintendent of Philadelphia's Central High School, and of the city's school system.

Even though there is no evidence that Bache was looking for such an opportunity, it would have been natural for him to leave the University of Philadelphia for a federal institution, perhaps in the nation's emerging national capital, Washington D.C. Our discussion of Bache's career has shown that his professorship provided him with a steady income and an opportunity to use some of his research in the classroom, but that he had focused on developing the more significant Franklin Institute. While the university post may have taken up most of Bache's time, the Franklin Institute's local and national significance stood out as his achievement. Its role in advising the federal government perhaps best represented the ambitions Bache shared with his family. But in the mid-1830s, few national scientific institutions existed. The very fact that the federal government had turned to the Franklin Institute in the case of steam boiler experiments indicated that this institution, prior to anything resembling a conscious national policy on developing scientific institutions, had been able to reach beyond its natural urban context to be as close to the federal level as was possible at the time.[7]

While Bache helped shape the Franklin Institute into a major American technology-oriented research institution, there emerged in Philadelphia another institution that promised to be of national significance. In 1831, French-born Philadelphia merchant and banker Stephen Girard (born in 1750) passed away and left the largest fortune ever amassed in the United States. Girard had arrived in Philadelphia in 1776. He had thrived in the Asia trade when he bought the first Bank of the United States in 1811 just after its charter had expired and, as it turned out, just before the War of 1812 assured that the federal government would need its services. Girard

6 Ibid.

7 While Dupree suggests that the amount of scientific activity at the federal level was surprising when compared to political rhetoric, these developments were only beginning to get underway and Bache was a part of them. There existed no federal institution that would have officially endorsed science. A. Hunter Dupree, *Science in the Federal Government: A History of Policies and Activities to 1940* (Baltimore: Johns Hopkins Univ. Press, 1986), 44–65.

was a widower and had no children. In his will, Girard stipulated that most of his seven million dollar bequest (roughly equivalent to the purchasing power of 173 million dollars in 2007[8]) be used to erect and operate a college for orphans.[9] The college was to house his collection of plate and furniture as well as his funeral casket. Girard gave precise instructions on where and how the buildings were to be set up. He provided a layout of the rooms and even set the strength of walls (at three feet). The buildings were to be of solid marble, the rooms in the center building to be adjacent to one another. The absence of a corridor and the shape of the ceiling (which Girard prescribed among many other details) would later prevent their use for teaching. In a fulfillment of a perhaps unconscious motive, the rooms in the massive central building are no longer used as classrooms and today host Girard's furniture and other collections within reach of the sarcophagus in which he is interred.[10] The building thus exudes the eerie feel of a mausoleum. Girard, Bache later observed, "has put himself in the place of a father to the orphan," and his college had become, post-mortem, a symbol of the fatherhood he had not been able to enjoy.[11]

These details may seem superfluous in discussing Bache's motives to join the project. It is important to detail those dimensions of the evolving college project because the school turned into a political symbol of leadership of disconnected political elites in the context of a strengthening democratic movement, and into a token of the ambitions and decline of an urban bourgeoisie in Philadelphia in which Bache's own family was heavily invested. Girard College could become such a symbol because of the ambitions of its most energetic proponent, Nicholas Biddle, and because of the political and cultural significance he chose to attach to it.

Even though Stephen Girard's minute specifications were a "technological and functional catastrophe, virtually unbuildable with the available skills of the day," the project's sheer size prompted a number of well-known architects to participate in a competition to design the college

8 MeasuringWorth, "Purchasing Power of Money in the United States from 1774 to 2007," http://measuringworth.com/calculators/ppowerus/.M.

9 Merle M. Odgers, *Alexander Dallas Bache: Scientist and Educator* (Philadelphia: Univ. of Pennsylvania Press, 1947), 28.

10 Girard had purchased a tract of land overlooking the city in the northwest. The buildings were to face north and south, and, for maximum protection from fire, no wood was to be used in their construction except for doors and windows.

11 Alexander Dallas Bache, *Report on Education in Europe: To the Trustees of the Girard College for Orphans* (Philadelphia: Printed by Lydia R. Bailey, 1839), 2.

buildings.[12] Nicholas Biddle was a former Federalist with a keen interest in European culture and Greece in particular, a country he had visited in the 1820s. Biddle assumed leadership of the board of trustees and of the building committee that had been instituted by the city of Philadelphia in 1832 to execute Girard's will. At this very moment Biddle's fight with President Andrew Jackson over the Second Bank of the United States (BUS) had entered a new stage, and the highly publicized and politically polarizing battle served as a lens through which the development of Girard College was perceived: After the charter of the first Bank of the United States had been allowed to lapse in 1811 (and after Girard had bought what remained of it), Alexander Dallas Bache's grandfather Alexander James Dallas, then secretary of the treasury, had championed the creation of the second BUS in 1816 (and written its charter) to provide the hard-pressed country with additional financial options. This bank was not a central bank in the modern sense but rather a federally chartered private bank with significant leverage in the financial market because the federal government deposited specie there, allowing the bank to extend or curtail credit to other banks. The BUS, however, was not responsible to the federal government but to its shareholders (which did include the federal government). Even though it was in the bank's interest, therefore, to use its significant financial leverage to guard the markets, it did pursue the interests of its shareholders as well.

Following the advice of Senator Henry Clay (who had presidential designs and was looking for a political fight with Andrew Jackson), Biddle decided in 1832 to apply for a renewal of its charter, which Congress granted. Bache's uncle George Mifflin Dallas, a Democrat but former Federalist foresaw the conflict between the bank and the president, reluctantly turned against his state's economic and financial interest, and introduced the bill in the Senate a few months prior to presidential elections.[13] He was

12 Michael J. Lewis, "Stephen Girard and his College," Bruce Laverty, Michael J. Lewis, and Michele Taillon Taylor, *Monument to Philanthropy: The Design and Building of Girard College, 1832–1848* (Philadelphia: Girard College, 1998), 23.

13 "The Pennsylvania Democracy was closely tied to the financial and transportation interests of the state, including the bank and internal improvements (canals and railroads). Both Dallas and {William} Wilkins {who was married to Dallas' sister} held bank directorships or stocks at one time. For many years there was a bitter rivalry between Pennsylvania and New York over these issues. The Keystone State began to falter politically in 1832, when it became increasingly evident that the president favored Martin van Buren," an old enemy of the Pennsylvania politicans. John M. Belohlavek, *George*

"fully aware of his awkward predicament" as his allegiance to Jackson was of enormous political significance and benefit to him.[14] The president took up the gauntlet, made the bank a matter of principle for asserting federal and executive authority over local economic interest groups, and vetoed the bill. The bank's supporters in Congress (including Dallas, who again voted in its favor) failed to overturn him, and so the charter was bound to lapse and the bank to expire as a federal institution in 1836. The conflict continued, however, when Jackson, reelected to the White House, ordered Secretary of the Treasury William J. Duane (married to Alexander Dallas Bache's aunt, a Philadelphian who had written Stephen Girard's will) to move all federal funds to smaller state institutions ("pet banks"). This was to make sure that Biddle would not use the financial leverage available to the bank to interfere in the 1836 presidential election in order to reverse the bank's fortunes. But Duane, the Philadelphian, refused to follow through with the plan, and he also refused to resign. Meanwhile, Biddle contracted credit in order to convince Americans, by way of hardship, of the continued relevance of the BUS. Jackson's opponents blamed the ensuing economic crisis on the president and his decision to remove the deposits. When businesses turned against the bank as the credit crunch tightened in 1833 and 1834, Biddle eventually gave in, restored credit, and began to disappear into political oblivion.[15]

It is in opposition to Jackson's "unconstitutional powers … 'dangerous to the liberties of the people'" that the so-called "Whigs," a coalition of diverse political interests, emerged as a new element in American party politics.[16] Slotten, in his essay on Bache's educational conceptions, used the urban bourgeois "culture" represented by this party as a reference to explain Bache's motives. Unlike John Quincy Adams who, as U.S. president, had proclaimed that Liberty *is* Power, Whigs declared that they defended liberty against it. In this way, some erstwhile Federalists, once skeptical of democracy, transformed themselves into a new democratic force opposing a Democratic president.[17] Jackson's anti-bank campaign had a strong anti-

Mifflin Dallas: Jacksonian Patrician (University Park and London: Pennsylvania State Univ. Press, 1977), 50.

14 Ibid., 38.

15 Sean Wilentz, *The Rise of American Democracy: Jefferson to Lincoln* (New York: Norton, 2005), 399–401; Daniel Walker Howe, *What Hath God Wrought: The Transformation of America, 1815–1848* (New York: Oxford Univ. Press, 2007), 373–86.

16 Wilentz, *Rise of American Democracy*, 398.

17 Ibid., 402.

statist element but Sean Wilentz has argued that this was part of an effort to free democratic government from "exclusive private business interests."[18] The administration's hard-money policy certainly ran counter to them.[19]

The bank war, in which Jackson and Biddle engaged from 1832, symbolized a struggle over broader issues that provide the context for Alexander Dallas Bache's decision to assume the presidency of Girard College in 1836 and of his work in Philadelphia during the following years. This significance was not restricted to the fact that Girard had taken over the first Bank of the United States in 1811, that Bache's grandfather had been instrumental in creating the Second Bank of the United States in 1816, and that his uncle George Mifflin Dallas was engaged in trying to save it in 1832. A key issue at stake was the nature of authority for federal and executive power in the context of a broadening American democracy represented by a growing voter base, and the diminution of Philadelphia's role in national politics. With the demise of the BUS in 1833 and 1834, Biddle's patrician outlook on American democracy was about to lose political significance. The Family Party, which had been directed by Bache's uncle George Mifflin Dallas, was failing politically because it was unable to disassociate itself from Biddle and his rigorous insistence on the bank's influence and on the kind of leadership for which it stood. Dallas had further destroyed his credentials as a state politician by proposing that the BUS not receive a state charter. He decided to accept a post as U.S. Envoy Extraordinary and Minister Plenipotentiary to Russia and thus opted for

18 Ibid., 438.

19 Sean Wilentz has stressed the sincerity of Jacksonian efforts to consolidate American democracy and federal power against the interests of commercial elites represented by Biddle, George Mifflin Dallas, and New School Republicans in Philadelphia. His aim, according to Wilentz, was not to establish an "imperial presidency" but to "head off {Henry Clay's pretensions to establishing an imperial Congress .... Jackson sought to sustain and enlarge the American presidency as an independent instrument of the popular will" (Ibid., 399). Wilentz points out that Jackson intended to replace the private BUS with a public institution similar to a modern central bank—an idea that remained absent, however, from Jackson's decisive 1832 veto message. Critics had pointed to Jackson's anti-developmental perspective by suggesting that this helped escalate the ensuing conflict as the president seemed set on destroying the bank without creating a substitute. See also Belohlavek, *George Mifflin Dallas*, 64. For a depiction of the consequences of the federal government's withdrawal of funds from the BUS, see John M. McFaul, *The Politics of Jacksonian Finance* (Ithaca: Cornell Univ. Press, 1972).

political exile.[20] At a time when Bache was helping to develop the Franklin Institute as a major intellectual resource for the technological development of the city's manufacturing interests, the bank war signaled that Philadelphia's former role in trade and finance, as well as national politics, was diminishing. And at that very moment, Girard College provided Nicholas Biddle with an opportunity to help erect a monument to the tradition of his kind of leadership with funds provided by fellow banker and Philadelphian Stephen Girard.

The Design and Ambition of Greek Revivalism

What was the intellectual and cultural trajectory of the project in which Bache chose to become involved? During the corner stone ceremony for the college in 1833, the chairman of its board of trustees, Nicholas Biddle, had this to say:

> {Education} of the people, which elsewhere is desirable or useful, becomes with us essential to the enjoyment, as well as to the safety of our institutions. Our general equality of rights would be unavailing without the intelligence to understand and to defend them—our general equality of power would be dangerous, if it enabled an ignorant mass to triumph by numerical force over the superior intelligence which it envied …. While, therefore, to be uneducated and ignorant, is in other countries a private misfortune, in ours it is a public wrong.[21]

Biddle found little encouragement for trusting America's general populace. He considered the college a tool for making America governable by providing the masses with an education sufficient for enlightened political participation. Politics was not a matter of negotiating diverse and pluralistic interests but of identifying the "correct" answer to a given problem. Greek Revivalism, the architectural style chosen by Biddle for the new college, complemented this political perspective in a particular way. Biddle connected the more general notion of republican nation-building that was associated with this style in the late eighteenth and early nineteenth centuries with a particularly skeptical and elitist outlook. Greek Revivalism had gained prominence in England after Greece and its cultural treasures had become increasingly accessible in the mid-eighteenth century. Even though this cultural reference obviously pointed to republican ideas, it was less

20 Belohlavek, *George Mifflin Dallas*, 64.
21 Nicholas Biddle, "The Address," *The North American Magazine*, August 1833, 216.

prominent in France than in Britain and in Germany (and in Berlin and Munich in particular). In the latter countries, the reference had an aristocratic, or at least an anti-French and nationalist dimension. In the United States, it was architect Benjamin Latrobe who introduced a variant of the style after he was invited by President Thomas Jefferson to become surveyor of public buildings in 1803.[22] The leading protagonist of the movement's second phase in the United States was Nicholas Biddle, for he was responsible for selecting this style for important public and private buildings. It was further associated with him because the Second Bank of the United States in downtown Philadelphia, the most prominent symbol of Jackson's opponents, had been conceived after a Greek temple. And in the context of the bank war, Biddle decided that Girard College was to become this style's most insistent manifestation. The buildings of that college, one art historian has pointed out, were "quickly recognized {after their opening in 1848} to be the fullest and most precise expression of Greek Revival architecture in America."[23] The style most obviously connected Biddle's bank and everything it stood for to the college Bache had chosen to join.

The architectural design competition for Girard College drew attention because of its sheer financial scale and this is why major American architects submitted plans even though Girard's will, which provided the rules for the competition, left little room to design anything other than the buildings' outer shell. A few amateurs participated; but so did, among other prominent architects, Edward Shaw, John Haviland (architect of the Eastern State Penitentiary and of the Franklin Institute in Philadelphia), both neoclassicists, and Isaiah Rogers, Alexander Jackson Davis, John Kutts, and William Strickland (architect of the BUS building). Several of Strickland's students submitted proposals, and it was twenty-eight-year-old Thomas Ustick Walter (1804–1887), one of the Franklin Institute's earliest drawing school pupils, whom the two branches of Philadelphia's city council chose to become architect of the college.[24]

22 See Roger G. Kennedy, *Greek Revival in America* (New York: Stewart, Tabori & Chang, 1989).

23 Laverty, Lewis, and Taylor, *Monument to Philanthropy*, 113.

24 Thomas U. Walter's Franklin Institute training (as well as his teaching activities there a few years later) is mentioned in Bruce Sinclair, *Philadelphia's Philosopher Mechanics: A History of the Franklin Institute* (Baltimore: Johns Hopkins Univ. Press, 1974), 236.

Walter's design, despite the columns along the main building's front, was French rather than Greek, inspired by the Place de la Concorde in Paris. In his proposal he had flanked the core building with two "extended horizontal blocks," thus breaking its monotonous weight.[25] His teacher Strickland was America's main proponent of Greek Revivalism at the time, and perhaps Walter's selection had to do with this background combined with an interest in an architect more malleable than the experienced Strickland. Once the decision for Walter had been made, Biddle worked hard, both in meetings with Walter and in committees, to change the latter's plans and to turn Girard College's main building into a full realization of a Greek temple with portico and peripteral colonnade.[26] In this way, he aligned Girard College architecturally with his BUS, which had been conceived after the Pantheon. Considering this obvious symbolic connection, political attacks against the bank were bound to target this second symbol of his leadership as well. In subsequent years, the college became a token of the conflict between Whigs and Jacksonian Democrats.

Four years after the building committee had been created in 1832, its members grew restless, sought to show progress, and decided to get work underway on an educational plan.[27] Perhaps because he knew Greece first-hand from his own travels there in the 1820s, Prussian émigré Francis Lieber was selected to advise the trustees on the college curriculum. With himself in mind, he suggested that a trip be undertaken to Europe to investigate schools there and to prepare a report for Girard College. Alexander Dallas Bache had been a trustee of the college since 1833, and in response to Lieber's proposals, it was Bache, not Lieber, who was selected to investigate the state of education in Europe as new president of Girard in 1836.[28] Lieber had prefaced his report to the trustees with a long list of titles on the subject, all of which were German.

25 I am drawing here on Michael J. Lewis, "The Architectural Competition for Girard College," in Laverty, Lewis, and Taylor, *Monument to Philanthropy*, 24–48, quotation 45.

26 Michele Taillon Taylor, "Nicholas Biddle and Greek Architecture in the Age of Jackson," ibid., 65–86, particularly 74 f.

27 Ibid., 96.

28 Nicholas Biddle to ADB, August 20, 1833, box 3, folder 3, Bache Papers, SIA. Bache had been appointed to fill the spot of John C. Stocker who had recently passed away.

Fig. 6. Girard College. Pictorial illustration to the Philadelphia Saturday Courier by John Caspar Wild, 1838.

*(Courtesy of The Library Company of Philadelphia, *W 155 [P.2058])*

The trustees chose to send "one of their own."[29] In July 1836, John Sergeant (1779–1852) commended the educational system installed at West Point, pointing out that it had been designed after a trip to Europe by the institution's head, Colonel Thayer.[30] Thayer's as well as Bache's journeys were of course part of a larger phenomenon of Americans studying European institutions.[31] Dennis Hart Mahan of the U.S. Army Corps of Engineers, for example, had travelled to Europe from 1826 to 1830, and Bache's Franklin Institute colleague Samuel V. Merrick had investigated gas works in Europe prior to their introduction in Philadelphia. Bache's

29 Frank Freidel, "A Plan for Modern Education in Early Philadelphia," *Pennsylvania History* 14, no. 3 (July 1947): 175–84.

30 John Sergeant to Nicholas Biddle, July 12, 1836, RH 2272, box 12, Rhees Collection, HL. Bache seems to have been insufficiently aware of the details of the College plan to be sent a copy of the Girard will by James Bayard later that month. Bayard to ADB, July 27, 1836, RH 906, box 12, Rhees Collection, HL.

31 Robert V. Bruce, *The Launching of Modern American Science, 1846–1876* (New York: Knopf, 1987), chap. 3.

scientific colleagues, Asa Gray, Elias Loomis, and Joseph Henry, among others, all considered their journeys significant in shaping their outlook on the field and of their careers. It was in line with Sergeant's observation that the trustees chose for such a trip Alexander Dallas Bache, a West Point graduate from a prominent Philadelphia family who was familiar with Thayer.[32]

Bache's European Trip and the Bache-Biddle Correspondence

Did Bache's involvement signal that he entertained a National Republican, even Federalist, outlook? How did his view coincide with, or differ from, that of Nicholas Biddle and George Mifflin Dallas? Is it sufficiently precise to subsume Bache under a "Whig culture" that set itself off from Jacksonian democracy in the wake of the Bank War even when it eventually aligned with it politically?

No letter remains in which Bache explained his motives for leaving a professorship to become president of a college that was years away from admitting its first orphan student. In addition to the European tour, there were other attractive features about the new commission: Bache's salary went up from 1,500 dollars per month, the amount he received as a professor at the University of Pennsylvania, to 4,000 dollars.[33] Eight years into their marriage, the Bache's still had no children and this suggests that none would perhaps be forthcoming at all. After his father had left, Bache, Sophia's oldest son, had financial obligations to support his family. At the time he left for Europe, his mother had moved in with him and his wife in Philadelphia.[34] The significance of the college project, furthermore, fell in

32 For the offer extended to Bache, see Nicholas Biddle to ADB, July 19, 1836, box 3, folder 5, Bache Papers, SIA. In response, Bache wrote to his friend Joseph Henry the next day, and prior to accepting the post, that "the Trustees of the Girard College have elected me Pres.[t] of the Institution, to proceed to Europe in the autumn." ADB to Joseph Henry, July 20, 1836, Joseph Henry, *The Papers of Joseph Henry* (Washington: Smithsonian Institution, 1972–2008), 3:79.

33 On Bache's salary as college president, see Cheesman Abiah Herrick, *History of Girard College* (Philadelphia: Girard College, 1935), 11.

34 Odgers, *Alexander Dallas Bache*, 32 f. Bache's later colleague Benjamin Apthorp Gould in his necrology of Bache reported an incident which must have taken place shortly before Bache accepted the Girard post. His mother Sophia Dallas had moved in with him and his wife Ency, and one evening, hearing an alarm, walked into Bache's room which "was held sacred to scientific investigation." She accidentally knocked over his elaborate ex-

line with Bache's earlier ambitions for developing national institutions from local ones. Finally, a two-year European tour, all expenses paid, was a significant opportunity in itself.[35] The thirty-year-old Bache chose to accept the restrictions attached to the new job (such as developing an institution that was to admit orphans only) and sail for Europe in the fall of 1836. In company with his wife Ency and her thirteen-year-old sister Maria, Bache left for England in September, and during the following two years, he also visited France, Holland, Belgium, Austria, Germany, Switzerland, and Italy.[36] His purpose, according to Biddle's letter of introduction for Bache, was to procure "information on the subject of the organization and management of Institutions of Education founded upon similar principles."[37] After his return in October 1838, Bache was able to report that he had seen 278 schools.[38]

While only few of Bache's letters have been preserved from the period prior to 1836, a significant bulk of his later correspondence remains. So many items exist today, in fact, and much of it in such precarious condition (both with respect to calligraphy and wear) that any attempt to conceive of Bache's intentions and perspective in a limited span of time will have to focus on selected documents. While I have consulted all of Bache's correspondence as well as his diary for this period, I will emphasize selected documents. Bache frequently wrote to his mother, keeping her informed of his progress. Bache's diaries for this period are detailed and meticulous but

---

perimental setup, "with Fourier's thermoscope of contact," which was destroyed completely. Bache, faced with the disaster, "stood white with emotion for a few moments; then turned away, only trusted himself to say that he would return soon, and hurried out of the house." Perhaps his mother had unconsciously signaled him how his experimental interests deviated from his broader mission. It may be significant also that this story, which was probably conveyed to Gould by Ency, was later remembered by Bache's colleagues. It may have symbolized to them Bache's particular relationship to his family and to his mother, and his somewhat muted experimental efforts after 1836 which paralleled his emerging success in institutional development. See Benjamin Apthorp Gould, "An Address in Commemoration of Alexander Dallas Bache," American Association for the Advancement of Science, *Proceedings* 17 (1868).

35 Coverage of expenses is mentioned by Herrick, *History of Girard College*, 12.

36 Henry, *Papers of Joseph Henry*, 3:367 n. 10.

37 Nicholas Biddle [to whom it may concern], September 16, 1836, box 6, folder 10, Bache Papers, SIA.

38 George V. Fagan, "Alexander Dallas Bache. Educator," *The Barnwell Bulletin* 18, no. 75 (April 1941): 17; Odgers, *Alexander Dallas Bache*, 76.

rather descriptive in style.[39] In the first section of this chapter, I will concentrate on the Bache-Biddle correspondence because it is of particular relevance.

Soon after his arrival in Europe, the dimension of his investigative project began to dawn on Bache. "To make anything more than a very superficial examination of the institutions of a place requires the cultivation of acquaintances," he wrote to Biddle from Glasgow. "[A]lthough I have had no difficulty in Ireland on this … still [more time] has been occupied than I had anticipated."[40] From his letters, it is evident that Bache considered as his object of investigation universities as well as schools. He paid particular attention to the education of teachers. When visiting one school that "intelligent men" had recommended to him as "among the best in this Kingdom," he wrote that the "English department of the high school engaged much of my attention." Without further commenting on or justifying why he moved on to this subject, he pointed out that the "university system of Scotland is here to be seen perhaps to the best advantage."[41] Bache considered all levels of education, therefore, trying to comprehend the entire system instead of just pieces of immediate application to the Girard College project. In that sense, he perceived his mission to be to explore educational institutions founded upon principles "similar" to those of his own college in the broadest terms.

Bache wrote to Biddle regularly, but the committee chairman was getting somewhat anxious for quicker results and ideas on how the new college may be organized. In October 1837, with Bache gone over a year, Biddle wrote that

> the College + your own mission are as you are aware objects of great solicitude with the whole community and the greatest difficulty which the Trustees have to encounter in the impatience of well disposed friends who are anxious to press forward the institution at some hazard of a premature commencement. That anxiety is so natural + so respectable that the Board of Trustees are very much inclined to meet it by every proper explanation of the progress of the enterprise. The mechanical part is going on very well. The two eastern outbuildings are in a great state of forwardness and this only stimulates the desire to know the progress of the

39 The only attempt at a somewhat coherent biography, Odgers, *Alexander Dallas Bache*, for this period mainly relies on Bache's diaries.

40 ADB to Nicholas Biddle, December [?], 1836, box 2, vol. 3, Bache Papers, SIA. All letters by Bache to Biddle in this collection are available only as copies that Bache had made while writing the letter.

41 ADB to Nicholas Biddle, March [9?], 1837, box 2, vol. 3, Bache Papers, SIA.

intellectual part of the work. It occurred therefore to the members of the Board, without any formal or official action by the body, that if you could from time to time give a report of your position, progress + general views it would be highly interesting. This need not be in great detail—it need not commit you to any premature expression of opinion—it would not be [designed] for publication, but it would bring the members of the Board into more immediate sympathy with the Institution + with you as the head of it, and would enable us to diffuse through our circle a knowledge that the work was going on steadily. In short, as is usual between friends at a distance from each other, we wish to hear more from you + of you [...][42]

Note Biddle's hint that more letters from Bache would put the Board into "more immediate sympathy" not only with the Girard College, but "with you as the head of it." Biddle stresses that he does not expect Bache to make concrete proposals on how the college should be organized but he does not avoid a possible implication of his letter: that the Board had reservations about Bache's presidency or remained in some other way to be convinced of his leadership, perhaps in view of the fact that Bache had been installed at a time when the date of the completion of the ambitious buildings and the opening of the institution remained vague. Biddle's letter, despite its rhetoric, effuses a sense of an aloof and demanding skepticism. This was different from the Franklin Institute where Bache had been on eye-level with other institutional leaders. Biddle, to whom Bache would later once refer as the "Golden Empower," took full charge of his role of chairman of the board of trustees, and Bache, though formerly a trustee as well, was a hired president and under obligation to perform.

When he wrote to Bache, Biddle had not yet received a letter that the college president had written from Locarno on September 28. In this official report, Bache did what he would just about to be asked—report his impressions and whereabouts. And Bache went further in that he began to develop ideas on educational policies and strategies by proposing that instructors be hired and sent to Europe immediately.[43] In the following months, Bache would write to Biddle repeatedly to outline and bolster this idea to have his future colleagues sent overseas for a few months. It was central to Bache and it warrants scrutiny. For this purpose, I will look at two letters in which Bache develops his proposal. In his letter from Locarno, Bache addressed Biddle as the chair of the board of trustees. In a

42 Nicholas Biddle to ADB, October 6, 1837, RH 930, box 12, Rhees Collection, HL.

43 ADB to Nicholas Biddle, September 28, 1837, box 2, vol. 3, Bache Papers, SIA.

subsequent communication from Stuttgart, Bache wrote to the Philadelphia banker privately. He could therefore be somewhat more direct in making his point.

Bache begins his first letter as follows:

> Locarno, Switzerland
> September 28th, 1837
>
> Dear Sir,
>
> Since my last report to the Committee of which you are chairman I have been principally employed in visiting the institutions for education in part of the [—] comprising the Swiss confederation. Since the [—] of [—], some school governments have been engaged actively in extending the opportunities for [—] instruction, and as they have increased the facilities for obtaining good instruction + their legislative branch have required by law that the citizens which they represent shall avail themselves of the means thus afforded.[44]

Bache had assumed the presidency of Girard College with the obligation to collect information about institutions of education similar to the one planned in Philadelphia. That his trip was to take two years suggests that the scope of his investigation was understood to be much more comprehensive and to include schools other than those for orphans. The beginning of this letter further suggests that his interests went far beyond the variety of institutional designs and that it encompassed the political and social setting for education.

It was common to refer to Switzerland as "Swiss confederation" and despite the many differences between that country and the United States, they shared democratic principles of which Bache was keenly aware. Perhaps it was no coincidence that Bache began to develop ideas for Girard College while he was traveling in Switzerland. The focus of Bache's initial comments extends beyond the immediate task of organizing Girard College as he considers an institution's role in the political nation-state. He highlights the political response to decisions by individual "school governments." Bache assumes a perspective in which change is initiated, not by state administrators from above, but by such "governments" from below. He implies that it is in response to their initiative that parliament made school attendance mandatory by law. Translated to the situation in which Bache finds himself at Girard College, this model would call for efforts to develop the institution into an ideal representation of a school for his

44 Formatting rules used for transcribing letters are in chap. 3, 58 n. 23, above.

country in order to provide either a state or the national congress with good reasons for making its attendance mandatory. There is nothing in Bache's letter that suggests that he wanted to pursue such a plan. It is the focus he provides in the opening section of this document that reflects his interest in developing education for the political community at large rather than just for orphans at Girard College.

The explicitness of Bache's phrase "their legislative branch have required that the citizens *which they represent*" (my italics, A.J.) emphasizes the material justification for introducing mandatory school attendance. This phrase could indicate, of course, that Bache points out that Swiss legislators made attendance mandatory only for those who were citizens in their jurisdiction and that they could therefore not be criticized for an abuse of power. But this is not his perspective at all. Bache's formulations "shall avail themselves" and "thus afforded" suggest, on the contrary, that he considered this law to be an expression of citizenship and a token of an attitude which considered this law, not a restriction of the individual's freedom, unwarranted abuse, or the result of effective lobbying by school administrators, but a sign of the sovereign's acknowledgment of successful efforts to provide the nation with proficient education. One could ask: If schools are implementing such education, why not let citizens decide for themselves by sending their children there? Bache's choice of words suggests that he considered such a law to shield children and teenagers from conflicting interests such as helping on the family farm or by supplementing the family income through paid labor, and that in the long run, Bache considered the benefit of education, to which the people had committed themselves, to outweigh what may have been perceived as a disadvantage.[45] Bache, in any event, viewed this measure as an opportunity rather than a restriction—an acknowledgment of common ambitions and standards.

Bache continued:

> As a powerful [means] of effecting rapid improvement in the schools they have established normal schools which are frequented not only by candidates for the [situation] of instructor but by the instructors of the common schools themselves. These schools have given me a number of interesting subjects of examination. Ef-

45 For the competition between education and farm work in rural areas, see Daniel Walker Howe, *What Hath God Wrought: The Transformation of America, 1815–1848* (New York: Oxford Univ. Press, 2007), 454.

forts to steer [—] the rural schools in which agricultural labour is combined in a greater or less degree with instruction have interested me.

In the last sentence of this paragraph, Bache points to an issue that could perhaps be anticipated in the context of his comments on laws introducing mandatory school attendance, i.e. the combination of education and "agricultural labour," though perhaps in a slightly different way. Rather than negotiating the value of education vis-à-vis supporting a family, he is interested in allowing students to improve their work through intellectual resources acquired in school.

In subsequent paragraphs of this letter, which I will not consider in detail, Bache argues that while schools were closed at the time he visited Geneva, he attended a meeting of the Helvetic Society of Public Utility where he was advised on other institutions. Bache provides Biddle with a list of schools he visited in Switzerland, pointing out an institution in Berne that experimented with "combining labour + instruction." Towards the middle of the second page of his letter, Bache moves on to a different topic, and the slightly indented placement of these paragraphs indicates that he wrote them at a different time. It is here that Bache introduces his idea that Girard College teachers be sent to Europe and he insists that the success of the college depended on the realization of this idea and that the trustees must lose no time in acting on it. But I will take a shortcut and rather than interpreting Bache's presentation of his idea in his letter from Locarno, I will move on to a letter Bache wrote from Stuttgart (Württemberg) as his idea comes into better focus here. As mentioned above, Bache wrote this letter to Biddle "privately" rather than to the chairman of the board of trustees and while Bache will have remained circumspect in making his case, it provided him with an opportunity to be somewhat more insistent, perhaps even more direct.

"My dear Sir," Bache wrote in this letter from Stuttgart,

In an official letter addressed to your committee I have expressed [certain] wishes in regard to what appears to the best way to devise the [most] [profit] for the Girard College from my mission to Europe. I am anxious [further] to [develop] them confidentially to you that you may use them as seems best to you for the welfare of the institution. Unless I [—] teacher[s] [—] which [—] the Trustees now intend shall be the case, + devote my time [to] the [very?/many?] elements (+ who assures that the attempt would be successful?) I shall be powerless in administering the beginnings of the institution unless provided with good teachers. The good + the bad schools which I have seen thus far in my [arrow], obviously are good or bad less from their organization, rules & regulations etc than from the spirit in which the whole school is carried on.[46]

Bache refers to his earlier letter to Biddle and the board of trustees. Bache's opening reflects his intention to acknowledge the chairman's influence which, in turn, underlines the addressee's responsibility to take seriously a proposal that promises to shape and improve the institution under his command.

Bache goes on to highlight the importance of the quality of teachers for his ability to administer the college. The first section of the second sentence is difficult to make out but it seems clear that with regard to getting the institution underway, this problem is a major, if not a decisive concern for Bache. This is his second letter to Biddle in which matters of administrative strategy are raised, and Bache begins this letter by making this point. He is not initially concerned with organizational questions but more diffusely with the type of teacher with whom he will have to cooperate as college president. In this context, he puts particular emphasis on the "beginnings of the institution." The quality of teachers will obviously matter beyond the institution's initial phase. What Bache implies is that he views the instructors as colleagues, whose decisions will have an impact on how the institutional framework evolves at the onset, setting precedents for faculty hired later. Bache has in view that such precedents will be set jointly, that his role as college president does not involve controlling all details of the institution's policies and development. As evidence for his argument, Bache, in the last sentence quoted here, points to European examples. He suggests that the "spirit" of an institution may not be controlled by the administration. He implicitly concedes to the entire faculty the responsibility of diffusing and sharing a common understanding of the principles according to which problems are to be resolved—including those for which no rules exist. By suggesting that without good teachers he would be "powerless" in his administration of the college implies that "governability" presupposes intellectual qualities. He does not conceive of power as an instrument to make people follow his will arbitrarily but as a dimension of an effective use of rational leadership. Bache hopes to be able to recreate this idea in a school in which the faculty shares this concern as he stresses cooperation rather than presidential command. He is

46 ADB to Nicholas Biddle, October 9, 1837, box 2, vol. 3, Bache Papers, SIA.

confident in his authority to provide leadership among equals and feels disinclined to insist on it formally.

Bache deduces these ideas from his European travels but they relate to the perspective he had assumed in developing Philadelphia's Franklin Institute. The Committee on Science and the Arts was premised on the idea that its members would cooperate as colleagues much in the way in which Bache wishes Girard College to function. If that is indeed the case, however, why would Bache point to the European example here? His reference reflects the prominence of European culture in America which provided the impetus for Bache's extensive trip in the first place. It serves as leverage vis-à-vis Nicholas Biddle and the other Girard College trustees who, by sending him abroad, had implicitly subscribed to this idea.

> Without regulations an individual spirit [frequently] [imparts] into a large establishment + gives life to the whole; [like in the contrary] good regulations tamely administered give [but] [an] [—] institution, after [all].

While only fragments of this sentence may be made out, the gist of Bache's comment is evident. Bache contrasts an institutional situation without regulations with another situation in which existing regulations are not adjusted to problems, or remain unenforced. He leaves out the obvious possibility of regulations having been put into place that are also well administered. This omission seems to make sense because Girard College was only starting out and Bache had to both create regulations and find ways of using them. But Bache could have also left out the second option of ill-managed rules. Why does he include it?

Bache contrasts the first option, which he favors, with this second possibility of undesirable consequences (suggested by "[like in the contrary]" and "[but]"). The latter is undesirable because, in the long run, even good regulations will have a tendency to be administered, not "tamely," but with rigor. The administration will assert its centralized control, and cooperation comes to mean obedience to rules rather than participating in the spirit in which they were conceived. Bache, in other words, seeks to avoid administering the college through regulations because he considers them as stifling spontaneous problem solving by other members of the institution. In line with his previous arguments, therefore, Bache perceives of Girard College as an organization whose ideals are imbued in and represented by every faculty member (and perhaps students as well) rather than being incorporated in a set of ossifying rules. In this way, Bache seeks to resolve the issue of how to set up an organization in which the authority to successfully solve problems emanates from the competence and sense of responsibility of its members. This contrasts with another model that Bache does not infer in which the authority of the members of the institution is "controlled" by its administration or president. Bache is concerned with the diffusion of rationality rather than with issues of establishing command for its own sake.

Bache next moves on to highlight the role of the "lower" levels of instruction, stressing that it would be difficult to find good teachers:

> If the college part of our establishment can be made to square with the requisition of the founder in regard to its age of leaving the institution, it will {be} [accordingly] easy to organize. The abilities required in its professors the Board can comprehend + with the lower departments in [good] order [it will] go on [well.] It is in these lower departments that I apprehend the most difficulty will be found, + the very lowest will probably be found the most difficult to supply with teachers + superintendents, even if we may doubt them more important in the whole college. While able professors are to be [had] every where, it is surprising how few really good elementary instructors there are. This task requires more devotion + more [practical] adaptation for it than that for professor.

According to Stephen Girard's will, students from the age of six to ten were to be taught at the school until they were between fourteen and eighteen years old, and they would then move on to practical training in areas such as "agriculture, navigation, arts, mechanical trades, and manufactures."[47] Bache considers this a limitation on the idea of establishing a distinct college within the larger school and he seemed to have in mind educating students beyond the age of eighteen and at the university level. Bache does not expect any problem recruiting professors for this upper branch of the school and his suggestion that the "Board can comprehend" their qualifications suggests that they were not particularly specialized and represented an educational canon familiar to educated citizens. He was writing to Nicholas Biddle, a prominent *hommes des lettres*, and he may not have wanted to offend but to charm him in order to more effectively convey his idea.

Why would it be more difficult to locate appropriate staff for instructing younger students entering the college at age six to ten? One obvious difference is that such students will need basic instruction and that they will act less like adults or as young colleagues than senior students. Bache

47 Girard's will is in Henry W. Arey, *The Girard College and Its Founder: Containing the Biography of Mr. Girard* (Philadelphia: Printed by C. Sherman & Sons, 1869), 57–85, quotation 74.

points out that elementary instructors need "more [practical] adaptation" and "devotion" than instructors of higher-level classes. But Americans in the 1830s were more likely to have elementary education and this implies that there probably were teachers, even if it may have been difficult to attract them to teach at Girard College. But Bache is less concerned with the job market than with the qualifications of teachers ("really good elementary instructors"). At a time when the idea of the normal school was emerging as a prominent model in Europe but remained largely absent from the United States, he is unwilling to compromise and to hire available untrained talent. Bache seeks to solve in principle the problem of providing good teachers for Girard College in the United States:

> In the [successful] attempts to reorganize public institutions in some of the republics of Switzerland, the chief agents have been supplied by the normal schools [the] establishment of which was made to begin the school reforms. With an elementary school in proper condition we may make of the Girard College a great establishment for training boys as mechanics, agriculturalists, teachers of different grades, [according] to the [best] of their dispositions. Any department of practice or of instruction may be [engrafted] with success if the [— is — of —] this impression of the importance of beginning aright.

Bache draws from his European experience the model for an American reform of teacher training in normal schools or, rather, their introduction.[48] To Bache, the establishment of normal schools is merely a beginning. From the idea of how school reforms were initiated in Switzerland, Bache moves straight on to Girard College and its opportunity to prepare boys (according to Girard's will, no girls were to be admitted) in the two main avenues of manual employment. Bache has in mind an informed way of solving problems in the areas of work for which students are prepared rather than continuing a tradition of dealing with such problems. The Girard College alumni, in Bache's perception, will not consist of blacksmiths and farmers but of "mechanics" and "agriculturalists." Schooling will make them aware of a general discourse in their respective fields, and they will be able to help develop and to translate findings in their areas. The third occupation mentioned by Bache, "teachers of different grades," concludes the trio and connects to his initial idea of making Girard a center for re-

48 The first American normal school would be created two years later in Lexington, Massachusetts. The first school so named was the École Normale Supérieur in Paris, founded in 1794.

forming American education. This is a crucial point for it relates to Bache's efforts at the Franklin Institute.

I am [now] really [anxious] that [some] at least of those who are to [act with] [me] should have an opportunity of seeing + perhaps of practicing for a [short] time in, the institutions which approach nearest to the model of what I think an elementary school ought to be.

Bache now arrives at what appears to be the occasion for writing this letter to Nicholas Biddle, his idea that his future colleagues, too, should be exposed to some of what he sees in Europe. He is "really anxious" that they participate in this experience and that by teaching there they immerse themselves in the European educational experience. The latter is not identical to a model that Bache has in mind for elementary education in America and perhaps for education in general. It merely comes "nearest to" that model. He consciously distinguishes between the two and does not surrender his American perspective on Europe while he concedes that elementary education in Europe is closest to its ideal realization at the time. Bache is not merely looking for practical answers to the immediate problem of developing the college's administrative and educational policy. He writes his letter to Biddle in the expectation that Girard College is to become a model institution that takes on the world's leading educational ideas in order to push them closer to perfection. In this sense, the college would not be a representation of European ideas in the United States or a model for that country; Bache aspires for the college to become a model for the world.

In connection with above observations, it is worth noting that Bache expects at least some of his colleagues to "act with" rather than "work under" him. He also assumes that the college project is sufficiently significant to consider using funds for sending abroad not only its president, but prospective teachers as well. By the standards of the time, it was unusual to send elementary teachers abroad for training, and yet Bache did not shy away from proposing drastic measures to launch the institution and informed education in America. Considering language barriers, he probably had in mind sending most colleagues to England even though the most progressive elementary schools, following work by Johann Heinrich Pestalozzi and Friedrich Froebel, were in Germany and Switzerland.

A gentleman of the requirements necessary to form an [impression] of the elementary class + afterwards to take part in the higher instruction, should be, [— part] of the agents chosen. For gentlemen to take part in [communicating] the elementary instruction + to devote themselves to the first [boys] admitted following them perhaps in the higher departments and succeeding the elementary[—], I think, [—] the other agents. The requirements should regulate somewhat the direction of their journey.

The sentences here are again difficult to make out but the direction in which Bache is taking his letter seems sufficiently clear. Bache assumes that unlike most elementary teachers at the time, those at Girard College would be male.[49] He develops his proposal of sending college teachers to Europe as "agents" for shuttling the European model to Philadelphia. The order in which teachers are to be sent ought to depend, in Bache's view, on the "requirements" of teaching duties at the college.

Supposing the first [named] to be appointed into [French] besides the [— —]: the normal-school {inserted:} + [Inp. class^{n}] of the Academy {end insertion} at Glasgow, Mr. [Hood's] school at Edinburgh, the model infant school at Edinb^{g}, [Arthur] Hill's school at Bruce Castle near London, the normal school at Versailles, the normal + [middle] scho[ols] at Lausanne + Zurich in Switzerland, M. Fellenberg's schools at Hofwyl near Berne are the principal establishments which I would designate as to be visited + for a longer time, [—], than I [would?] spend at them.[50] These with perhaps [a few] other points which I may yet [meet] would occupy six or seven months. The [under] teachers should probably remain longer in [particular] schools in England + four or five months should suffice for their [—]. Setting out next spring all could be accomplished by the autumn and thus could return [— —] for the opening of the institution. I hope myself to return by the [— — — — —] early in the autumn + [to] see a beginning [made] before winter sets in.

A year after he had left Philadelphia, Bache had developed a good sense of which schools in Europe he considered worthy of providing a model for education at Girard College as he lists at least one institution which is today considered to have been among the most progressive of its time, Philipp Emmanuel von Fellenberg's pauper school in Hofwyl.[51] Fellenberg had initially cooperated with Johann Heinrich Pestalozzi. The school in Hofwyl was known for its integration of education and work so as to

49 Opponents of the College would later argue that Stephen Girard had in mind employing female teachers as was customary at the time. *Report of the Special Committee Appointed by the Common Council on a Communication from the Board of Trustees of the Girard College* (Philadelphia: s.n., 1840), 26.

50 Names of schools and individuals in this sentence were compared to Bache, *Report on Education in Europe*, 305, 409.

51 *Encyclopedia Britannica, 2005 Ultimate Reference Suite DVD*, s.v. "History of education."

provide students with resources for their prospective occupation. Bache also referred to normal schools, a recent development in European education.

Bache would like to send prospective teachers at Girard College to Europe for several months, and to do so soon. He does not hesitate to call for immediate action and suggests that teachers spend in Europe six or seven months. Nor does he refrain from the general remark that he may later identify additional places for these Girard College teachers to visit. While he had stressed that Biddle should use his ideas as he sees fit, Bache outlines his plan firmly.

One issue Bache has not addressed is that if he is to return in the fall of 1838, he cannot be involved in selecting proper candidates. He has not mentioned any specific names and summarily referred to the group of teachers he wants to go to Europe, and this reflects the fact that teachers had not yet been hired.

In fact, Bache turned to these matters in the following section of his letter which, unfortunately, is on a third page in worse condition and very difficult if not impossible to decipher. What can be made out is that Bache discusses three candidates for filling posts at Girard College, considering their qualifications, a Professor [Hoar] whom Bache had known for several years and to whom he attests "patience" and "conciliating manners," John [Fa—], a University of Pennsylvania graduate, as well as [John McKirby] and his brother. Bache advises Biddle how to contact them and closes his letter by stressing that

> my anxiety daily increases. Without freely communicating with you I should not be satisfied for so much depends upon the right beginning to the ultimate success of the Girard College that those who have it in hand cannot confer too often or too [intimately].

More on Bache's European Tour

Bache again wrote to Biddle about his idea from Holland on November 20, but had not yet received a response to his proposal, and neither had he heard anything when he anxiously wrote to Biddle from Hamburg on November 27, 1837. In that letter, Bache felt prompted to explain why he had expressed himself so openly, not merely reporting about his trip but suggesting ideas for developing Girard College as an educational facility.

But he repeated his proposal, stressing that sending teachers to Europe would be the "foundation" on which "we shall be successful."[52] A month later, still not having heard, Bache wrote that at times he "felt discouraged at hearing nothing of the [progress] of the material parts of the establishment, or of the receipt of communications," but that he had "successfully struggled against this." His idea of sending teachers abroad he advertised as providing an opportunity to "catch their spirit if they did not [catch] the details." He added that "in coming here even without the language [they can] in less than three months be fitted to derive profit" from such a trip.[53]

Biddle finally responded to his idea in mid-February, 1838, conveying the "general idea of the board" that following Bache's return "in midsummer or early in the autumn" (rather than in the fall as the college president had suggested), "the business of [instruction] [commences] in the outbuildings, if in the opinion of the council, it can be done under the will."[54] At the time of Bache's departure it had remained unclear whether the college, under Stephen Girard's legal stipulations, would be allowed to open for instruction before all of the buildings had been fully completed. Bache obviously hoped to be able to begin work upon his return. Biddle made clear that this was not certain at all. He rejected Bache's proposal "to send out some tutors to perfect themselves at different schools" on two grounds. Biddle argued that the board of trustees was not qualified to select the personnel Bache wanted to have sent abroad because the college president was to submit ideas on how to organize education at the college after his return. It was unclear to the trustees what subjects were to be taught and what qualifications the college teachers were to supply. To hire teachers, Biddle pointed out, and to send them to Europe, would involve additional expense, and this would draw criticism from those insisting on Stephen Girard's provision that such costs be incurred only after the college buildings had been completed. "I wish all parties to think," Biddle wrote to Bache, "that your mission is sufficient that you can do all that is necessary to be done without sending others + that if any thing else is

52 ADB to Nicholas Biddle, November 27, 1837, box 2, vol. 3, Bache Papers, SIA.

53 ADB to Nicholas Biddle, December 26, 1837, box 2, vol. 3, Bache Papers, SIA. Bache refers to this topic in at least this one other letter he wrote to Biddle: January [31], 1838, box 2, vol. 3, Bache Papers, SIA. In his October 9 letter quoted above, Bache had written that it would take four or five months for teachers to benefit from their trip to Europe.

54 Nicholas Biddle to ADB, February 15, 1838, RH 931, box 12, Rhees Collection, HL.

wanted hereafter, you can send the proper persons to the proper places with proper instructions." Biddle did not rule out realizing Bache's plan at a later time but he declined to pursue it immediately.

Considering the political difficulties in which the trustees would find themselves soon after Bache returned from Europe in 1838, Biddle's concerns about keeping a low expense profile turned out to be well-founded. Opening the college would prove to be more difficult than the trustees had anticipated. While Bache had taken for granted that the college would at least be partially staffed by academics (he had referred to professors), Biddle refers to the college teachers as "tutors" who may need additional refinement but no outright infusion of the European spirit. While Bache harbored ambitious plans for the institution, aiming to connect it to an academic discourse, Biddle put less emphasis on this aspect and considered the educational work in traditional terms and on par with existing schools. The chairman's ambitions, in other words, were focused on the symbolic significance of the school's impressive architecture and size, while Bache emphasized the institution's model role for developing the country's cognitive resources on the basis of collegiate cooperation. But Bache had been a trustee of Girard College for several years prior to his appointment as its president and he must have been aware of at least some of the political problems which Biddle pointed out to him. It is surprising that Bache's ideas were very much at odds with the intentions of Biddle and the other trustees. This suggests that Bache, even though he had been a member of that committee since 1833, had remained aloof from its work and that in choosing to accept the post of president in 1836, and in putting forth his unusual ideas of sending abroad other college faculty, he was expecting Biddle and the other trustees to politically shield his ambitions. Bache had apparently accepted his nomination with the expectation of support from the trustees for a wide range of ideas and measures in line with his interest in developing leading American institutions, and this helps explain his decision to leave the University of Pennsylvania professorship for the new position.

What is striking today, of course, is the degree of informality involved in a college project of unprecedented financial size. Bache expected to have been hired to develop the administrative and educational framework for creating a model institution, and he could be hired and depart for a two-year trip to Europe with views which turned out to diverge significantly from those of the trustees and the board's chairman. In line with the municipal basis of his earlier efforts at the Franklin Institute in Philadelphia, Bache, as president of Girard College, relied on Nicholas Biddle and his standing in that city as a political framework for stabilizing his intellectual and cultural efforts on behalf of Philadelphia and the United States on a new scale. In late summer 1838, however, it began to dawn on Bache that the basis for these ambitions was eroding.

"I feel that I have discharged my duty," Bache responded to the chairman from Vienna, "which I owed to the [trustees] and, + to the Girard College. I can but regret that my imperfect [introduction of my ideal] on the matter has placed it in a different light before your committee than that in which it appeared to me."[55] Bache did not mention the matter again and neither did he give up on his plans to administer the college. Just as he was ready to return to the United States after his two-year absence, Bache summarized and reviewed his trip in a letter to Biddle. "In reviewing the tour which I have made," he wrote on August 6, 1838,

> it is interesting to find the rights of orphans so little considered. Objects of justice [are] private charity {underlined by Bache}. The leading consideration in [educating] them is here to do so at the lowest expense. It is time that … an institution where each individual receives an education in preparation to + in adaptation with his capacity + with the [bent] of his [mind], exists as far [as] my information [reveals]. Such an institution we have to supply then to the world and [Mr.] Girard has left us the means to furnish it. If I understand his intention he would have us consider the institution as the head of a large family all the members of which are to be [perfected] for just such a station in life as the qualities of their mind enables them to fill. We are to educate {Bache's underline} in the true sense of the word.[56]

While his initial idea on teacher training by way of European immersion was rejected, Bache sought to reclaim the initiative by declaring in this letter to Nicholas Biddle that the college was to avoid "making machines rather than mechanics." He boldly asserted that he hoped "to find the Trustees prepared to carry through" with the views he laid out in his letter because it would "determine peculiarities in the arrangement of the material as well as of the intellectual part of our institution." Bache thus sought to assert his presidential leadership and while his more ambitious administrative and educational plans were rejected, he insisted on the idea of aiding each child in developing its specific talents and capacities.

---

55 ADB to Nicholas Biddle, March 27, 1838, box 2, vol. 3, Bache Papers, SIA.
56 ADB to Nicholas Biddle, August 6, 1838, box 2, vol. 3, Bache Papers, SIA.

In his letters to his mother, however, Bache disclosed his growing disappointment with the college. In January 1838, Bache wrote from Halle (Prussia) and reported that he feared that the trustees had sent him

away from home [without] supposing that I meant to labour in earnest. If so it is [badly] conceived. I feel deeply the [responsibility] which I have assumed [+] the [debts] in[curred] for this journey, but if petty considerations were to be in the way of rightly organizing the Girard College + I am rejected [—] yield to [—], to claim a good salary as President, to avail myself and sacrifice my trust[,] the obligation will be cancelled + I shall [feel] myself [free]. I try not to charge when so far away, but the very expression of feelings sometimes relieves them. I feel that in many ways the Trustees [of] the Girard College [—] the wrong man in me. Time will show.[57]

Bache's disappointment is grounded in the suspicion that the trustees were unwilling to honor his broader ambitions for the college and did not share his view of the institution as an opportunity for testing and developing new educational ideas. He rejects the idea of being a mere figurehead to implement a standard policy because this obviously violates Bache's developmental goals and the responsibility he feels in living up to the pledge made by accepting to go on a two-year sojourn to Europe. This does indeed raise the question what intellectual justification the trustees considered Bache's trip to have. Bache's experience suggests that their heeding Francis Lieber's advice in sending abroad the future college president in order to study European schools was not associated with any radical intentions to modernize the American system. This unprogressive elitism paralleled Biddle's architectonic choice.

The significance of the failure of Bache's plan to have teachers sent to Europe should not be overemphasized. It remained a perhaps minor episode in the development of the college, and much more significant problems would soon arise. But it remains significant as a token of ideas and ambitions Bache associated with the project: an opportunity to develop a model institution by using leading European ideas as a propellant for American development towards world leadership. In this sense, Bache's interests were not restricted to professionalizing education; education was a dimension of his broader agenda of pressing forward the culture of the young American republic.[58] This helps explain why, subsequent to his

57 ADB to [Sophia Dallas], January 8, 1838, box 2, vol. 2, Bache Papers, SIA.

58 While Odgers mentions Bache's interest in teacher training, he does not explicate their broader motivational context and meaning. See Odgers, *Alexander Dallas Bache*, 74.

return to the United States in 1838, Bache continued to support the improvement of education in his native state but refrained from pursuing it as a career outside of the period under discussion. Bache did not become a Pennsylvania version of educator Henry Barnard who would go on a similar European tour a few years later. Barnard was central to the development of education in the United States by helping create the Massachusetts board of education and the first teachers' institute there, founding the *Common School Journal and Annals of Education* in 1838 and the American Association for the Advancement of Education in 1851.[59] Bache would be a member of that organization and even its president in 1855 (succeeding his friend Joseph Henry) but its scope was too narrow and merely embedded in his broader cultural ambitions which contained, but did not focus on, education. It was from the vantage point of an already self-conscious scientific community that Bache perceived of both the trajectory of American cultural development and of the role of education within it.

While I have strategically confined my discussion of Bache's trip to Europe to questions pertaining to his attitude towards education and to his ideas on the role of Girard College, historians of science have usually emphasized the significance of Bache's trip for the development of his views on the role of the profession in America. The college president's European travels coincided with that of physicist Joseph Henry, a close friend who had been awarded a year abroad by Princeton University. (Henry would soon be the first secretary of the Smithsonian Institution.) Bache's and Henry's correspondence during and after their tour is commonly viewed as evidence of a significant initiation. They compared conditions in Europe to their own country's scientific situation, and readied themselves for implementing higher and independent standards for their colleagues, ideas which they promoted by helping found and direct the American Association for the Advancement of Science in 1848 and the National Academy of Sciences in 1863.[60] The selection of Bache's letters to Biddle discussed above suggests that the ideals and values represented by the scientific community provided Bache with a platform for engaging in his work as college president, providing orientation and principles for his educational ideas and for his itinerary and recorded impressions of Europe and of its scientific scene.

59 *Encyclopedia Britannica, 2005 Ultimate Reference Suite DVD*, s.v. "Barnard, Henry."

60 Bache's and Henry's trips to Europe are discussed in Bruce, *Launching of Modern American Science*, 15–19.

As an indication of this perspective, Bache would frequently sign his letters as "Prof. Nat. Philos. University of Pennsylvania" rather than "President of Girard College," when corresponding with European colleagues.[61] While Robert V. Bruce has reminded us that Bache's and Henry's European tour "helped shape {their} vision of what American science should be," it seems in order to stress that Bache was thoroughly grounded in a scientific community and that his tour merely reinforced his belief in the development of American cultural potentials.[62] From his letters to Biddle, to his mother, and to scientific peers at home and abroad, it becomes evident that Bache, who had been decisive in orienting the Franklin Institute towards science and technology, viewed the institutional development of Girard College and of education as an aspect of the broader development of American culture. To Bache, science was the relevant arena for a future competition with Europe. The effect of the European trip was not to establish but to intensify his citizen's sense of pride in American achievements. In addition to serving as an opportunity for connecting to colleagues abroad, therefore, the trip served as a booster rather than a starter.

When he left for Europe, Bache was no longer a student of science but thought of himself as a representative of relevant scientific work being undertaken in the United States. As an entrée into European scientific circles, Bache referred to work by fellow Americans or to his own study on steam boiler explosions. In London, for example, Bache wrote to geologist John Phillips, a prominent member of the British Association for the Advancement of Science, asking for a "rational account of the late disastrous steamboat explosion at Hale." From what he had heard, the explanations of possible causes of that disaster were off the mark, and he suggested consideration of the Franklin Institute's research on the subject.[63] Bache sought to promote American science in Europe and not only was he occasionally annoyed that his colleagues in the United States wrote infrequently, but that his colleagues did little to enhance an independent American sci-

61 See, for example, ADB to "Monsieur," November 24, 1837, box 2, vol. 2, Bache Papers, SIA. Thomas Biddle, second cousin of Nicholas Biddle, had sent Bache "fifty pounds from a few of us who wish you to buy any thing useful more particularly to extend our acquaintance in mining + manufacturing iron, copper, lead + knowledge of mill + steam mill machinery" {Biddle's emphasis}. Thomas Biddle to ADB, September 19, 1836, RH 932, Rhees Collection, HL.

62 Bruce, *Launching of Modern American Science*, 18.

63 ADB to Prof. John Phillips, June 12, 1837, box 2, vol. 2, Bache Papers, SIA.

entific culture where they clearly had the option to do so.[64] In June 1838, Bache complained to English-born chemist John W. Draper, who had earlier written to Bache about his research, that European scientists were receiving information about American developments via England. "We ought to make an effort to change this state of things," he wrote and regretted that Draper had not provided him with information on his work "as in moving about I endeavour by conversation to make known what circulates most slowly in Europe."[65] And a year later, after he had returned to Philadelphia, Bache complained of Draper's appeal to a "foreign tribunal as calculated to foster this same narrow spirit" as relying too much on scientific opinion from abroad.[66] The European tour, in other words, intensified Bache's sense that the American scientific profession was insufficiently self-reliant.

Reminiscent perhaps of his own aspirations, Bache developed a particular admiration for the French mathematician, astronomer, and physicist François Arago. That French scientist, he wrote to his mother from Paris, "combines in a most surprising way science, politics, + municipal affairs."[67] In comparison, he seemed impressed, but not quite as awed, by Michael Faraday. At the "Royal Institution in London, Bache observed the famous physicist", "the genius loci," to be "exceedingly busy in saying a word to every body. … He devotes himself in fact to this institution & is its atlas."[68] On Christmas Day 1837, Bache wrote a long letter to his mother in which he described Alexander von Humboldt, whom he had met in Berlin a few days earlier, as reminding him of an "old gentleman who visited at Grandpa's" but stressing his "vitality … and polyglot character of his mind."[69] To Bache, Humboldt was an impressive figure for his knowledge and versatility and yet he felt that there was something outdated about him. Perhaps Humboldt stood for a mode of investigation that provided the basis for Bache's own work but symbolized heroic individual

64 ADB to John Vaughan, November 1, 1837, box 2, vol. 2, Bache Papers, SIA.

65 ADB to John W. Draper, June 23, 1838, box 2, vol. 3, Bache Papers, SIA.

66 ADB to John W. Draper, February 13, 1839, box 2, vol. 1, Bache Papers, SIA.

67 ADB to Sophia Dallas, August 1, 1837, box 2, vol. 2, Bache Papers, SIA. In a letter to his mother on July 28, 1838, Bache reported on his second meeting with Arago, an "acquaintance that was soon more agreeable than the first one." Box 2, vol. 3, Bache Papers, SIA.

68 Alexander Dallas Bache's European Diary, [April 7, 1836], Henry, *Papers of Joseph Henry*, 3:245–46.

69 ADB to [Sophia Dallas], December 25, 1837, box 2, vol. 2, Bache Papers, SIA.

investigation and achievement rather than the coordinated effort in which Bache sought to become involved and which he helped shape in the United States. In general, when comparing European countries and their achievements, Bache was particularly interested in the German educational system and, for the reasons explicated above, impressed by developments in Switzerland.[70] Bache was particularly interested in the canton Vaud with its capital Lausanne because of that region's progressive politics. "There are many things in which it might be followed by us fine republicans," he suggested, putting his own country's perceived leadership in perspective. He had apparently taken a close look at the canton's constitution and detailed it in a letter to his mother (which suggests that she took a great interest in her son's intellectual experience and in this particular topic). Bache pointed out that the "sovereignty resides in the people" and "all Vendois are equal in the eye of the law, there being neither restriction of locality, birth, patrons or families," which sheds light, of course, on limitations Bache took to be relevant in his own country. He went on to connect his summary of Vaud's political framework with the stipulations for the education of citizens. Every citizen is trusted to participate in the political process "for every child in the canton must be educated from the age of seven until sixteen" and that "there is an absolute necessity by the law that every child shall be sent to school, either to a public or a private school." He then goes on to describe the consequences, fines and imprisonment, for parents choosing to keep their child away from school.[71] After his return to the United States, Bache would revisit this topic in a speech in which he suggested that "the institutions of Switzerland are in harmony with the free spirit of the people," which contrasts with his perception of English aristocracy and its achievements, feeling that Eaton Hall, impressive as it might be, should not belong to one man.[72] Bache was no socialist *avant la lettre*, as Odgers implies; rather, his views were an expression of an avant-garde republicanism grounded in the idea of creating public institutions as an expression of the community's dedication to universalistic ideals represented by art and science.[73] In line with his reaction to his educational ideas being turned down by the trustees, he strongly felt the

70 Odgers points to Bache's lecture on Switzerland but does not explicate the significance that country had for Bache. See Odgers, *Alexander Dallas Bache*, 68–72.

71 ADB to Sophia Dallas, September [5], 1838, box 2, Bache Papers, SIA.

72 Odgers, *Alexander Dallas Bache*, 54, 71.

73 Ibid., 54.

obligation to live up to the standards implied in his remuneration, and after his return to the United States, he would do so in various ways and in difficult institutional and political circumstances.

Central High School

These difficulties were caused by the political symbolism and conflicting interests attached to Girard College. The trustees had been confident of being able to get teaching underway before the imposing and technically challenging structure would be completed. The city of Philadelphia, however, chose to adhere strictly to what it considered Girard's intent as stated in his will, and president Bache returned to a campus with five buildings, two of them under roof and one, Founder's Hall, with arches for three stories and four out of thirty-four columns standing. Architect Thomas Ustick Walter had gone to Europe (where he had met Bache) to study the construction of marble roofs in order to design one for the main building (which was required by Girard's will but had never been done in America). It would obviously be several years before the building would be completed and teaching commence at Girard College. Bache, a college president without a college, used his time to write a 660-page *Report on Education in Europe* in which he provided details about the continent's educational system, including descriptions of hundreds of schools and universities he had visited abroad. The book became a standard for educational reformers such as Henry Barnard.[74] In a survey of American schools, the latter considered the work by Bache and others who had quickly developed the Philadelphia's school system, "a proud monument of disinterested zeal, intelligence and fidelity of men who have been entrusted with its administration."[75] During this time, Bache also became instrumental in designing the educational program of the rather unusual Central High School in Philadelphia that was designed as the crown of the state's reformed public school system. The school building on Juniper and Market Street had been designed by Thomas Ustick Walter but unlike Girard College, it was in operation when Bache returned from Europe.

---

74 Bache, *Report on Education in Europe*; Odgers, *Alexander Dallas Bache*, 87 f.

75 Henry Barnard, "Report on a System of Common Schools, For Cities and Large Villages," *The District* 3, no. 2 (August 1, 1842): 28.

Pennsylvania had instituted public education in Philadelphia in 1818, providing education dominated by the Lancasterian tutor system in which older students helped teach their younger peers, but only paupers were admitted as everyone else was considered to have the option of attending private schools. This changed in 1836, when the state legislature passed a bill creating a public school system that was open to the middle classes, no longer operated on the Lancasterian system, and, in addition to elementary and middle schools, established in Philadelphia "one central high school for the full education of such pupils of the public schools of the First School District as may possess the required qualifications."[76] During the same year the educational system was reformed, President Andrew Jackson ordered federal funds in the Second Bank of the United States to be transferred to the several states, and Pennsylvania drew on this windfall to build a high school across from today's Philadelphia City Hall. Teaching got underway in October 1838. Bache had heard of the school while he was travelling in Europe. After his return, he agreed, or successfully proposed, to submit recommendations for the school's organization that were accepted in late 1839. Bache, with permission by the trustees of Girard College and remaining on the college payroll, became the director of the new downtown institution. It was here that Bache, during the following two years, put to use his European insights. The school's rationale ideally suited his educational ideas.

The school's raison d'être differed significantly from that of Girard College. While the latter had been conceived and was restricted by the will and the whims of its founder who could no longer be consulted, the high school was an ambitious public project. High school students would be older than those at Girard College, as students were at least twelve years of age.[77] The school served students who would otherwise have attended private Latin schools, or none at all.

Many of the policies suggested by Alexander Dallas Bache for Central High School in 1839 reflected ideas he had conveyed in his letters from Europe, such as the argument that rules of discipline "should be as few and general as possible; in this form they are convenient as guides to both teacher and pupil, but detailed regulations, which trammel the teacher, and exercise the inventive ingenuity of the pupil in their evasion, are perni-

76 Quoted in Fagan, "Alexander Dallas Bache," 22.

77 Edmonds, *History of the Central High School*, 64.

cious."[78] Despite initial difficulties cooperating with a faculty that had not been supervised for years, the "professors" at Central continued to enjoy a relatively large degree of "robust self-governance" after Bache was installed as director.[79] Bache stipulated that the principal and the professors were to meet as a board, and that as principal, he considered himself to be the faculty spokesman.[80] While conflicts were not absent from Central's faculty, Bache did indeed seek to follow the model he had in mind when writing to Nicholas Biddle from Europe in 1837.

More significantly, Bache turned Central, which was becoming just another Latin grammar school (so designed by the initial faculty of 1838), into a modern high school offering instruction on a wide variety of subjects including the natural sciences. This reflected Bache's earlier experience at the Franklin Institute whose school one author considers to have been Central's model.[81] Bache arranged two educational programs (a brief two-year course of "higher elementary instruction" and a "classical course") around a core of a four-year program in preparation for "commerce, manufactures, and the useful arts," which he considered to be the basis for the other two. "The phrasing of Bache's proposal," David Labaree has argued, "suggests that he designed the curriculum around the principal course and that he included the classical course reluctantly because of the demand from middle-class families who wanted to send boys to college."[82] The focus was a preparation for work rather than college and the school was thus able to disconnect itself from the demands of college entry exams in Latin and Greek to focus on modern languages and science. While Bache did not abandon the "classical" and "English" courses of studies, the school's focus on the "principal" course indicates that he was able to provide leadership for the idea that schools be dedicated to helping develop the student's abilities in line with a more immediate potential for

78 Fagan, "Alexander Dallas Bache," 21–31, quotation 29; Labaree, *Making of an American High School*, 10 f.

79 Ibid., 113.

80 Edmonds, *History of the Central High School*, 64.

81 Allen, "The Franklin Institute of the State of Pennsylvania," 277.

82 Labaree, *Making of an American High School*, 14 f. Labaree goes on to suggest that in eliminating the elementary and classical option over the years, the school more fully realized its rationale of putting out a culturally coherent set of students, "to mold a community of citizens" (p. 16). He does not address the question of how Bache and his colleagues at the school sought to respond to the pragmatic needs of their day. He also does not discuss the ideals which Bache had formulated or used in documents such as those discussed above.

practical application than a "classical" course of studies could provide. The principal course consisted of four years of English, French and Spanish, geography, history, mathematics, mechanical and natural philosophy, natural history, writing, and drawing, and (unlike Girard College) "moral lessons" such as "evidences of Christianity." It contained no Latin or Greek. As its name suggests, Central High School was designed as the ambitious apex to a city-wide school system. "The teachers of the grammar schools watch most anxiously its result, as involving in a measure the characters of their respective schools," the Controllers later reported, "and it is no exaggeration to say, that from the time of admission into the grammar schools, the pupil is trained with a view to his successful application for admission in the high school."[83] Competition was keen, and Controllers were eager to point out that the design of the entrance exams forestalled any "opportunity for partiality and favoritism" through written tests which were evaluated anonymously. Considering the employment opportunities taken up by the school's graduates, Central doubled as a normal school and became the kind of institution Bache had in mind when writing home from Switzerland: A public school dedicated to high standards provided by professors dedicated to science who would instruct, among others, students who would go on to teach at other school levels.[84] In 1841, Bache expanded these educational interests by assuming, in addition to his work at Central, the duties of general superintendent of all public schools in Philadelphia. By this time, the system had grown significantly. In 1842, it comprised 62 primary schools (9,342 students, supposedly between four and eight), 11 secondary schools (2,597 students between seven and ten), 19 grammar schools (8,445 students between ten and thirteen), and Central High School (307 students between thirteen and seventeen). All of the students were boys.[85]

What distinguished Bache from other school reformers at the time was that he tied vocational training to scientific research. As the most prominent sign of this orientation, the Central High School had installed a telescope (which had been ordered from Munich, Germany) in an astronomical observatory on top of its building, one of only four such observatories

83 Barnard, "Report," 29.

84 Concerning the idea for a normal school, see ibid.

85 In 1840, Bache developed recommendations for a school for girls. Alexander Dallas Bache, *Report on the Organization of a High School for Girls, and Seminary for Female Teachers* (Philadelphia: J. Crissy, printer, 1840); Barnard, "Report," 28.

in the United States. Bache wholeheartedly supported and organized its use by students.[86] He also attempted to connect Girard School to scientific work, first by discussing and seeking support for an "experimental school"[87], then by convincing the trustees to erect a magnetic observatory on college grounds. Bache had begun to work on geomagnetism and the recording of magnetic declination in the early 1830s.[88] He had then recorded declinations at home, and his wife had supported him as an assistant. In his necrology of Bache, Benjamin Apthorp Gould later remembered that his colleague, who would spend most of his career supervising the triangulation and charting of the American coastline, remained particularly interested in terrestrial magnetism.[89] The observational scope of this project was similar, of course, to Bache's idea of using the army for recording meteor showers. After his return from Europe, Bache had enthusiastically accepted the suggestion that he provide geomagnetic observers who would contribute to an international project to establish charts of magnetic deviation.[90] Central High School students began to operate both the magnetic observatory at Girard College and the astronomical observatory at their own school. Under Bache's supervision, the results of the magnetic observatory were published as a U.S. Senate document in 1847.[91] At a time when President John Quincy Adams' 1825 suggestion that "light-houses in the skies" be built in America was still remembered, by the Jacksonian Democracy, as a ridiculous phrase reflecting Adams' aloof elitism, leading Philadelphia citizens chose to go ahead and implement the idea.[92] Three years after Bache had joined Central, Henry Barnard

86 Labaree, *Making of an American High School*, 21 f.

87 ADB to "H.", February 3, 1839, box 2, vol. 1, Bache Papers, SIA. The school is also mentioned in ADB to [Fowley], February 16, 1839, box 2, vol. 1, Bache Papers, SIA.

88 Alexander Dallas Bache, "Observations on the disturbance in the direction of the horizontal needle, during the occurence of the Aurora of July 10th, 1833," *Journal of the Franklin Institute* 17, no. 1 (January 1834): 1–9.

89 Gould, "An Address in Commemoration of Alexander Dallas Bache," 9.

90 ADB to Prof. W. Lloyd, July 29, 1839, box 2, vol. 1, Bache Papers, SIA. Bache had written that "you may rely upon me to join in your great project." Edward Sabine in England was the instigator of the worldwide project.

91 Girard College Magnetic and Meteorological Observatory, Alexander Dallas Bache, and United States Army Corps of Engineers, *Observations at the Magnetic and Meteorological Observatory, at the Girard College, Philadelphia, Made Under the Direction of A. D. Bache, LL. D. and With Funds Supplied by the Members of the American Philosophical Society, and by the Topographical Bureau of the United States* (Washington: Gales and Seaton, printers, 1847). Observations are listed on over one thousand pages.

92 Edmonds, *History of the Central High School*, 82.

referred to Bache's work in Philadelphia as a model for his work in Massachusetts. In a broad review of institutions in his own state as well as in Rhode Island, Ohio, Kentucky, Maine, Connecticut, and New York City, Barnard underlined the significance of developments in Philadelphia. He pointed to the influence Central High School would have on the entire system by creating an incentive to other schools to train their students so as to muster Central's strict entrance exams.[93] But Barnard did not comment on the scientific work that was being pursued at that school.

It is obvious, therefore, that Bache continued to identify with his scientific peers between 1836 and 1842, a period when he had little opportunity to actively contribute to the advancement of knowledge. While he conceived of the Girard College project as falling within the scope of advancing his project of embedding in America institutions the will to attach the country to a universalistic discourse, and while the directorship provided Bache with an attractive income, he felt keenly that his scientific ambitions had been put on hold. Upon his return to Paris in July 1838, Bache had reported to his mother that while he enjoyed the sensation of again visiting what seemed a familiar place, it was also an ambivalent one. His esteem among colleagues there had not risen, he reported, "for in regard to science my [course] has been not even stationary, but retrograde."[94] While he was working on his *Report on Education in Europe*, Joseph Henry urged his friend to put much effort into it as "this is your first essay before the public in connection with Girard College" and because it would be of "much importance to yourself and to the cause of education in the United States." He added that there was "no little ill feeling in reference to the college among some of the clergy." He reminded Bache that he was no longer a professor at the University of Pennsylvania and could not rely on the cache that came with that post.[95] In his response, Bache rejected a notion he felt was implicit in Henry's advice—that he ought to make his *Report* malleable to the country's religious interests. Where "truth is at stake [I] will not flinch for the fear of the lash," he wrote. He rejected Henry's idea of publishing his report as a commercial matter and asserted that he was himself

93 Bache explained how students were examined in a letter to an unknown addressee, December 26, [1842], N.R. II, reel L, Reingold Papers, 1952–1991, Record Unit 7470, Smithsonian Institution Archives (hereafter cited as "Reingold Papers, SIA"). This item was copied from the Griscom Correspondence, New York Public Library. See also Barnard, "Report," 28.

94 ADB to [Sophia Dallas], July 9, 1838, box 2, vol. 3, Bache Papers, SIA.

95 Joseph Henry to ADB, May 20, 1839, Henry, *Papers of Joseph Henry*, 4:218.

critical of Girard's stipulation that no religious instruction was to take place at the college: "The natural supporters of the institution are those with thoroughly worldly views & principles, with them I have no affinity."

Fig. 7. Philadelphia's Central High School for Boys, South Juniper Street, facing Penn Square below.
Detail of 1853 photograph by Frederick Richards.

(Courtesy of The Library Company of Philadelphia, (6)1322.F.115c)

For our purposes, however, it is significant that Bache stressed that he had "no idea of relinquishing a scientific career" and that despite his work for Girard College, Central High School, and the public schools in Philadelphia, he endeavored "to extend, not merely to diffuse, the science of the day."[96] One may dismiss this assertion as a sign of Bache's defensiveness in a situation in which he was confronted with the restrictions on his scientific efforts but it remains noteworthy as an indication that he had not

96 ADB to Joseph Henry, May 28 [1839], ibid., 4:224–26. See also the editorial comment in the introduction to that volume, 4:xxi.

given up a dedication to a discourse and a mode of inquiry that informed and guided his institutional work. During his tenure there, Bache did indeed seek to implement scientific projects at Girard College and Central High School, efforts that shed light on his Humboldtian motivation to connect education with science.

Bache Ejected

Bache lost his hold on his educational efforts in two successive steps. The first involved developments at Girard College that had long been in the making. The institution had never been able to shake off the difficulties associated with the grandiose architectural designs by Nicholas Biddle and the delay in finishing the buildings. The trustees attempted to gain permission to open a preliminary school but they were rejected by lawyers who argued that this would violate Stephen Girard's will. In 1842, Philadelphia city councils, which supervised the college trustees and again responded to their wish to open a preliminary school, voted to investigate the work of the board and its expenses, including those of Alexander Dallas Bache. The committee turned down the application for an early opening of the institution and, in addition, found it extravagant that there was a president but no college.[97] Even before the Committee had begun its investigation, Bache had declined further compensation and had moved into the employment of the Controllers of the Public Schools in Philadelphia.[98] When the committee recommended in September 1840 that both the trustees and the

97 Bache had in fact been without salary from the college since December 1840 when he chose to forsake it in order to preclude the kind of criticism the City Councils later formulated. "The relinquishment of my salary was {as} a volunteer on my part, but the Council would try to make it appear otherwise. I was intending to allow the Trustees to make their fight entirely irrespective of pecuniary considerations, or those of a personal kind." For details, see Henry, *Papers of Joseph Henry*, 5:128–30 n. 1–8. The quotation is from ADB to John Vaughan, February [3?], 1841, box 2, vol. 1, Bache Papers, SIA.

98 ADB to Elias Loomis, October 22, 1840, N.R. II, roll S, Reingold Papers, SIA (copied from Elias Loomis family papers, Manuscripts and Archives, Yale University). The Common Council "have taken it into their hands to stop further progress" at the college, Bache wrote.

president should be dismissed, Bache resigned.[99] In its report, the committee were critical of the fact that

> the effort to obtain a plan of instruction and government, through their assistance, has proved, and is likely to prove abortive. For after having procured all the information that America and Europe could supply, they are not yet in possession of sufficient materials out of which a satisfactory system can be formed.[100]

This attitude towards the school's role was in opposition to Bache's much more ambitious plans. The committee had in mind available modes of running a college such as employing female teachers to instruct the orphans rather than the professional male staff Bache had referred to in his letters.[101] The committee found it "'hard to believe that this country does not afford more than one individual who is fully qualified to preside over an institution for educating boys,'" thus implying that Bache's appointment had been the result of favoritism and denying that the college could have a model role. After Bache had left, the board of trustees continued to operate until December 1841 when it was abandoned.[102]

The city council's decision marked the end of Nicholas Biddle's leadership in Philadelphia and, following his removal from the Second Bank of the United States, the closing of his public career. It was a significant popular success for the Jacksonian movement and the political press associated with it in Pennsylvania. It marked the end of a type of political and social leadership, and the Whigs, which had evolved in opposition to "King Andrew," adopted the popular political strategies introduced by the Jacksonians.

The December 23, 1841 resolution that severed Bache's connection with Girard left him in charge of the observatory and instruments there. "Should the movement take effect," Bache wrote to Joseph Henry, "I intend to embrace the first opportunity of returning to science *as a business* {his emphasis}." He added that the "magnetic observations were treated {by the Common Council's committee} as pretty amusements for grown children." With the help of John F. Frazer and students at Central High

99 "Report of the Special Committee Appointed by the Common Council on a Communication from the Board of Trustees of the Girard College," 51 f.; Herrick, *History of Girard College*, 27 f.

100 "Report of the Special Committee Appointed by the Common Council on a Communication from the Board of Trustees of the Girard College," 39.

101 Ibid., 26.

102 Herrick, *History of Girard College*, 28.

School, Bache continued to operate the observatory until 1845.[103] "Perhaps when we meet again we may weep together over the green grave of the first president of Girard College," Bache wrote to Henry.[104]

But Bache still retained his connections to the Central High School and the city's school system. "With an opportunity thus of doing much good to my native city," Bache wrote to mathematician Elias Loomis, "I do not complain if in some degree, for the present, removed from the opportunities of cultivating science as continuously as I desire to do."[105] But Bache's standing at the High School soon became precarious as well. The school was successful in the eyes of reformers but the idea of public schools that were not designed to provide education as alms for the poor remained new and untried in Pennsylvania. Critics focused on the school's "aristocratic" character or opposed the school because they found their own children not eligible or too far away to attend. Others had no children and rejected the idea of having their taxes support institution that seemed to them of little benefit; or, if they had children, chose to send them to private grammar schools and disliked the idea of paying "double."[106] In the report quoted by Henry Barnard, the Controllers of Philadelphia's public schools conceded that the school's "expense must necessarily increase." But they asked rhetorically: "What patriot would wish them diminished, to return to the scanty provisions for pauper instruction, or to the precarious, expensive

---

103 As mentioned above, they were published in 1847: Girard College Magnetic and Meteorological Observatory ..., *Observations at the Magnetic and Meteorological Observatory, at the Girard College*.

104 ADB to Joseph Henry, December 1, 1841, Henry, *Papers of Joseph Henry*, 5:128–30. The support by the APS "relieves me from the necessity of begging for funds for the purpose," Bache wrote (his emphasis). ADB to Loomis, September 17, 1841, N.R. II, roll S, Reingold Papers, SIA (item copied from Elias Loomis family papers, Manuscripts and Archives, Yale University).

105 ADB to Elias Loomis, October 22, 1840, N.R. II, S, Reingold Papers, SIA (item copied from Elias Loomis family papers, Manuscripts and Archives, Yale University).

106 Reese, *Origins of the American High School*, 59 ff. See also Fagan, "Alexander Dallas Bache," 31–34. Ironically, such criticism reverberates through some educational literature today. Reese takes the side of such critics, and so does sociologist David Labaree who uses Central High School as a sample case to show how reformers (including Alexander Dallas Bache) used it as a vehicle to implement their "particular social interests" (Labaree, *Making of an American High School*, 41). The merit-system introduced by Bache, Labaree argues, was used as a marketing tool to appeal to a middle-class which had sent its children to private schools but now was attracted to this new institution in downtown Philadelphia—an institution supported by taxes rather than tuition (ibid., 38 f.).

and exclusive system of private schools?"[107] The consequence of abolishing tax-funded public schools, they implied, was to return to a system that had stigmatized students and parents as poor and deserving, thereby strengthening private institutions such as grammar schools that had no connection to the public welfare.[108] In his annual reports, Bache refuted charges of social exclusiveness by providing charts showing the varied background of students.[109] But the school continued to be under pressure as it was considered as not being "'for the benefit of all'" and for serving the exclusive clientele which many found to be symbolically represented by Girard College and its Greek Revivalist design.[110]

Bache's connection to the Central High School, however, was dissolved for reasons diametrically opposed to such criticism. In his recommendations for the school curriculum and organization in 1839, Bache had argued against Latin and Greek as components of the core curriculum, and these languages had been omitted from the Central High School's "principal" course of studies. Bache had argued that "the Latin and Greek languages have no place in the foregoing scheme of instruction" because the "pupils begin them too late, and end the study of them too early."[111] Bache pursued this idea against strong opposition from the humanists among the school's faculty who did not share his modern interest in the physical sciences as a basis for vocational training. While the political details cannot be reconstructed from available documents, it seems certain that the humanists ejected Bache.[112] While Bache's role at Girard College had become precarious because, among other technical reasons, the institution was considered too aristocratic, it was "aristocrats" who assailed Bache at the Central High School. Girard College was attacked by Jacksonian Democrats upset with elitist aspirations associated with Nicholas Biddle and Greek Revivalism; at the Central High School, those representing what Laurence Veysey would later call "Liberal Culture," a traditionalistic elite

107 Barnard, "Report," 29.

108 For a detailed consideration of the attacks on the Central High School, see newspaper clippings in "Record of the Wars of 1842, 1845, and 1856," John S. Hart Papers, 1826–1875, Manuscripts Division, Princeton University Library. Labaree, *Making of an American High School*, 35.

109 Ibid., 13.

110 Ibid., 35. Quotation from newspaper clipping in Odgers, *Alexander Dallas Bache*, 132.

111 Alexander Dallas Bache, "Report to the High School Committee," 1839, quoted by Fagan, "Alexander Dallas Bache," 30.

112 Odgers, *Alexander Dallas Bache*, 133–36; Fagan, "Alexander Dallas Bache," 39–41.

sometimes attached to pre-revolutionary standards, disfavored Bache's interest in preparing students to solve the republic's problems with the aid of modern physical sciences.[113]

In a letter he wrote to Central High School professor of French, John A. Deloutte, Bache merely hints at the difficulties which had evolved among the Central High School faculty. The University of Pennsylvania had apparently inquired whether he would be interested in returning to a professorship there.[114] "I was in no haste to send in my resignation" to the High School, Bache wrote to Deloutte, "until it appeared entirely essential to do so from the necessity of accepting the appointment at the University."[115] After veiled references to his enemies inside the school faculty, Bache added that one

> service of good to the school which I supposed might possibly accrue from my removal was that it would afford a pretext to those who had opposed it from ignorance for yielding + again those who from political or other feeling against me had opposed it would cease to do so. But I have always supposed that there were two sides to this, + that the school might lose as much from the warm support of my friends as it would gain on the other side.

Bache had made the decision to leave the Central High School and in his letter to Deloutte he was explaining his decision. He had considered for some time whether his departure may have the effect of uniting two distinct groups which had opposed the school for different reasons: One group disliked the school out of "ignorance," and perhaps Bache had in mind the populist impulse aligned with the Jacksonian movement that

---

113 Laurence R. Veysey, *The Emergence of the American University* (Chicago: Univ. of Chicago Press, 1965).

114 In late July 1842, Henry considered advising Bache to open a private school, prospects of a chair at a university not being good (draft of a letter by Henry to ADB, [late July 1842], Henry, *Papers of Joseph Henry*, 5:254.) When he was offered the post at the university, Bache wrote to Henry that his financial situation was too precarious to decline it. "I have been reluctantly, I confess, brought to the conviction that my present position is too precarious to allow me to decline the University offer" (ADB to Joseph Henry, [July 29, 1842], ibid., 5:255.). Bache applied for a post at the University of Virginia, so as to be able to negotiate about his salary at the University of Pennsylvania where he received 2300 dollars (ADB to Henry, August 22, 1842, ibid., 5:258 f.).

115 ADB to John A. Deloutte, August 13, 1842, Gratz Collection, Historical Society of Pennsylvania. Odgers quotes from this letter and his transcription of some words differs from mine. See Odgers, *Alexander Dallas Bache*, 135 f. The letter is also referred to in Fagan, "Alexander Dallas Bache," 38 f. Neither Odgers nor Fagan take a closer look at this letter.

identified the school as "aristocratic." The other was opposed to it because of a "political or other feeling" against Bache himself, a reference which may point to former Federalists or Whigs who may have considered the principal to be a token of his family's stake in Jacksonian Democracy. This reading of Bache's letter aligns very well with all of our other observations, including the following: Bache writes that he had long balanced the political perspectives of both camps, whose political alignments and interests he considers to constitute "the other side," with the "warm support of my friends." Who are these "friends"? Against the backdrop of everything we have laid out in this chapter, Bache is most likely referring to his scientific peers in the United States and abroad. Bache implies that he conceived of such support as a unique opportunity not easily provided by someone else. This corresponds to our assessment that Bache stood out from the group of school reformers by his conscious devotion to science. It would not be surprising should he fear that his removal from the school would result in its politicization, i.e. in the loss of the idea that a republican elite in America should derive its legitimacy through rational problem-solving and practical achievements for the common welfare.

Bache was writing from Newport where he was likely visiting his in-laws. After suggesting that it was a promising sign that the Central High School succeeded in developing an institutional sense of integrity and an esprit de corps (by not allowing "[all] who were courted by the latter institution {i.e. the University of Pennsylvania} to be drawn away"), he concluded his letter to Deloutte by adding that he had

> just returned from a military review at the fortification where I made my first forays in practical life, the plan of which once put upon paper has been carried out by very various persons, by a succession of officers;—the work is now complete + no record is preserved of who built the different parts; the present generation cares nothing about <u>that</u>. In walking over the fortification with the young officer now in charge I found myself more interested in tracing its <u>history</u>, than in admiring its present finished condition.

Bache's somber comments on his visit to Fort Adams, where he had worked as an engineer from 1826 to 1828, illustrate his general theme of how individual efforts and achievements are valued and remembered. The disappointment of leaving the Central High School made Bache reminisce that his work, of which he was proud, would likely not be acknowledged by future generations. His comments point to Bache's ability to develop motivational momentum in the absence of any such promise by reassuring

Deloutte of the continued relevance of the Central High School and of public schools in general. He contrasts the younger generation's lack of historical consciousness with the sudden realization of his own interest in how different generations approach and tackle their respective opportunities. Bache's perspective is that of a community which provides the focus for such efforts, and the context for their appreciation and memory.

In September 1842, Bache returned to a professorship at the University of Pennsylvania. Slotten suggests that Bache believed that a man of science would also "take an active interest in other social and cultural pursuits." Bache's educational work between 1836 and 1842 shows that he was less keen on performing as a scientist than to use a scientific discourse, whose relevance he took for granted, for developing institutions that facilitated the cultivation of a political community.[116] Bache had become a protagonist for efforts to establish a competitive, research-oriented school system as a public project, not a private one, but unlike other educational reformers such as Henry Barnard, he did so with the aim of connecting these institutions to the universalistic discourse of science. The following months would provide him with an unforeseen opportunity to take his ideas to the federal level in Washington D.C.

116 Slotten, *Patronage, Practice, and the Culture of American Science*, 10.

Chapter 6

Bache's Program for National Consolidation I

Bache's 1842 Address on "American Manufactures"

1842 was a year of transition for Alexander Dallas Bache. In the fall, he had just returned to his professorship at the University of Pennsylvania, when the superintendent of the U.S. Coast Survey, Ferdinand Hassler, died. This provided Bache's colleagues with an opportunity to lobby effectively for him to become Hassler's successor in Washington D.C. Considering Bache's previous work in Philadelphia, the 1863 founding of the National Academy assumes the quality of a "cumulative" event towards the end of his life that naturally follows from his early ambitions and ideas. For this reason, I will discuss in detail Bache's 1863 rationale for founding it in a later chapter. Before I do so, I will test the evolving hypothesis on Bache's perception of the role of science in the emerging United States by slightly shifting the analysis, and focusing even more thoroughly on important documents, and commenting on institutional contexts only in passing. Bache's subsequent work for the Coast Survey and his professional involvement, such as helping found and direct the American Association for the Advancement of Science (AAAS), unsuccessful attempts to found a national university in Albany, and public fights over the directorship of an observatory, are much better known and accessible than his earlier work in Philadelphia.[1]

---

1 Important guides include: Robert V. Bruce, *The Launching of Modern American Science, 1846–1876* (New York: Knopf, 1987); Sally Gregory Kohlstedt, *The Formation of the American Scientific Community: The American Association for the Advancement of Science, 1848–1860* (Urbana: Univ. of Illinois Press, 1976); Hugh R. Slotten, *Patronage, Practice, and the Culture of American Science: Alexander Dallas Bache and the U.S. Coast Survey* (Cambridge: Cambridge Univ. Press, 1994); A. Hunter Dupree, *Science in the Federal Government: A History of Policies and Activities to 1940* (Baltimore: Johns Hopkins Univ. Press, 1986); Mary

In the following chapters, I will take a close look at speeches Bache gave in 1842, 1844, and 1851. Historians of science have long considered two of them to encapsulate Bache's program for developing science as a profession, a program he was able to pull off by using his political and intra-professional leverage as head of a small but influential group that called itself the "Lazzaroni." His group dominated science organizations such as the AAAS. My focus on these speeches will be complemented by my analysis of an 1854 exchange of letters between Bache and Harvard mathematician Benjamin Peirce, fellow Lazzarone and instigator of the National Academy of Sciences during the Civil War. The discussion of these documents will serve as a test for the interpretation as it has evolved so far and as an opportunity to adjust, refine, and expand the findings.

American Mythology

The first document to receive scrutiny is the address on "American Manufactures" which Bache gave at the close of an exhibition by the Franklin Institute, an institution he had helped shape.[2] Bache gave this speech in October 1842, just a few months after he had left his post at the Central High School for his old professorship at the University of Pennsylvania, and weeks before he could anticipate the vacancy of the post of superintendent of the U.S. Coast Survey. What does this speech reveal about Bache's view of science and of the Franklin Institute, now that he had spent several years as a director of educational institutions?

Bache's published Philadelphia speech begins as follows:

> The traveler in the deserts of Syria, resting at one of those few favored spots where the turf shows the presence of the refreshing well, and the date palm gives him shade, finds himself amid the ruins of a great city.

Ann James, *Elites in Conflict: The Antebellum Clash over the Dudley Observatory* (New Brunswick: Rutgers Univ. Press, 1987).

2 Alexander Dallas Bache, "Address delivered at the close of the Twelfth Exhibition of American Manufactures, held by the Franklin Institute of the State of Pennsylvania, for the Promotion of the Mechanic Arts, October 1842," *Journal of the Franklin Institute* (1842): 379–94. The Franklin Institute also published the address separately. It is available at the New York Public Library: Alexander Dallas Bache, *Address delivered at the close of the twelfth exhibition of American manufactures: held by the Franklin Institute of the state of Pennsylvania, for the promotion of the mechanic arts, October, 1842* ([Philadelphia]: [The Institute], n.d.). References below are to the latter edition.

The speech concluded the Franklin Institute's exhibition and allowed Bache to review the event and the public's response. Even though Bache had been a member of the Franklin Institute, his former affiliation is not mentioned. He is presented as a "Professor of Natural Philosophy and Chemistry in the University of Pennsylvania."

Instead of focusing on the exhibition itself or on its cultural or political context, Bache invokes a literary and generalized traveler in the Syrian Desert. He invites his audience to imagine this traveler arriving in a hospitable environment after a tiring journey in difficult conditions. There is shade, and the traveler may now refresh himself with water and dates. The trekker assumes a more leisurely outlook and realizes that he has stumbled upon the remnants of a great civilization. The ruins are evidence for eminent cultural achievements in a hostile environment. The adjective "great" implies that this bygone culture was not merely able to successfully deal with infrastructural problems in the desert but to live up to significant aesthetic aspirations.

In the context of an exhibition of manufactures, Bache's opening provides an analogy to the United States in 1842. The traveler's leisurely attention is directed to the remnants of a past civilization that was able to master the challenges of its environment. The technology on display at the Franklin Institute's exhibition on manufactures served a similar purpose. It provided the United States with tools to master its surroundings and to live up to the country's self-imposed cultural challenge of "settling the continent."

> Broken columns—architraves, and fragments of pediments half imbedded in the sand—heaps of ruins, indicating the former existence of massive structures, and deluding him with the idea that even now he may trace the extent and form of the space once occupied by the dwellings of men—all speak of the magnificence, the grandeur, and the vastness, of a great commercial capital.

Bache is speaking of Palmyra, ancient trading city in the Syrian Desert. Architraves (as well as pediments) were typical for classical Roman architecture. Palmyra's wealth stemmed from its important role in trading between east (Euphrates River) and west (Mediterranean Sea) that matches Bache's reference to "commercial capital." Even today one may find among the city's ruins (also called Tadmur or Tadmor) the graves of wealthy traders. Palmyra was excavated in the 1840s and then received renewed cultural attention. Bache is aware of such developments, but he does not mention the city's name here merely hints at its identity. His clues are unambiguous. No other place in the Syrian Desert matches his coordinates. Bache draws his audience's attention by providing it with a cultured riddle. He assumes that the Franklin Institute, with its exhibition of manufactures, appealed to the city's educated classes.

Bache further delineates the traveler's impressions of the city's ruins and suggests that he is "deluded" by them, as he seems unable to reconstruct the settlement's size and dimension. Palmyra's true size and significance, in other words, cannot be established and remain mysterious, and the beholder of these archeological remains is in awe in the face of this *Kulturlandschaft* as his imagination recomposes the city's majesty. It is the role of the traveler's imagination that is striking here. Bache's topic is the aesthetic presence of culture and of its past. By referring to "men," Bache does not restrict his perspective to a specific culture and instead has in view the provenance of all of humanity. This matches his use of the indefinite article in "a great commercial capital" as Palmyra is thus viewed as an instance in the global development of such centers, a particularly striking example of the cultural potential of civilization in general. In the context of an exhibition on manufactures, Bache focuses his attention on cultural achievements in a challenging environment that presuppose infrastructural accomplishments ("magnificence, grandeur and vastness"). In this way, Bache implicitly raises the question of what his own country's past will once look like: What will America's cultural achievements have been when its culture has vanished?

> He is amid the ruins of Tadmor of the wilderness, Palmyra, the great commercial emporium of former days—now part of the greater desert.

Bache resolves his riddle and reveals what he had been speaking of. He highlights the settlement's adverse circumstances and this provides an additional clue that the city's setting was an important aspect in choosing to speak of Palmyra in the first place. The definite article in "the great commercial emporium" singles it out as the leading merchant capital of its time. Bache's use of the term "wilderness" further suggests that he considers the city on a cultural trajectory that includes the United States in 1842. The concrete historical circumstances and particular timing ("of former days") are less important than the city's historic world leadership.

> Here was once the entrepot of the commerce of the East and West, and here arose a city—as it were one vast temple to that commerce which linked together the far East and West.

Bache highlights Palmyra's role as a point of exchange between the Euphrates and Asia in the east and the Roman Empire and the Mediterranean in the west. He explicitly connects the city's cultural achievements to its trading role and even suggests that the city's culture was *dedicated* to this "intercultural" function ("as it were one vast temple"). What comes to mind here is the role that early American leaders such as Thomas Jefferson envisioned for the United States: to become a trading hub between Europe and Asia.[3] Palmyra becomes a role model for America's ambition and future.

> Amid the lagunes and marshes at the head of the Adriatic, the gorgeous fane and splendid palace are reared, and the varied ornaments of a florid architecture are lavished to decorate the homes of the merchant nobles.

Bache moves on to another instance of a successful trading city: He provides his audience with clues (such as "Adriatic" and "merchant nobles") to guess that he is speaking of Venice. The contrast to Palmyra lies in that city's setting: While the Romans were able to erect Palmyra in the midst of a desert, the builders of Venice were challenged by water. There is no apparent cultural connection between the two cities, between the Roman Empire and Venice merchants. Considering his choice of cities here, it is unlikely that Bache was not aware of a Romantic discourse in which both places were associated with decay and desolation. While Bache highlights Palmyra's aesthetic appeal, however, he emphasizes both cities' technological achievements among adverse conditions.

> The very difficulties of the site are made to contribute to luxury; no noise of wheels disturbs the quiet of home, or the hum of business on the Rialto, but the luxurious gondola glides silently through the vast canals which connect the distant quarters of the queen of the sea. Commerce has been again at her work.

Bache continues in the same vein. He still does not mention Venice by name but now clearly identifies the city by mentioning the Rialto Bridge that crosses over the city's Grand Canal. Bache continues to be interested in the challenges of, and the successful adaption to, the specific challenges to a given settlement's locale. Not only did Venetian architects deal with the difficult conditions in which they erected that city but they aptly used them to their advantage. In contrast to the sandy roads of Palmyra (and the

3 Henry Nash Smith, *Virgin Land: The American West as Symbol and Myth* (Cambridge: Harvard Univ. Press, 1950), 15–50.

dusty unpaved roads of contemporary American summers), Venetian canals provide for a seemingly effortless and a quiet mode of transportation over a vast urban area ("distant quarters"). In the given context of an exhibition of manufactures and against the backdrop of contemporary American ideas of settling an "untamed" continent, Bache continues to focus on predecessors of successful settlement and cultural achievement in a challenging environment.

Through this comparison, Bache implicitly refers to a potential future role of his own country. It is connected to its predecessors through "commerce," an almost god-like entity seemingly independent of time and space. In addition, Palmyra and Venice provide a standard for the emergence and success of American culture. The fact that Bache makes such a comparison by focusing on cities suggests that he was still very much rooted in Philadelphia as a hub of his intellectual and professional development even though his emphasis on taming the natural environment is in line with an American perspective.

Civilization has advanced westward; and while Tadmor is crumbling, and the sands of the desert are gathering over its ruins, Venice is rising from the waters, the new entrepot of commerce between the East and West.

The conclusion to this initial paragraph adds few ideas. Its main function consists in solving the riddle by providing the audience with the cities' names. In addition, Bache highlights the fact that Venice lies west of Palmyra, proposing that "Civilization" (as an extension of commerce) not merely moves, but "advance[s] westward." The line of development thus extends to the United States, and this confirms that Bache draws on the examples of Palmyra (Tadmor) and Venice as predecessors to an anticipated American role. He seeks to answer the question: What will the United States have left behind when its commercial and cultural development has run its course? At the same time, he takes for granted that the country's future legacy will be compared to Palmyra's and Venice's outstanding cultural achievements, confirming Bache's view of his country's development as self-assured and ambitious. Even though he does not mention his country here, Bache's comments presuppose that the United States aspires to assume cultural leadership of the western world.[4] If Bache

4 Bache adopts for his purpose long-standing notions of *translatio imperii*, the westward movement of empires in history, but he adjusts it by stressing the role of trade and the natural challenges for settlement. The concept of *translatio imperii* is described by

feels compelled to invoke such mythological questions at all (the rise and fall of cultures in the past) his country's emerging role was not yet buffered and legitimized by an independent and a uniquely American past. In line with a discourse on the nature of the American empire, Bache scanned world history for precursors to his country's situation, i.e. eagerness for cultural achievement under adverse settling conditions based on successful commerce. Palmyra and Venice came to mind. He connected the dots and the emerging historic diagram pointed to his own country's anticipated role.

> A new route is discovered, by which the products of the agriculture and arts of India are conveyed to Europe; commerce departs with prosperity in her train, and Venice is given over to the destroyer.

We had anticipated that Venice would also be viewed as moribund and Bache now indeed makes this point.[5] He implicitly raises the question of what other city (or country) along a "new route" will become the next trading capital. Bache does not have in view settlements that exist beyond their economic and cultural peak. He presents an all-or-nothing situation in which a city either prospers or is abandoned. The continued existence of both Tadmor and of Venice suggests otherwise but Bache is interested in cultural pinnacles.

> In the early periods of history these changes were few, their progress was gradual, like the slow changes of the scenes of a diorama; ages elapsed before the tide ceased to set through Palmyra. In modern times the changes are like those of the kaleidoscope, sudden and striking. Agriculture changes its objects or its methods—manufactures spring up and flourish, or decay—the arts find new seats and new subjects for their exercise—commerce, which connects the producer and the consumer, runs in new channels. Cities greater than Tadmor or Venice spring up, the creations of a new civilization.

Antonello Gerbi, *La disputa del Nuovo Mondo: Storia di una polemia, 1750–1900* (Milano: Riccardo Ricciardi Editore, 1955), translated by Jeremy Moyle, *The Dispute of the New World: The History of a Polemic, 1750–1900* (Pittsburg: Univ. of Pittsburg Press, 2010), 130–45. See also Loren Baritz, "The Idea of the West," *American Historical Review* 66, no. 3 (April 1961): 618–40.

5 While travelling in Europe, Bache had visited Venice and in a letter to his mother, his observations matched those of contemporary travel guides. "There is nothing in ruins in Venice," he had written, "but everything in decay [—] that looks as if centuries in neglect have accomplished it + as if centuries more would be required." ADB to Sophia Bache, April 8, 1838, box 2, vol. 2, Bache Papers, SIA.

While Bache continues to discuss his two samples, his initial riddle is completed, and he shifts his focus. He has not mentioned his own country but in the context of an exhibition on American manufactures, his reframing of the Romantic reference to Palmyra and Venice as places of decay suggests that Bache took it as a point of comparison for future American achievements in technology and culture. Bache emphasizes the likelihood and the altered pace of change. By explicitly shifting his attention to Tadmor's and Venice's "successor civilizations," Bache continues to address his own country's cultural potential and ambition. The stepped-up tempo of progress, he seems to suggest, allows his listeners to anticipate American cultural leadership within their life span.

> Increased production, whether in agriculture or in manufactures, is so obvious and powerful a source of prosperity to a country, that we naturally look with interest upon every circumstance which may effect it, endeavoring as far as may be, to understand, that we may aid. While all are agreed as to the necessity for cherishing agriculture, manufactures, the mechanic arts, and commerce, as the essential elements of national wealth, few agree as to the means of protection.

Bache now zooms in on questions of how best to live up to American ambitions for cultural development and leadership. His focus clearly is on "national wealth" and development as he discusses national (and not municipal) policy.

Bache speaks of "means of protection" rather than "support" or "development." He seems to take for granted that agriculture, manufacturing, engineering, and trade will strongly evolve without any support except for protection. Is Bache referring to the contentious political issue of tariffs? This does not seem to be the case as he is speaking of a difference in opinion about the *means* of protection while Southern economic interests opposed tariffs to protect American interests rather than demanding a different way of protecting them. Bache uses the term "protection" in a more basic sense. He implies that the different areas of American development will develop naturally as long as they are "protected" in the sense that nothing impedes their rational development.

> One would think that by this time facts enough had been accumulated to settle all doubts, and to establish a science whose principles should be as well ascertained as those of the philosophy of nature. But the passions, prejudices, and interests of men must be overcome before they desire to find the truth; and then all the difficulties remain of interpreting the results of complex experiments, and of assigning the just influence to each of their numerous and varied attendant circumstances.

The problem, according to Bache, is not that there existed a dearth of ideas on how to develop American efficiency; it is their implementation that remained difficult. He takes for granted that the principles of the philosophy of nature, the principles of natural science, had been settled, and he uses them as a test for evaluating what would later evolve into the "social sciences:" understanding the forces shaping economic, social, and cultural development. Bache is optimistic about the potential of such research for he doubts his country's belief in the relevance of such principles rather than the principles themselves. He takes for granted the validity of scientific principles and seeks to install them in the United States where "passions, prejudices, and interests of men" are yet to be "overcome." In a first step, cultural progress will necessitate a "desire to find the truth," a willingness to let go of cherished beliefs. On this basis, the analysis of "influences" may proceed. Bache assumes that such a readjustment and the ensuing discussion will open up the prospect of arriving at widely accepted results but stresses that such analysis remained difficult.

> It is conceded in every civilized community that the products of its agriculture, manufactures, and arts, should be brought as nearly as possible to perfection, and that improvement is the necessary consequence of the increased intelligence of those who follow the various callings connected with them. Avoiding, then, debated and debatable ground, and planting ourselves upon that which is fully and fairly our own, it may be profitable for us to consider *the means employed in different countries for the promotion of manufactures and the mechanic arts, and of the intellectual improvement of their cultivators.*

Bache rolls out the standard against which American self-civilizing efforts are to be measured: to strive for improvement and perfection in all areas of occupation. Bache focuses on the individual and this corresponds to his earlier criticism of "passions, prejudices, and interests of men." For Bache, the individual is responsible for helping to increase the community's intelligence.

What does Bache mean by "debated and debatable ground?" Considering his focus on individuals engaged in the areas of "agriculture, manufactures, and arts," Bache avoids a discussion of these branches in a more comprehensive, organization-focused, perhaps even state-oriented and political fashion. Bache suggests that to discuss the improvement of agriculture, manufactures, and arts without a focus on the individual is to depart from ground "which is fully and fairly our own." This reflects, of course, a contemporary American penchant for what used to be called "individualism," for independence from political interference, an idea strongly reflected in contemporary American politics. His comment suggests that Bache may well consider this perspective to be part of the "prejudices" he had criticized above. Bache does not seek to address this matter. He is aware of the limitations imposed by contemporary American culture and chooses a non-confrontational approach by reassuring his audience that he is putting himself squarely into the American camp. At the end of this paragraph, Bache outlines his topic, in italics, as a presentation of such ideas but he concludes by stressing the relevance of these ideas for individual "cultivators." In his negotiation of a perceived cultural resistance against ideas that did not originate within the United States, Bache assumes the role of the cultivator of America, cautiously suggesting that his fellow citizens adapt an intelligent approach to solving problems. What is striking here is the combination of Bache's ambition for his own culture, his acknowledgement of his country's limitations in developing it, and his almost pedagogically humble attempt to translate his international experience.

> From this general survey, we may derive materials for a comparative estimate of our own efforts—encouragement it may be, or stimulus to increased exertion;—hints of new lines of usefulness, or assurance that perseverance in those in which our efforts are already directed, will ultimately be crowned with success. In a country like this, where public opinion makes, alters, or repeals, the laws, there is always reason to hope for the success of what is right. It may not come this moment, nor the next, but as sure as darkness of night heralds the approach of dawn, which certifies the coming noonday, so surely will truth finally prevail where public opinion rules.

Bache now continues as one may expect him to. Other countries' insights become a standard for American development, and the United States was still trying to catch up to their achievements. Bache considers himself to be intellectually on a par with developments elsewhere, and it is from this vantage point that he turns back to consider his own country's situation and needs. In the wake of his experience as president of Girard College, furthermore, Bache is keenly aware of his country's emerging mass democracy. Bache suggests that public opinion has a very decisive role by proposing that Congress merely reacted to it so that, in his view, it "makes, alters, or repeals, the laws." In line with his role at the Franklin Institute's committee on steam boiler experiments, Bache would prefer Congress to assume a more decisive role in seeking support for rational legislation. But the time for enlightened legislation had not yet arrived. Bache assumes that politicians could or would not assume a leadership role akin to that of the intellectual, and yet his assessment differed from Nicholas Biddles's in that Bache took no condescending view of the American masses.[6] He agrees that education was required but expected rational policy to eventually evolve.

Prospects for Consolidating the American Nation

We can easily connect many aspects of Bache's introduction to his experience during the preceding years. The issue of political legitimacy for education and for rational political action had been a central concern of his work at the Central High School, a public institution that continued to be exposed to political attacks. Bache's engagement in his home town of Philadelphia, furthermore, is reflected in his looking back to other cities as a precedent to the American experience. His address suggests that he was using it as an opportunity to make sense of the political context of his work during the preceding years.

This becomes particularly evident in subsequent passages of Bache's speech. The historic dimension of the American experience that he had highlighted in his introduction provides a backdrop for a more thorough investigation of contemporary problems and opportunities. Bache takes for granted that his compatriots hesitate to acknowledge other countries' cultural achievements because of a lack of self-confidence. I will now proceed at a quicker pace, summarizing Bache's speech, and focusing on selected passages that promise to test or adjust the interpretation.

Bache restricts his survey of European countries and their efforts to support the different branches of "the arts" by briefly addressing his interpretation of the situation in Austria, Prussia, France, and England. In his depiction of the situation in each of these countries, Bache is highly selective. His samples strengthen his broader argument that public support *as such* was an important element for cultural development, a fact that, in his opinion, Americans would eventually need to accept. Bache consistently argues in favor of the public support of cultural institutions but he does not do so in order to implement his particular profession's interest for "support without strings." He insists on the importance of fully developing

6 See a brief discussion of Nicholas Biddle's views in the preceding chapter.

a uniquely American potential for innovation, which, in turn, he considers the basis for developing American aspirations for a global and historic significance through artistic, scientific, and technological achievement.

At the same time, however, Bache's speech suggests that he remains tied to an urban and Philadelphian perspective on American developments. While discussing the future role of an American "sovereign," Bache continues to assume a perspective associated with his city and its institutions. This was in line with the particular context of Bache's speech at the Franklin Institution. At the same time, however, it is striking how Bache intuitively assumes a "we" that is at home in the city rather than the state or the country. He is able to make out the future of the American body politic but Bache remains a child of his time.

Bache begins his brief international survey in Austria, highlighting how this country's museum exhibits include manufactures from abroad. But Bache holds on to a pluralistic view of how to support industry, taking for granted that there is not one correct model. Austria, he writes, "has *her* way of encouraging manufactures." Bache highlights Prussia's role in "uniting the States of Germany … in a commercial league" and his attention to this matter suggests that he had an eye for models of integrating the American states. At the same time, however, Bache considered the American political situation to have distinct intellectual benefits.

> We may admire the efforts of the Austrian and Prussian commissions, but after all, the plodding spirit of routine which clogs the limbs of activity in these countries, renders the measure of success of the plans *there*, no scale of what would be accomplished where the load of despotism was not to be borne forward.[7]

By contrasting the "clogging routine" in Austria and Prussia to the American condition, Bache implies that he was interested in its reverse, in "crisis" and problem solving and its potential for expanding knowledge by providing unforeseen solutions. By speaking of activities in these European countries as "limbs," Bache considers them as organic units. In Bache's depiction, Austria's and Prussia's potentially innovative branches of industry and the arts were constricted by a political system they were designed to support, a system that ignored the citizen's concrete interests and the innovator's pursuit of new ideas. For the time being, however, they remained models for the emerging United States whose innovative potential, rooted

7 Bache, *Address delivered at the close of the twelfth exhibition of American manufactures*, 5.

in republican and democratic principles and the premise that the individual could radically pursue his or her own interest, was yet to unfold.

In his brief tour of alternate models of supporting public institutions, Bache touches upon the French sample only briefly. He suggests that "France has halted in her scientific career since the youth of the nation have drunk so deeply of the excitements of political life" and that the country relied on "acquisitions of a past day."[8] He then moves on to England, a case he obviously considers more relevant. If the United States had the advantage of deducing the legitimacy for its political institutions from the interests of its citizens, why not take England as an example and abandon the idea of improvement through public institutions? Did not England provide an example for a successful implementation of individual liberty while not providing for public institutions in education and elsewhere? Why should individuals surrender their autonomy by succumbing to public institutions? Bache accentuates a possible conclusion from the English sample—that no promotion of the arts was necessary for that "which can and will take care of itself." If England was successful, why should America not also be successful without publicly supporting the development of its manufactures?

In anticipation of such arguments, Bache criticizes conditions in Great Britain which he does not wish to be a model for the United States. "Keen competition," Bache stresses, "keeps men much asunder." For fear of compromising their ideas to others, the famous workshops (such as James Watt's) were closed to visitors. The focus on economic benefits strongly obstructed an exchange of ideas. This is not only in contrast to Bache's efforts, in the 1830s, of helping shape the Franklin Institute into a platform for open intellectual exchange. He does not give up his idea that the United States, in contrast to Great Britain, had an opportunity to combine individual aspiration and effort with public support and to provide the individual with the cultural wherewithal to make contributions that are innovative, not just for himself or for his immediate community, but with respect to the state of culture in all areas including manufacturing. In his speech, Bache does not have in mind the scientific profession alone. His interest is not confined to expanding his profession's role but to arguing in favor of strategies that maximize the potential for cultural innovation in all areas of science and industry.

8 Ibid.

While Bache does not consider English inventors as providing a model for America, he suggests that British associations should perhaps be emulated. "The collision of mind with mind that takes place in these numerous associations," he argues,

> is of high importance; the tendency is to make men aware of their own deficiencies and to furnish a motive to supply them, to liberalize the feelings, to promote mutual confidence, and to produce esprit de corps.[9]

Bache's underlying question: What type of organization will prompt men to engage in an exchange of ideas and to recognize each other's arguments as potentially better than their own? His core expectation is not that the individual expects to take away from such an exchange an idea that he may then turn into a profit. Bache instead takes for granted a desire to be accepted as someone with intellectual gusto. Bache implies that members of associations (such as British ones) are interested in being recognized and accepted by one another in their mutual concern for a discussion focused on providing valid arguments. He also assumes a willingness to accept, however grudgingly, if someone else provides such arguments. How would this "liberalize the feelings," as Bache suggests? Without opportunities for exchange (Bache seems to consider this to be the American standard) the individual will have to rely on publications, if available, or other means of exploring ideas and of checking them. Such an "exchange" of ideas will be driven solely by the intellectual interest of one investigator and his commitment to a book or a journal. An interchange of ideas among people interested and versed in similar subject matter, however, will force both discussants to respond to ideas that, had they been reading a book, they may have glanced over or ignored. Discussants insist on the relevance of ideas in ways a book will not, and responses will be spontaneous. Bache suggests that discussions result in an awareness of the resourcefulness of others, and in an acknowledgment of the plurality, not of standards, but of ideas. It is interesting that Bache speaks of "feelings" here. He takes for granted that such experiences shape emotional dispositions. Finally, what is the basis for "mutual confidence" and an "esprit de corps" that may emerge over time? One could argue that a "collision of mind with mind" may lead to the exact opposite, to competitive disintegration. Bache has something else in mind. When the "collision of mind with mind" serves to make participants aware of their own "deficiencies," discussants have made

9 Ibid., 7.

a decision to expose such deficiencies to others; it requires confidence to trust in others not to turn this deficiency against oneself.

In this one sentence, Bache provides a formula for professional integrity and cohesion, and he uses it as a model for cultural advancement in general: Conflict and debate are constitutive for intellectual advancement. In order for such advancement to take place, participants need to open up and to expose their own ideas and convictions to the review and scrutiny of others. They share a dedication to accepting the better argument with respect to a given problem, regardless of who provides it. This implies a readiness to face the uncertainty of debate and a readiness to abandon ideas and convictions in favor of more effective and convincing ones. To open up in such a way is possible only within a protected environment of "mutual confidence" in which the logic of competition is disconnected from practical consequences. An "esprit de corps" is a natural concomitant because members of such an association share this readiness to expose their ideas and beliefs to such mutual testing. They share a commitment to expose themselves to "crises" within the protection of their association. This aesthetic, non-practical crisis-orientation produces a feeling of community because the members of this "corps" realize that their way of looking at things, their readiness to expose themselves to uncertainty, sets them apart. Bache's description adds little to our understanding of the nature of the scientific profession but it is remarkable that he is able to sketch the logic of professional organizations so clearly. The way he speaks of this process of professional community building suggests that he considered this a process that had not yet taken place in the United States. He could have pointed to the example of the Franklin Institute where Bache had helped install modes of cooperation that were close to the model he provides here. The U.S. Military Academy at West Point also provided a model for Bache's thinking here, a model which he had rejuvenated for his idea that the army be used for meteor observations. But Bache speaks of such associations as though they did not exist in America. This was certainly true with respect to professional scientific associations. No equivalent to the British Association for the Advancement of Science existed in 1842 but we can see that Bache anticipated it.

Bache continued his discussion of England by addressing the economic circumstances and prerequisites for intellectual "improvement." In this passage, there is further evidence to sustain the argument that Bache intuitively connected the development of science to that of education, not merely because he had found employment in both areas, but because he considered both essential in helping advance and shape an emerging American intellectual culture. With respect to England, Bache points out that low wages "keep the mass of [operatives] from intellectual improvement." He suggests that their

> youth is passed before they can judge of the necessity for culture, and when manhood is reached, the cares of providing food and maintenance for themselves, and usually for a family besides, press upon them … heavily.[10]

What is the common denominator to the workers' low wages and adulthood? It is the pressure that prevents them from pursuing intellectual "improvement." In keeping with his coordinates for describing scientific associations, Bache takes for granted that intellectual advancement and culture require leisure and absence of immediate financial pressure, and both are incompatible with financial hardship and the pressures of adulthood. Bache's sentence here implies that if adulthood was reached later and if "operatives" were given more time during adolescence, they might themselves come to understand the pertinence of education not offered by their milieu. In the absence of such freedom, however, one possible solution would be to mandate early education by law, an idea Bache had been impressed with while traveling through Switzerland. Bache moves on to suggest that until "the means of life are more uniformly distributed, the mass of the mechanical population of Europe cannot become intellectual." He argues that intellectual refinement requires leisure and time away from dull labor. Bache is equally concerned with the common worker's refinement as with that of professional organizations such as the British Association. It would be a mistake, therefore, to consider his address solely as a document concerned with the development of the scientific profession. It is apparent that he takes a scientific community for granted. But he seeks to extend the logic of professional discourse to other areas of life in America.

In his consideration of the British model, he is certain that American conditions better facilitate the cultivation of the worker because American workers receive higher wages. He assumes that "intellectualization" was a concern widely shared among American workers if only conditions allowed them to live up to it. Again, in comparing U.S. conditions with European ones, Bache looks for clues for how the country, in the long run, could live

10 Ibid.

up to its ambition of pulling ahead. Bache's perspective is republican in that he considers his country's political system to provide the basis for a broad elevation of cultural standards and education. The "inestimable rights" of American workers as citizens

> put him on a par as a citizen with every other citizen, he occupies a different place in the social scale—may, by education in school and out of school, put himself on an equality with any other citizen—and may have comfort and competence for himself and his family. Thus relieved from the grinding pressure of want, wo {sic} to him if he forget that he has a mind and soul as well as a body—an intellectual and moral as well as a physical nature![11]

"This Most August Sovereign"

This brings Bache to an evaluation of American conditions and to the main point of his speech. The individual American worker, he points out, has sufficient means to engage in intellectual cultivation. The potential for creativity of the American population, Bache argues, is higher than in Europe, an important advantage considering the ambitions for cultural leadership that Bache laid out at the beginning of his speech. What avenues are open to an American citizen seeking "self-improvement"? What has been done in the United States to facilitate such interests? What, according to Bache, should be done?

In his review of possibilities for education and cultivation, Bache focuses on the Franklin Institute and its efforts during the preceding decade. His outline serves to highlight his core argument that institutions such as the Franklin Institute should be supported, not by individual groups of citizens, but by "our great sovereign, the people." Bache hopes that "the eyes of this most august sovereign might be opened to the importance of fostering these institutions!"[12] Why would he suggest that the public instead of private initiative support associations? Had not the Franklin Institute's success proven that private efforts could make a significant difference? Most of Bache's address circles around this question, and (perhaps surprising today) he comes out strongly in favor of public support of all branches of education and research. This, indeed, is the main point that Bache considers from different angles and perspectives. Demon-

11 Ibid.
12 Ibid., 8.

strating a boldness he would abandon after moving to Washington as superintendent of the Coast Survey the following year, Bache highlights the relevance of cultural development against the political developments of his age. His appeal to "this most august sovereign" is followed by these remarks:

> If for the improvement of the mass, he {this most august sovereign} would but contribute a little of what he lavishes in raising up the political princes of the land! In the olden time, the commons of England gave every ninth sheep and every ninth fleece to their ruler, to enable him to wage war; now a large portion of our commons devote at least the ninth penny to king Party; to enable him to carry on the strife political. Would that they would spare the ninth part of this to put down ignorance and elevate virtue![13]

Bache distinguishes between an elite and the masses. In the given context of his speech this distinction points to the difference between those who are concerned with the direction the community is taking versus those who, because of limited education and exposure to a broader consideration of their role within society, consider politics mainly in terms of advancing their immediate personal interests. The difference lies in a broader versus a narrower outlook and contextualization of one's interests, of assuming or disregarding the view of a "generalized other." Bache takes for granted that the masses require "improvement." He faces a dilemma: The masses are part of the "sovereign," and the decision to redirect funds to support cultural institutions for their own benefit requires their support.

There are other implications. Bache speaks of "raising up the political princes," suggesting that party politics in the American republic brought about results beneficial to a small group of leaders. He ironically suggests that these interests of republican politicians were not dissimilar from those of European aristocrats. Taken literally, the potential of American independence had not been fulfilled because republican ideals remained compromised. Bache's observation comes full circle when he suggests that the relevance of waging war had been replaced, in the United States, by supporting a political caste which was not addressing the political community's problem of putting "down ignorance." Finally, Bache implicitly compares waging war to party battles, suggesting that they are comparable in terms of integrating and mobilizing a community. The difference between the two, of course, is that war is waged against a common outside enemy while

13 Ibid.

political fights are directed against political enemies within the community. Bache suggests as a "moral equivalent to war" the fight against "ignorance" in order to increase the likelihood for cultural innovation. In the absence of an external enemy against whom to unite as Americans, Bache relies on the hope of American culture pulling itself up by its bootstraps. His hope rests on those who are not part of that "large portion of our commons" which devotes its attention to party strife.

This prompts Bache to move on to discuss the Franklin Institute's efforts that he puts squarely into the camp of those seeking to elevate "the mass." In doing so, he assumes a regional perspective and considers the Institute to represent Philadelphia and its vicinities. When Bache asks what "*we*" have done "to advance the progress of the useful arts," his ensuing discussion of the Institute suggests that his audience, and the focus of identification for cultural advancement in 1842, was distinctly urban. Elites in cities such as Boston, New York, and Philadelphia, Bache suggests, each help to solve the common problem by advancing culture in their own specific way. Through exhibitions such as the one in Philadelphia, the "taste of the public is improved [...]—the taste of the artist is elevated."[14] In a paragraph, Bache sets out his perception of the relationship between market and culture. "It is obvious, then," he writes with respect to public taste,

> that there are reasons why exhibitions should contribute to *aid* that which requires other causes to support. If they neither form the foundation of the building, nor yet its superstructure, they serve to determine its shape and the arrangement and distribution of its parts.[15]

Exhibitions of manufactures, Bache concedes, provide no rationale for developing machines exhibited in Philadelphia and at shows elsewhere. Innovation relies on profits realized in a market or alternative sources of income as "other causes of support." (Bache could have written: "other means of support;" his formulation here suggests that public taste in turn needs mobilization on behalf of goals other than those provided by an exhibition.) But Bache conceives of exhibitions as either a "foundation" or a "superstructure" for public refinement. As showcases of innovation, exhibitions reflect standards of cultivation. These standards, Bache implies, have not yet become the "foundation" for American public culture ("the

14 Ibid.
15 Ibid., 9.

building"). They are upheld by a small segment of society represented by the Franklin Institute. But Bache assumes that such standards of cultivation will become ("yet") the aspiration of American culture in general, its "superstructure." In this sense, it is a matter of time and a possible result of efforts to guide American culture. Meantime, exhibitions and the opportunities for comparison and the cultural guidance they provide support this process by providing standards for promising areas of innovation. As an extension of this argument, Bache moves on to stress that the economic value of assessments made by the Franklin Institute should not be underestimated. Some "would value in the same way expressions of esteem as so much breath," but Bache points out that "the opinions of the judges, expressed at the exhibitions, become the guides of many and many purchasers" and have an impact on prizes and on the market.[16]

In order to facilitate the development of American ingenuity, Bache moves well beyond the immediate idea of exhibitions organized by the Franklin Institute. He focuses on a more reliable institutional setting so that exhibitions of manufactures can be made permanent. "Why should not provision be made in the ordinary and regular working of the Institution," Bache asked his audience, "for a constant exhibition?" And it is here that Bache picks up his earlier train of thought, that projects on such a scale could hardly be realized by private initiative. "Voluntary association may do much," he suggests with reference to his own experience as one of the Institute's leaders, "but not everything."[17] In this context, it may be worthwhile asking whether Bache's interest was pecuniary. Why did he point out that only public support could guarantee continuous evaluation? One dimension of his critique, of course, has indeed to do with the financial possibilities of the public purse versus that of private organizations. Against the backdrop of Bache's overall argument, however, the financial dimension seems nothing but an extension of the underlying interest in ongoing cultural development. This is in keeping with the observation that Bache, in the same paragraph in which he recommends a "constant exhibition," points out that the Institute would be unable to organize it.

He then moves on to more general observations:

> We already begin to have a history. Already the obelisk is raised, upon the base of which the names of the useful, zealous, and able, among the members of the

16 Ibid.
17 Ibid.

Franklin Institute, are to be inscribed at death—that tablet bears even now the names of {William H.} Keating and of {James} Ronaldson {two founders of the Franklin Institute}.[18]

Bache's implicit question again concerns the future of American cultural development. When would the Institute, a unique institution in America, "begin to have a history?" When would it begin to accumulate a record? This is hardly a matter relevant for developing the American market economy but of significance with respect to evaluating whether intellectual progress had been made in areas such as manufacturing. Not to have a "history" implied that the country lacked the means of comparing ideas that were considered to be innovative to the precursors of these ideas, and to chart its progress. If there was no historic record of American achievements, there was no index against which to measure novelty, except standards provided by others. The country, in short, lacked "cultural sovereignty," an implication of Bache's initial reference to Palmyra and Venice. While this was a common theme among American intellectuals, Bache's address, rather than contemplating the point in the abstract, points to the pragmatic restrictions imposed by such shortcomings, as well as to possible solutions.

As Bache's comments make clear, this was a relevant matter, not only because it was tied to issues of American development but because it had consequences for a "system" of gratification guiding individual ambitions. In his address, this had been Bache's theme all along. Bache had highlighted the relevance of "esteem" provided by awards for innovative machines. By suggesting that names of the Institute's founders were inscribed in an "obelisk" Bache invoked the historic dimensions of Palmyra and Venice. Certainly an obelisk, just like its Egyptian models, would stand the test of time and remain as a permanent testimony to successful American efforts once the United States (or Philadelphia as it were) had run its historic course. The names of the Institute's prominent members are to be "inscribed at death." The Franklin Institute, in lieu of a similar organization of national stature not yet in existence, provided the means to honor successful engagement on behalf of the (urban) community. The Institute, however, was a "voluntary association" with all of its limitations. It was not an official or legal representative of the political community. The Institute's claim rested on the initial expectation that its leadership could muster sup-

18 Ibid., 10.

port because of its record in other areas and, even if it had not remained unchallenged, this expectation had since been verified.

In the context of these considerations, Bache's concern for the *public* support of culture appears in a new light. Money is part of the issue. But at the heart of the matter is the lack of adequate republican legitimacy. Bache had identified the "sovereign's" immaturity as a cause of this deficiency; we may add that his national perspective remained programmatic and his intuition regional. Even though Bache was just a few months away from being appointed superintendent of the U.S. Coast Survey in Washington D.C., an important national scientific post, his regional perspective stands out. Bache sought to raise awareness and political support for local cultural institutions that presupposed an "enlightened sovereign;" but he had to shift to the next level and mobilize such awareness nationally. It was here that America's ambitions would ultimately have to be expressed for despite his regional identification, he projected his ambitions onto the nation-state. Only a national institution, an institution created by the federal government, would eventually provide an arena for recognizing and honoring relevant American contributions to human knowledge.

Elites in the American Republic

In discussing Bache's 1842 address, we must turn to his conclusion. The prominent observation here is that Bache bundled a bouquet of ideas and suggestions that all point in the direction of public support for cultural institutions. He returns to the financial issue. While the "sovereign has now awakened to the advantages of supporting public schools" (and Bache is referring to Philadelphia's Central High School) that "sovereign" was "not yet fully prepared to push the principles upon which they are based to their utmost limit."[19] Again, Bache does not doubt that the "sovereign" will eventually catch up ("yet") and that the development of public institutions will take place. He implies that politics will evolve in the direction of the radical ideas ("utmost limit") he anticipates. He moves on to illustrate his point further by asking whether "public opinion {will} ever so far ripen as to furnish these means" of supporting a variety of schools such as the ones he had visited in Europe. "I have heard it remarked by more than one

19 Ibid.

person conversant with the minutiæ of the institutions in Philadelphia," Bache wrote,

> that all the enterprises for the diffusion of knowledge are supported by a small portion of our population; and yet they are intended for the ultimate good of *all*, and should be supported by the *whole community*. There was no doubt a time when the idea of paying for the support of a fire department would have seemed preposterous, and now we quietly pay for insuring our houses, and then in addition, a portion of our taxes goes to furnish the means and appliances for extinguishing fires. What would we think now of supporting a fire department entirely by voluntary contributions? … *Shall the principle be that, what is for the good of the whole, shall be supported by the whole?*[20]

Bache boldly insists that "{we} agree that education shall be put upon a *truly republican* basis." In his view, this implied that the "public purse" should finance schools of all levels: "one wide system {with} all the institutions of every name for the promotion of knowledge."[21]

In the light of the arguments Bache had been developing in his speech, his conclusions hardly need comment. "The scheme is not so Utopian as, at first sight, it might appear," he assured his listeners, confirming that he was laying out what must have seemed an impossible plan at the time. "Let me commend to your thoughts," he continued,

> the idea of forming *a system* from these various parts, not centralized, but like our own political union, each independent, while all are united, *a great system of public instruction, worthy the patronage and support of a free and enlightened people.*[22]

Bache then wrapped up his address by warning his audience that

> without intelligence, virtue is comparatively powerless; without virtue and intelligence, liberty degenerates into licentiousness, independence into brutality. Liberty and independence exist but in name. When virtue, liberty and independence fail, the commonwealth which has chosen them as her watchwords, and has emblazoned them with the emblem of agriculture, commerce and the arts upon her arms, will cease to have a being.[23]

Education and culture were matters that could make or break the American national project ("cease to have a being"). By raising these issues,

---

20 The criticism quoted by Bache is of course reminiscent of political campaigns against Central High School. Ibid., 14.

21 Ibid., 15.

22 Ibid.

23 Ibid., 16.

Bache returns to his initial theme of his country's larger historical perspective and the essential conditions for its survival and development. While in earlier chapters, we could infer from Bache's family background and from his early career that he intuitively looked for ways in which he could help provide a basis for his country's cultural development, his 1842 "Address on Manufactures" and its conclusion are additional evidence that Bache, while identifying with science as a model for a rational and universalistic discourse, assumed a holistic view of his country's historic opportunities and of the difficulties in living up to them.

In his concluding sentences, Bache looks at the role of institutions that represent a rational discourse (such as science and education) from the perspective of "virtue," of moral values embedded in and represented by a community. He connects these terms and the social practices for which they stand in a cascading fashion: "Liberty" and "independence" are central, "virtue" follows next, and "intelligence" lies on a third level. All three are part of a "commonwealth," i.e. they constitute America's mythological core. Bache thus infers the rationale of the country's founding and the primary value of personal independence to be balanced by "virtue," i.e. the dedication to a spirit of cooperation and the common welfare. This distinction corresponds to our observation (of his biography as well as the reference to Palmyra and Venice in his speech) that Bache identified with a national American mission but that he viewed this mission from an urban, Philadelphian perspective. What he is trying to do here is to attach the idea of a rational discourse (represented by science, education, etc.) to his country's mythological self-conception. He is arguing that the institutional implementation of such a discourse is essential to the preservation and success of America, and as we have seen, this idea was indeed central to his efforts before 1842. Finally, Bache's imagery (he speaks of "emblems" and "arms") invokes for the American nation a knightly quest, ambitiously using "agriculture, commerce and the arts" as weapons to assert her virtuous ideals. This spirit relates, of course, to Bache's West Point experience and the sense of mission attached to its national role. In this concluding passage of his speech, in which Bache is obviously trying to distill the essence both of his talk and of his Philadelphia experience between 1836 and 1842, he extrapolates from the familial basis of his own membership in that exclusive national institution the model for building the American nation.

Chapter 7

Bache's Program for National Consolidation II

The United States Coast Survey

When Alexander Dallas Bache arrived in Washington D.C. in 1843, federal science policy remained in its infancy. The Franklin Institute's steam boiler experiments had assumed a national dimension because they had been sponsored by the federal government. But science had been touched on only sporadically in national politics. President Thomas Jefferson's initiative, following the 1803 Louisiana Purchase, to send explorers west along the Missouri River, ostensibly for commercial and military purposes, remained an exception. Jefferson had had plans for a national university but these failed in 1807. The University of Virginia had been founded instead. Of lasting service, of course, was the U.S. Military Academy at West Point, another military measure with scientific implications.[1] Other than federally sponsored explorations (directed by the War Department) and West Point, by 1842 the U.S. Coast Survey had begun to emerge as the most significant and largest science-related federal institution before the Civil War. The Survey had been created by Congress in 1807. Four years later, Ferdinand Hassler, its first superintendent, was sent to Europe to investigate procedures and to acquire instruments. But in 1818, Congress effectively discontinued the survey by limiting its staff to military and naval officers, excluding its civilian superintendent.[2] It was not until 1832 that the law of

1 A. Hunter Dupree, *Science in the Federal Government: A History of Policies and Activities to 1940* (Baltimore: Johns Hopkins Univ. Press, 1986), 26, 24, 14–16.

2 Florian Cajori, *The Chequered Career of Ferdinand Rudolf Hassler* (1929; repr., New York: Arno Press, 1980), 87.

1807 was put into operation once again, so when Bache arrived in Washington, Hassler had had ten years to shape the organization.[3]

After Hassler had fallen seriously ill in 1842, John K. Kane, once a Girard College trustee and a supporter of its observatory, wrote to Joseph Henry in November 1843, suggesting that Alexander Dallas Bache be recommended for the post. Bache, he argued, had administrative talent, and as the

> descendant of Franklin, a democrat always and of the best sort, allied to our Pennsylvania democrats Dallas and Walker and Wilkins and Irwin[4], a graduate of West Point and therefore in favour with the army, professor of our University, president of the Girard College while it had the promise of vitality, the leading mind of our Mechanic Institute and Philosophical Society, in a word, the nucleus around which science gathers in Philadelphia.[5]

Much like Secretary of War James Barbour had written to Alexander Dallas Bache's mother about her son's performance at West Point, Kane used the definite article to point to Bache as "*[t]he* descendant of Franklin."[6] Kane's use of the first person plural suggests that he considered himself to represent Philadelphia vis-à-vis Henry whom he saw as "almost the only man except Bache whose appointment would secure the confidence of scientific men." Following Kane's suggestion, Bache and his friends indeed sought to mobilize the scientific community after Hassler had passed away.[7] As part of this concerted campaign, Bache asked Elias Loomis for a letter of recommendation to be sent to the president, pointing out in follow-up correspondence that "if there is any object above all others which I shall aim at, it is to promote science as far as my position enables me." This he sought to do by "calling to the aid of the survey the real talent of the

---

3 Charles Henry Davis, *Life of Charles Henry Davis, Rear Admiral, 1807–1877* (Boston and New York: Houghton Mifflin, 1899), 75–77; Cajori, *Chequered Career*, 161–232.

4 Kane was of course referring to Bache's uncle George Mifflin Dallas and his grandfather Alexander James Dallas; Robert John Walker was Bache's brother-in-law who in 1845 became secretary of the treasury; his aunt Matilda Dallas was married to Congressman William Wilkins; William Wallace Irwin, also a U.S. Congressman, was married to one of Bache's sisters. See previous chapters and notes to Kane's letter that was printed in Joseph Henry, *The Papers of Joseph Henry* (Washington: Smithsonian Institution, 1972–2008), 5:450 f.

5 John Kent to Joseph Henry, November 20, 1843, ibid., 5:450.

6 My emphasis. See James Barbour's letter quoted in chap. 4 above.

7 See Hugh R. Slotten, *Patronage, Practice, and the Culture of American Science: Alexander Dallas Bache and the U.S. Coast Survey* (Cambridge: Cambridge Univ. Press), 68–75.

country," inferring that Hassler had let this opportunity pass.[8] One historian has concluded that in supporting Bache for the post of superintendent, the "American scientific community had united, possibly for the first time, behind one of their own."[9] This endorsement was not unanimous, of course, but the community did support Bache for perhaps the most prominent national science post, and he it did so against competing interests from within the federal bureaucracy. Following the death of Ferdinand Hassler, therefore, the new position provided Bache not only with a higher income when compared to his professorship; it also matched his long-held ambitions for the development of federal science policy. With the support of his colleagues, Bache was appointed by President John Tyler. With no other prominent scientist within the federal administration at this time, Bache effectively became his profession's representative in Washington D.C.[10]

Bache held his new post until his death in 1867, and during the twenty-five years of his tenure, he significantly expanded the survey by sending out several teams organized to work in different sections along the Atlantic Coast, on the Gulf of Mexico, and later along the Pacific Coast. Relying on civilian engineers as well as army and naval officers, triangulation work on land was complemented by hydrographic work offshore. By 1858, the Survey had published 250 completed and preliminary harbor and hydrographic charts. Tens of thousands of copies were distributed free to newspapers, institutions, societies, federal and state administrations, and commercial interests all over the country.[11] Under Bache's direction, the Survey became active in other scientific areas: Magnetic variation had long been Bache's interest, and during the 1850s the Survey set up more than 100 stations along the American coastline. It worked out tide patterns for the Gulf Coast and helped establish the country's border with Mexico. The

---

8 ADB to Elias Loomis November 24, 1843, and December 13, 1843, both N.R. II, roll S, Record Unit 7470, Reingold Papers, SIA (photocopies of original at Yale University [Elias Loomis family papers, Manuscripts and Archives?]).

9 John L. Martin II, "The Congressional Struggle over the Coast Survey, 1848–1851" (M.A. thesis, Univ. of Hawaii, 1988), 28. See also Slotten, *Patronage, Practice, and the Culture of American Science*, 68–75.

10 Bache was simultaneously appointed head of the Office of Weights and Measures. "United States Standards of Weights and Measures: Their Creation and Creators," Arthur H. Frazier, *Smithsonian Studies in History and Technology* 40 (1978), http://www.sil.si.edu/smithsoniancontributions/HistoryTechnology/sc_RecordSingle.cfm?filename=SSHT-0040, 13 f.

11 Ibid., 99 f.

Survey, in the 1850s, supported work by Harvard biologist Louis Agassiz on organic matter collected along the Florida coast. Bache inherited Hassler's post of superintendent of weights and measures and, just as Bache had proposed in the 1830s, the Survey conducted experiments to establish the accuracy of standards. Astronomical observations were critical for surveying, and Bache's organization established close connections to Harvard University where mathematician Benjamin Peirce became its advisor on this matter. This established a strong Washington-Cambridge "axis" of which more will be said later.[12]

Fig. 8. The U.S. Coast Survey on New Jersey Avenue near the Capitol occupied several buildings, including the one on the left used as a hotel when this photo was taken in ca. 1920. For a map of the organization's offices in 1836, see Frazier, "United States Standards of Weights and Measures," 10.

(Courtesy of the Historical Society of Washington D.C., CHS-03444)

12 For this paragraph, I am drawing on Dupree, *Science in the Federal Government*, 100–05. On the role of the Coast Survey in establishing borders, and in relation to the Mexican border commission in the 1850s, see, for example, ADB to his brother-in-law Major William H. Emory, April 1, 1853, as well as other correspondence by Bache here, N.R. II, roll T, Reingold Papers, SIA (copied from the Emory Papers, Yale University). Concerning Bache's relationship with Peirce, see chap. 9 below.

Coast Office triangulation staff usually spent six to eight months per year in the field and the rest of the time computing results at the Washington office or at home.[13] Bache frequently worked in the field himself. While Congress was in session, it devolved upon him to secure appropriations for the surveying work, a task at which he proved notably successful. When he first arrived, he found the work in a state of disorganization and Congress skeptical of it.[14] Significant political interests and pressures were put on an organization with hundreds of jobs, significant expenses, and information crucial for navigation and trade. In March 1844, Bache wrote to Joseph Henry that politicians had managed successfully "to obtain an independent footing" while Hassler had been in command, and these forces, with Bache's arrival in Washington, now saw "their independent powers wrested from them at once. … They have literally *conspired* to unhorse me." One issue, of course, was appropriations. "The meddling with Congress is a delicate affair because it might happen that in crushing the Philistines, the whole might be destroyed, a self sacrifice."[15] Work in the field and in Washington kept him tied up. Again, his post left Bache little room for investigative work.[16] While Bache published occasional papers, most of his writing outside of his vast correspondence now focused on the Coast Survey's annual reports that included information on the work as well as scientific papers and charts prepared by his staff. During the 1850s and with the support of his scientific peers later organized under the umbrella of the American Association for the Advancement of Science, Bache used his growing political influence and standing in Washington D.C., as well as his insights into the necessities and opportunities of research tied to the interests of an expanding and industrializing country, to vastly expand Coast Survey operations. He turned the Coast Survey into the foremost scientific organization in the United States with more than 700 employees in 1855.[17] With close to half a million dollars available to it in 1854, its appropriations significantly exceeded the Smithsonian Institution's budget.

13 ADB to Stephen Alexander, October 25, 1844, roll 3, RG23, M642, Coast Survey Records, National Archives.

14 ADB to Joseph Henry, December 26, 1843, Henry, *Papers of Joseph Henry*, 5:475 f.

15 ADB to Joseph Henry, March 3[–4], 1844, ibid., 6:46 f.

16 His friend Henry wrote to him a year after he had arrived in Washington: "I fear your time is now so much occupied that you can have scarcely any to spare for friendship or science." Joseph Henry to ADB, April 16, 1844, ibid., 6:75.

17 Slotten, *Patronage, Practice, and the Culture of American Science*, 99.

Dupree concludes that the Coast Survey was "in reality the general scientific agency of the government."[18]

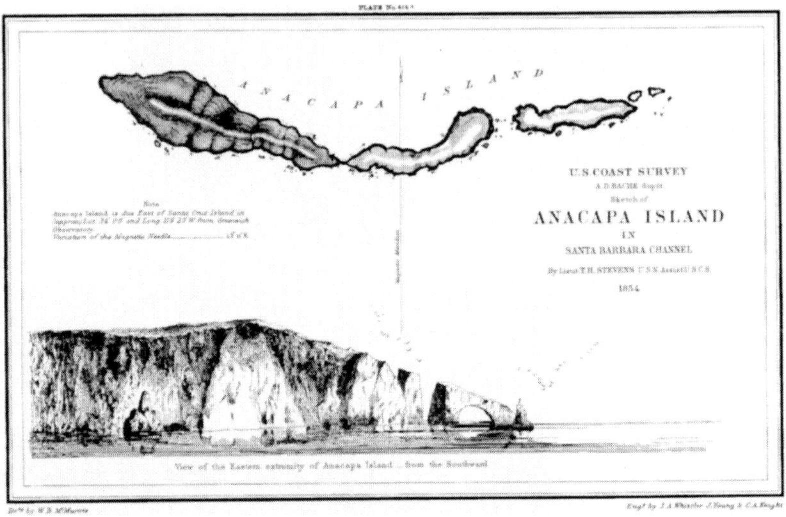

Fig. 9. Drawing of Anacapa Island by James McNeill Whistler who worked for the Coast Survey in 1854 and 1855.

(Image Archives of the Historical Map & Chart Collection, Courtesy of Office of Coast Survey/National Ocean Service/NOAA)

But even if no one in 1843 expected the survey of the expanding U.S. coastline to be finished any time soon, it was a pragmatic response to national expansion and not designed to be permanent.[19] In his annual reports to Congress, Bache responded to this tacit expectation by charting the remaining work. If "your Coast Survey is to expand with the practice of annexation and the enlargement of objects for its exercise," one of Bache's correspondents wondered, "where will it end?"[20] In 1858, Bache explained that given the length of the nation's coastline, the survey would need at

18 Dupree, *Science in the Federal Government*, 104.

19 Dupree makes this point. He argues that the Coast Survey had a "tendency toward permanence" but its temporary character remained. Dupree, *Science in the Federal Government*, 100, 102, respectively.

20 Joseph Reed Ingersoll to ADB, November 21, 1853, RH 1582, box 15, Rhees Collection, HL.

least another ten to twelve years to be completed.[21] While the Coast Survey was the most significant scientific agency supported by the federal government, therefore, its existence did not signify that the American public or the federal government had agreed, in principle, to establish a national scientific organization.

Bache had been concerned with the issue of political endorsement of educational and scientific work for some time. At the Franklin Institute, Bache had helped draft proposals for weights and measures he preferred to implement on the national rather than the state level. His committee had argued that the federal government had the Constitutional wherewithal to proceed in such a way and national politicians such as John Quincy Adams quite obviously shared this perspective. But even on the state and the local level, it had remained difficult to support science or education. In 1843, things were moving forward as offices such as the Naval Observatory were created "by underlings in the executive branch of the government in the very shadow of congressional disapproval"—but the matter remained unresolved politically.[22] This paralleled many scientists' assessment of the prospects for developing scientific organizations. During the 1830s, the idea of a national scientific organization such as an American version of the British Association for the Advancement of Science was discussed privately. After his return from Europe in 1838, Bache rejected such an idea by arguing that "We have half a hundred persons engaged in diffusing science for one who is occupied with research."[23] These two developments (a reconsideration of national science policy and of the federal government's Constitutional powers, on the one hand, and the development of professional scientific organizations, on the other) began to enter a new phase just after Bache had moved to the nation's young capital.

The second speech I will discuss in more detail relates to this period. The background to the events that gave occasion for Bache's talk had begun to unfold in 1836 when an obscure Englishman, James Smithson, died and left 500,000 dollars to the United States for the support of science. Since then, ideas had been discussed on whether Congress could accept the money and how it was to be used. The founding of both the

---

21 U.S. Coast Survey, *Annual Report of the Superintendent of the Coast Survey Showing the Progress of that Work during the Year Ending November, 1851* (35th Cong., 2d Sess., Exec. Doc. 6 (Ser. 980), 1852), 6.

22 Dupree, *Science in the Federal Government*, 62.

23 ADB to Prof. Lloyd, November 30, 1838, box 2, vol. 1, Bache Papers, SIA.

National Institute and of the Association of American Geologists in 1840 was part of a general development of national scientific institutions. But while the latter grew out of practical scientific interests of coordinating state surveys among geologists, the National Institute, according to Sally Kohlstedt, sought an "on-going and direct relationship to the federal government."[24] It was at an 1844 meeting of the National Institute that Bache gave his speech, and his involvement in this organization requires a comment or two.

A variety of ideas had evolved after Smithson's surprising 1836 bequest had been accepted by Congress—such as a national university or a school of another type.[25] One such initiative had been launched by Joel R. Poinsett who had organized the United States Exploring Expedition as secretary of war under President Martin van Buren, a federally sponsored exploration of the Pacific Ocean from 1838 to 1842. Poinsett was interested in finding a repository for material brought back by this expedition headed by Captain Charles Wilkes. In 1840, Poinsett gathered a group to found the National Institution for the Promotion of Science. There was a West Point and an engineering connection through the membership of Colonel John James Abert, chief of the corps of topographical engineers, and through Bache's former mentor, Colonel Joseph G. Totten, of the Corps of Engineers. This was a local initiative in a city lacking the thriving culture of Boston or Philadelphia, but because Washington D.C. happened to be the capital this group could attach to its organization national ambitions, hence its change of name to "National Institute" two years later.[26] In its 1842 constitution, the Institute included among its directors, "with their consent," the "Secretaries of State, Treasury, War, and Navy, and the

24 Sally Kohlstedt, "A Step toward Scientific Self-Identity in the United States: The Failure of the National Institute, 1844," *Isis* 62, no. 3 (Autumn 1971): 340. For an account of the National Institute, see Robert V. Bruce, *The Launching of Modern American Science, 1846–1876* (New York: Knopf, 1987), 252 f.

25 For a general overview over these developments, Dupree, *Science in the Federal Government*, 66–90, remains very useful and I am drawing on it here. The conflict between proponents of classical education and of an education of "things, not words," which Bache had become familiar with in Philadelphia, was evident in these proposals (p. 68).

26 When Bache had decided to leave his chair at the University of Pennsylvania in 1842, a colleague insisted: "You are made for social labor, and a city is your proper home." He implied that Washington City did not meet the former's metropolitan standards. Robert Maskell Patterson to ADB, February 5, [1844], RH 951, box 13, Rhees Collection, HL. The date seems indeed to be 1844 even though Bache left Philadelphia in 1842/1843. I was unable to further verify the date.

Attorney General, and Postmaster General of the United States." The president of the United States or the vice president was to "preside at all meetings of the Institute" (which, in fact, never happened).[27] The Institute sought to mobilize political support by incorporating the political capital's elite.

Because of these developments, the Institute could quite obviously not be expected to represent the interests of the scientific community. The Institute's handling of the expedition's artifacts, its personnel decisions, and its ever-increasing ambitions indicated to the scientific community that the Institute was incompatible with its expectations. This became evident when the Institute invited to its 1844 Washington meeting members of other scientific organizations, including those of the recently founded Association of American Geologists who were planning to meet in the same city a few months later.[28] That the National Institute's meeting followed political protocol (with President John Tyler giving a brief opening address) further points to its disengagement from the scientific profession. These issues came to a head at the 1844 meeting. "The key question to be resolved" between proponents of the National Institute and scientists who criticized it, Sally Kohlstedt points out, "was the role which politics and the national government should play in any institution designed to promote science."[29]

The National Institute

As a second sample for deducing Alexander Dallas Bache's perception of the role of science in the emerging United States, I will focus on his 1844 speech, "What are the wants of science in the United States," a handwritten manuscript among the Bache Papers at the Smithsonian Institution Archives. The situation in which Bache gave this speech was ambivalent with respect to a federal support of science, and it shows how he positioned himself in the emerging national scientific context.

27 The constitution may be found in the National Institute for the Promotion of Science, *Bulletin of the National Institute for the Promotion of Science* (Washington: Gales and Seaton, 1841), 389 f. Includes later publications by the Institute.

28 I am following Kohlstedt's account here. Kohlstedt, "A Step toward Scientific Self-Identity in the United States," 342–53.

29 Ibid., 340.

The title of this speech indicates a significant shift away from issues that Bache had addressed in Philadelphia. In his 1842 speech at the Franklin Institute, Bache focused on the development and legitimacy of cultural institutions *en gros*; in his 1844 speech, he proposed to address the role of science in particular. Because Bache had moved to the nation's capital, and because he had assumed one of the most prominent scientific posts in the nation, we may assume that this shift reflected his more general engagement in matters of the nation's professional policy after he had arrived in Washington D.C.

Bache had been elected a Corresponding Member of the National Institute in 1840.[30] He was also a member of the organizing committee for the April 1844 event that sent out a "circular" in March 1844, inviting "friends" of the National Institute as well as "Presidents and Officers of Colleges and Universities; the Members of Scientific and Literary Societies of the United States, &c." The committee explained that the Institute looked "for essential and permanent support to the constituted authorities of the Nation."[31] The core idea was to provide an organization that would hold, on behalf of the national government, collections of scientific objects in Washington D.C. The circular, which was signed by at least five politicians and Alexander Dallas Bache as the only science-related government administrator, was vague on the meeting's purpose, suggesting that such matters could be discussed later. "Resting on the basis of a popular government," Joseph R. Ingersoll had written for the committee,

> the Institute is designed literally for the people. ... It is obvious that there can be no limit to the number of its associates except that which may be imposed by individual reluctance to do good in this especial form. The door is open wide to the friends of the advancement of the power, knowledge, and happiness of mankind: for the ability to be useful extends throughout them all.

Ingersoll tacitly suggested that whoever was not in favor of the Institute might perhaps be against it, or at least "reluctant to do good." The circular's tone hinted at the author's elitism disguised as nationalism. "Officers

30 Francis [Markoe] to ADB, August 22, 1840, box 6, folder 12, Bache Papers, SIA. Martin writes that the National Institute had been influential in the reorganization of the Coast Survey in 1842 and 1843" which may help explain why Bache would be careful and try to have a hand in the Institute's affairs. Martin II, "The Congressional Struggle over the Coast Survey, 1848–1851," 34.

31 Joseph Reed Ingersoll, *Third Circular of the Committee, Relating to the National Institute* (s.n, 1844). A copy of this document is at the American Philosophical Society.

of the association," Ingersoll proposed, "are thus representatives of the nation" because "it acts only as the agent and organ of the Government." In the absence of active interference by politicians, in other words, the Institute could be controlled by its leadership but members outside of Washington had little opportunity to benefit from the organization or to influence it. The Institute was a decidedly local affair but unlike institutions in Philadelphia, it claimed to represent national science. "The seat of political duty is necessarily here," the Institute's leadership explained. "Why should not the seat of learning and the arts be here also?"[32] Or, as the committee had found during an earlier meeting: "this Institute affords an opportunity which ought not to be neglected of concentrating the genius and learning of our country at a common centre, from which the beams of intelligence will radiate to gladden and bless the land."[33]

Bache signed this circular, and he agreed to deliver a paper "On the state of science in the U.S. + Europe."[34] Two weeks after the meeting had taken place, Bache's friend Joseph Henry wondered what was "to be done with the mass of diluvium which the Institute has drawn down on itself in an avalanche of pseudo-science?" Only because Bache participated in the Institute's affairs, he wrote to his friend, did he not "despair of the future existence of the scientific character of our country."[35] Henry thus took the Institute's intentions seriously and feared their impact on American science. Bache has been perceived by historians as having shared these reservations but to have participated in order to retain influence and to "'direct … the host of pseudo-savants … into a proper course,'" as Henry put it.[36]

A letter Bache wrote to mathematician and astronomer Elias Loomis in March 1844 provides strong evidence for this view.[37] Bache was ambiva-

32 "Preamble to the Proceedings of the Board of Management of the National Institute," ibid., [2].

33 "Board of Management of the National Institute," December 23, 1843, ibid., [3].

34 Francis Markoe, "Preliminary letter of invitation to address a meeting of the National Institution," February 1, 1844, RH 457, box 9, Rhees Collection, HL.

35 Henry to ADB, April 16, 1844, Henry, *Papers of Joseph Henry*, 6:76.

36 Joseph Henry to ADB, April 16, 1844, as quoted in George H. Daniels, "The Process of Professionalization in American Science: The Emergent Period, 1820–1860," *Isis* 58, no. 2 (1967), 157. The letter is in Henry, *Papers of Joseph Henry*, 6:77. For the most detailed discussion of the National Institute and its 1844 meeting, see Kohlstedt, "A Step toward Scientific Self-Identity in the United States."

37 ADB to Elias Loomis, March 7, 1844, Reingold Papers, SIA (copied from the Elias Loomis family papers, Manuscripts and Archives, Yale University?).

lent towards the meeting. He explained that the Institute's leadership "would like to have the Smithsonian fund to draw upon" but that to

> obtain money directly or indirectly Congress must be satisfied that they have the confidence of men of science. At first I doubt not the active movers in the matter of this meeting, had the British Assoc. {for the Advancement of Science} in view, but the reception which their scientific pretensions met with from men of science did not encourage that idea.

Men of science were skeptical of the National Institute's plans, and their endorsement was necessary just as it had been for Bache's appointment to the Coast Survey post. But the American scientific profession was not represented institutionally. While there had, of course, been a national scientific community before 1844, and even though it had occasionally influenced federal decisions, its consideration of matters that were to be decided by the national government was a relatively recent development. The very fact that the National Institute, a local organization, was able to try to mobilize the community's support on its behalf attests to the novelty of such a perspective. There existed only one national scientific organization, the American Association of Geologists, founded during a meeting held at the Franklin Institute in 1840 (renamed Association of American Geologists and Naturalists in 1842). Because it sought to establish the claim to speak for science, the National Institute could not ignore the Association as the only "official" organ of professional science in America. The British Association for the Advancement of Science had come to represent science in England. In America, too, "there could be only one such authority in a nation."[38] Once a national perspective was opening up in the early 1840s, in other words, any initiative toward filling this unique position had to be taken seriously for it could irreversibly shape the nation's scientific culture and institutions. In his letter to Loomis, Bache had written that he had discouraged "the idea of a great scientific meeting {by the National Institute}, to represent American science, or any gathering which is to pin the eyes of the men of science of Europe upon itself as a type of science of our country."

---

38 Kohlstedt, "A Step toward Scientific Self-Identity in the United States," 351.

Bache's Speech at the 1844 Meeting of the National Institute

But the meeting did take place, and a detailed reading of Bache's speech, which Sally Kohlstedt considers a "subtle plea for specialized science at a meeting of amateurs," will provide an opportunity to better grasp his understanding of the state of the scientific profession in 1844, and the consequences he sees for contemporary efforts by the National Institute.[39] One question to be pursued is whether Bache's speech reflected a contemporary "impetus" for an "organization designed to stimulate professional development and to publish reports of significant scientific activity … and to the self-evident financial necessity of popular appropriation of science in a democratic society," as Kohlstedt suggests.[40] Bache rejected the idea of government control of scientific organizations, an idea around which politicians had built the National Institute. A reconsideration of Bache's speech will show the nuances and the implications of his critique and, most significantly, the professional ideal from which he deduced his arguments.

Bache begins with a question:

> What are the wants of science in the United States?[41]

Compared with Bache's 1842 speech in Philadelphia, there is a shift of emphasis. In his earlier statements, Bache had discussed the relationship between culture and the political community, and science had been one aspect among others. With an eye on the Franklin Institute's development, he had pointed to the limits of private support of cultural institutions and he had recommended public support instead. His question here suggests that the relevance of science could be taken for granted, that there existed a consensus that science ought to be supported. The frame of reference is the nation-state: The National Institute's initiative was premised on the idea that scientific organizations could be developed in the context of the federal state, not on a local or international basis. By raising this question, Bache focuses on science policy from the perspective of those engaged in science for they might be the ones qualified to answer this question but he does so in a way that would make its answer relevant to any American

39 Ibid., 389 n. 106.

40 Ibid., 341.

41 The manuscript in box 5, folder 36, Bache Papers, SIA, contains no title, nor is Bache's name mentioned. It is, for the most part, not even in Bache's hand. His authorship is deduced from the paper's present location.

citizen (or politician) concerned with advancing science in the United States.

In the context of the National Institute's ambitions, Bache's programmatic opening sentence suggests that he either sought to bolster the Institute's plans by stressing their relevance for the development of American science, or that he felt that the project was moving along too quickly and that he consciously returned to a consideration of the premises of the plans the Institute wished to have discussed. Given the above introduction, the latter is of course more likely.

When and where can this question be better agitated than in a meeting of the friends of science!

Bache uses the verb "to agitate" which implies that the idea to support science was not as settled as his opening question seemed to suggest. If this question had commonly been accepted as relevant, it would hardly have needed such propagation. The meeting at which Bache spoke was designed to provide an opportunity to discuss concrete proposals. Bache ignores these ideas and instead raises more fundamental issues. He suggests that it may be insufficiently clear why science should be supported in the first place. Bache addresses the audience as "friends of science." There were scientists among the audience, even if their number was small, and Bache would of course include himself in that group.[42] Because he addresses "friends of science" only, he assumes a perspective in which he speaks on behalf of his colleagues.

To press such a question to a hasty decision by any scientific tribunal regularly or irregularly constituted, however composed, however numerous, would be in the highest degree improper.

In his initial question, Bache had suggested that American science was in need of support but he continues to shift his focus by insisting on thoughtfulness and caution. He puts into perspective the National Institute's meeting by comparing it to others ("*any* scientific tribunal," my emphasis). By speaking of "regularly or irregularly constituted" Bache even hints at doubts concerning the meeting's legitimacy and implicitly rejects the meeting's authority. His tone is self-assured ("in the highest degree") and

42 Among the 52 who accepted the invitation, most were "local amateurs and statesmen." Among the scientists were John W. Draper, Matthew F. Maury, James Espy, and Elias Loomis. Kohlstedt, "A Step toward Scientific Self-Identity in the United States," 358.

he leaves no doubt that he represents the norms according to which a decision on the future of science in America will have to be made.

> The attempt would be as vain as improper. To throw facts bearing immediately or remotely *upon it* before men of science and the friends of science, to discuss temperately all matt[ers] ~~bearing upon the question~~ *relating to the question* must always be in season, especially so, on an occasion like the present.[43]

Bache continues along the same lines. By repeating the adjective "improper" he warns his listeners against immature decisions, and he suggests that at least some of them might be driven by conceit. Why else would he mention vanity? By highlighting that there should always be room for critical discussion, Bache again qualifies the significance of a meeting he considers a mere "occasion."

Bache now explicitly distinguishes "friends of science" from "men of science." He continues to address his audience as the scientific community's spokesman or representative, suggesting to everyone else that they are mere "friends of science." By calling for a calm discussion of the issues, Bache implies that his listeners may have opposing views but share a strong interest in science, and that he represents a discourse in which ideas are considered, not on the basis of personal preferences, but coolly and with reason.

Bache's initial remarks thus confirm that he remained aloof from the National Institute's project. In addition, they provide a glimpse of the role he had assumed in Washington and among the nation's scientists.

> I would then respectfully invite your attention to some remarks upon the causes which have produced the present condition of science in Europe & in the United States & as ~~on~~ a commentary upon them, would offer some suggestions as to the manner in which science may be advanced among us.

Bache moves on to introduce his topic. The cultural circumstances of science stretched out over both Europe and the United States—he speaks of it in the singular ("condition"). He also sets out to compare conditions in the United States to those in Europe, instead of comparing them to individual European countries. Bache levels national differences and considers Europe en bloc. In a certain sense, of course, the transnational character of science corresponds to Bache's view of science in the western

43 I am using *italics* for an insertation by the author, [brackets] for guessed words. {Curved brackets} contain my own comments. In general, I have tried to reproduce all original formatting (~~deletions~~ etc.). See also chap. 3, 58 n. 23.

world, but this does not explain why he would not distinguish between separate conditions within Europe.

Europe is so parcelled out into independent states, with a Babel like confusion of tongues reigning among them, that to take a minute view of the influences producing ~~that~~ *the* condition *of science* would require the exhibition of a mass of detail which could by no means be compressed within a moderate compass.

Bache responds to anticipated criticism that he was looking at Europe as though it was a homogenous entity. But do his comments here really change that perception? Bache justifies the brevity of his remarks on Europe with that region's diversity. His phrase "parcelled out into independent states" implies that Europe really is a unit, and that it was fragmented only later. Even though Europe shares a common Christian culture, however, the continent has never been unified politically. Bache compares Europe to the historic Babel. His metaphor turns European diversity, the plurality of languages, into a curse. Europe's discrete political entities Bache views as mini-states, their "lack of integration" obstructs her development. When Bache speaks of "confusion," he overlooks how the diversity of languages in Europe corresponds to a diversity of coherent cultures and political sovereignties. He implies that Europe really should be integrated and that it should share assimilated conditions for science. In this way, Bache conceives of Europe on distinctly American terms. According to this perspective, the continent's diversity should be enfolded in one state and in one overarching culture. Bache's implicit assessment of the American condition is favorable: one state corresponds to one language while Europe remains cacophonic.

Bache's more limited intention as a speaker, of course, is to elucidate the status of his comments and to explain why he cannot easily summarize the European "confusion." He moves on to raise methodological issues:

It might then on a cursory view of the matter be supposed that a stranger could form no opinions worth promulgating.

As a hypothetical objection, Bache introduces the idea that a stranger may be unable to grasp the complexity of European conditions. His assessment reflects back on Bache's perceived distance to a coherent European continent that he introduces as a term of comparison for plans to develop science in America.

In the manuscript, there follows a boxed-in paragraph:

~~A trial will show the reverse.~~ *Further consideration may show that this is not necessarily the case.* ~~Have you ever analyzed~~ *If we analyze* the process by which ~~your~~ ~~judgement~~ *impressions* of places is formed ~~in visiting a strange place you~~ *we shall find it something like this. At first we* are struck with the great features of nature and of art, ~~and~~ *& we* visit the most interesting localities. *Let us* Chronicle these first impressions for they are wonderfully fleeting. Minor objects which at first were unnoticed come out into relief, and as ~~you~~ *we* draw closer and closer, as it were, to the place, by longer studying its details, ~~you~~ *we* are bewildered with the endless variety which now presents itself. ~~Work~~ *If we work* now diligently, to master the ~~details~~ *minutiae* + *to* draw ~~y~~our conclusions, & note them: Work on—and note anew—~~and~~ *and* at last—after changing ~~y~~our views from day to day *if we* look back at the first bold sketch ~~and~~ *we* find it true to nature.

Parts of the next paragraph were set off in the same way but the line was then crossed out. Bache had probably deleted the sentence below when he became aware that he could economize even more by taking out the passage above:

But It is not the daguerreotype which putting before us with equal and entire accuracy every line of the face produces the true likeness. {Deletion undone:] Denner's miniatures of old men and old women may be examined with the microscope and are found as to number and distinct~~ion~~*ness* and size of wrinkles faultless, but we do not see the lilkeness of our ~~old dear~~ *dear old* friends in them, the wrinkles do not express the soul.{End of cancelled deletion.}

Bache chose to go with a more graphic version of his argument and to disregard his outright methodological considerations. What had prompted Bache to feel that it would be worth while continuing to discuss methodological issues instead of moving on to discuss his topic? Among scientific peers, a reconsideration of methodological issues would hardly be necessary. In addressing a public audience, Bache turns his talk into a demonstration of how scientists solve problems by cautiously describing and analyzing their subject matter.

Bache could omit one of his two paragraphs because they essentially convey the same idea in different ways: Even though it might seem difficult to comprehend a phenomenon with many different aspects, such understanding is possible by looking for underlying principles. Bache distinguishes between type and token, and in line with his reference to Babel, he is interested in the common cultural denominators of Europe's "endless variety." Bache likens his own role as a "social scientists" to that of the painter who, unlike the photographer, conveys a person's "generative idea." Bache thus takes the side of the artist over mere technical innovation.[44] And what does he make of the European "soul"? Perhaps Bache implies that the continent's cultural principle consists in discrete cultures being organized into states, each with its own history.

As we judge of the material so of the [—] ~~of the~~ intellectual. While a stranger sees all things in the general, *the results of* his judgment may *nevertheless* be ~~true ones~~ *accurate*. We will then confine our attention in speaking of the influences which seem to preside over European science as it is, chiefly to the prominent features.

Bache speaks of "influences" that affect contemporary European science. In the context of his speech, institutional developments are of particular relevance, but these take place within the particular contexts of European nation-states. Bache is interested in a broad transnational assessment of science because he is looking for answers to develop science on a continental American scale. Another solution would have been to compare the United States to just one European country but this is not what Bache announces to do.

The initial sections of Bache's speech thus reveal a gap between his American ambitions for a continental state and that state's political and cultural legitimacy. He compares Europe to the United States but ignores that Europe consisted of many different national states. Bache could overlook this difference because, instead of perceiving his own country in terms of a coherent nation-state, he viewed it as a federation. This federation needed guidance by elites anticipating a national sovereign. Bache's speech at the meeting of the National Institute suggests that these elites were engaged in discussions about the principles for developing the federal state. In line with his 1842 speech in Philadelphia, Bache presumed that the United States aimed at surpassing Europe by adopting its most effective cultural principles. Bache was interested in the "influences which seem to preside over European science" because he was looking at the development of scientific institutions in America from the perspective of a future political sovereign.

44 Incidentally, the oldest surviving photograph taken in the United States is a faded image of Central High School in Philadelphia taken in 1839. The photograph was taken by Joseph Saxton who became chief mechanic in the Office of Weights and Measures (U.S. Coast Survey) in 1844. Arthur H. Frazier, "Joseph Saxton and his Contributions to the Medal Ruling and Photographic Arts," *Smithsonian Studies in History and Technology* 32 (1975), 9–13, http://www.sil.si.edu/smithsoniancontributions/HistoryTechnology/pdf_hi/SSHT-0032.pdf; Hugh R. Slotten, "The Dilemmas of Science in the Unites States. Alexander Dallas Bache and the U.S. Coast Survey," *Isis*, no. 84 (1993): 30.

European Conditions

While Bache announced that he would look at European conditions, he moves on to discuss the conditions in the separate European national states. For brevity's sake I will only discuss a few sections of Bache's text and include his closing statements because there he offers concrete suggestions for developing the National Institute.

England is Bache's first example. He not only describes scientific organizations and institutions in that country but includes in his characterization colleges and universities. In this way, he considers options for pursuing a career in science: "channels through which the great body of talent of the country seeks a more or less immediate entrance into active life." He stresses that all grammar school students are evaluated against the same standards of merit. In a passage he later erased, Bache conceived of a particular opponent for a debate about education in America:

> But it is objected that ~~a mass of~~ *many* idle and careless young men enter the universities and graduate. ~~True~~ *This is so* but what influence have they upon the society of which they form a part. Little, indeed, compared with that *of the* hardy spirits which competition has trained as intellectual gladiators. Whatever be the action of such a system upon the mass of a nation the effect upon the few who have the power to use above from the root of competition, is it likely that the fruit will be that of any other plant?

Bache takes for granted that the English educational system may be considered a model for American schools, and the criticism aimed at British institutions affects their American counterparts as well. Bache refers to critics who use the standard of income work to ridicule the leisure available to students as excessive luxury promoting laziness and irresponsibility. Even though Bache later erased this passage, he takes such criticism seriously. He had become familiar with these arguments, of course, in public debates about the role of the Central High School in Philadelphia. Bache's metaphorical reference to "intellectual gladiators" counters such criticism by insisting on the students' dedication to intellectual competition and achievement. In ancient Rome, the fate of prisoners of war, slaves, or convicts who became gladiators depended on their aptitude in a fight in which their lives were at stake. Bache's use of this metaphor is ironic because he dramatizes a romantic and adamant dedication to an intellectual cause as a matter of life and death. The fight is closely followed by spectators in the arena but of little consequence to them. Bache considers the gladiators' attitude relevant for the entire university; accordingly, he feels that the outlook of those who did not dedicate their leisure time to a cause was negligible as long as it provided others with the opportunity to do so.

Bache is interested in the benefits of such education for society. In his somewhat difficult sentence at the end of this paragraph, he suggests that it is students dedicated to engaging in intellectual fights who will have a cultural impact, not their fellow students who shirk such conflicts. Intellectual competition at universities provides a basis for developing an appreciation and achievement in other areas. This assessment falls in line with the observation in previous chapters that Bache was interested in providing in his country a discourse relevant for intellectual engagement.

Bache continues as follows:

> ~~But this is not a question requiring discussion, it may be answered by~~ *There are objections to this view but rather than to discuss them let us make together a tour of* observation. *Let us* ~~F~~*f*requent the meetings of the Royal Society of London of the Astronomical Society, *listen to* the spirited discussions of the Geological Society, go to the observatories of England, attend the reunions of the British association~~s~~, and the~~n~~*re* ~~you will~~ *we shall* find the men of Cambridge & Oxford, and there ~~you~~ *we* may learn something of the spirit which animates the elite of their graduates.

Bache was indeed moving from a review of English schools to a discussion of English science. True to the question he set out to answer, he is interested in the institutional and intellectual prerequisites of successfully implementing and enhancing national science.

Bache assumes that unlike colleges and universities, institutions such as the Royal Society were admired by his Washington audience, for otherwise his overall strategy here would not work. He stresses that former college and university students become members in organizations such as the Royal Society and that their education is thereby validated. Bache's comments suggest that public opinion contradicts itself when it appreciates the role of national scientific institutions such as the Royal Society and disregards educational ones such as colleges and universities. In the previous section of his speech, Bache had presumed that his listeners had little interest in the role of educational institutions in shaping a particular scientific work ethic. He also presumed that among his audience, institutions such as the Royal Society and the British Association nevertheless yielded significant national prestige. In line with his initial reference to "vanity," Bache tells his listeners that their interest in enhancing the country's cultural prestige through national institutions was propelled by self-importance if such efforts were detached from a true understanding of the prerequisites of science including education. Bache is not interested here in providing financial support for scientists. Neither the Royal Society nor the British Association provided incomes for their members. He instead seeks to point to the intellectual and "ethical" preconditions for science.

In the following paragraph, which was later taken out, Bache had initially developed his ideas on the role of education:

> Then ask yourself if this is not leaven enough. Again, a new provincial university is to be founded, where are the teachers drawn? A new central university is to be established for conferring degrees, whence are its fellows [—] *taken*? Colleges are to be founded, whence are their Professors? Schools are to be organized, where are the principals obtained? And these all connected with the immediate movements of the day?

Bache's comments leave little doubt that he was less concerned with creating scientific institutions that could contribute to a universal scientific discourse in the short run than with developing the institutional arrangements for sustained cultural development on a national plane. This confirms observations in earlier chapters that Bache sought to connect institutions on all levels to the "immediate movements of the day," i.e. to a universalistic rational discourse. His implicit goal is to create in his country a system of education that will allow students to understand contemporary research problems. This makes sense only if Bache also assumes that Americans will not only keep up with these developments but also play a significant role in them.

Bache actually erased his comments here but our observations fall in line with others. He deleted for economy rather than content. Bache skipped his comments on American universities because he could shorten his text by developing his ideas with reference to their esteemed British counterparts:

> There is a tone which exalts English science far above all other. You ~~feel it~~ *recognize it at once + always,* when in contact with its votaries. Their characters are not only exalted by great intellect, but chastened by true religion. This spirit they owe to the Universities, and where can be found its substitute. Not to the Universities ~~ties as connected the~~ *in their peculiar relation to a* national church ~~peculiarly~~, but as christian institutions, in which the christian spirit must be fully recognized, and where it must necessarily be the leading ostensible motive of action.

As a metaphor for good students and their readiness to engage in an "intellectual fight" without practical, extra-academic relevance, Bache had introduced the image of the "intellectual gladiator." In this passage, he speaks of students as "exalted … but chastened." Bache adjusted his raw metaphor with a more refined and restrained characterization of intellectual rigor. He now takes a different, humbler look at the British university system by stressing education's moral dividend. He even mentions that there may be institutions other than universities that effectively convey such a spirit but he does not explicitly say what he has in mind. Religious communities? Autodidacts? Bache qualifies his initial, more direct appreciation of the university's role in preparing students for the tougher experience of intellectual fights.

There is another level on which Bache makes concessions to his audience. By looking at British universities as institutions conveying a particular moral mindset and attitude, Bache stresses that they are able to assume this role despite their specific Anglican context ("national church"). He anticipates the argument that universities of the British type (or perhaps universities in general) have little use for his country because they do not sufficiently adhere to the ideal of a given congregation's independence. This is odd because such a view would equate a university's dedication to a universalistic body of expanding knowledge with a lack of self-sufficiency cherished by American tradition. But Bache seems to have in mind such criticism when he argues that universities as "christian institutions" should be taken seriously in their moral capabilities even if their institutional setting seems to conflict with it.

Bache's reassuring tone and his corrections (replacing his initial romantic gesture with a more humble assertion of Christian ideals) points to a hiatus between his own motives and those of his audience. Why else would he have adjusted his comments on the university's role? While Bache is critical of ambitions for national scientific institutions without a solid educational foundation, he tactically concedes to religious communities their prominence in shaping American life. While Bache assumes a secularized perspective on the development of educational and scientific institutions in the United States he cannot ignore the authority of religious communities.

Developing American Science

Before he shifts his focus to the American situation, Bache continues his tour of European science, of Britain, Germany, and France. He is interested in those countries' leading scientific institutions and in their ability to support research science.

Bache considers both the Royal Institution of Great Britain (founded in 1799) and the British Association for the Advancement of Science (BAAS, founded in 1842) of limited use for the development of research science. He reports that the BAAS provides the logistics to coordinate larger projects among its members, but he also points out that the organization's annual meetings are open to the public and that the limited quality of papers presented at these meetings is frequently overlooked. Papers are written for effect rather than substance. Such an organization, Bache suggests, "may do more harm than good." Bache is interested in strengthening cooperation among the comparatively small number of scientists trained at universities or having assumed a research-oriented focus in a different way, and his considerations continue to be independent from questions of financial support. He suggests that science does not yet have the critical mass and autonomous institutions to convey its results to a wider public. With respect to scientific organizations in Britain, Bache writes that "the machine has so little friction that perhaps it may go too fast." He is afraid that initial American efforts to create professional institutions will be overwhelmed by public interests incompatible with science.

Bache views conditions in France to be similar to those in England. There is no easier way to control science and to limit the impact of charlatans than to create an Institut Français by government decree, but one "of the evil fruits of associations like the French Institute … is that they put the possession of the place of member in lieu of activity in the pursuit of science." The Institute, Bache suggests, was unable to stimulate its members to continue to pursue scientific questions once they had become members. Bache's view of the Institute's success may be disputed; but he was interested, not in organizational acknowledgment, but in finding organizational means to stimulate the development of new ideas (such as an "intellectual gladiator" would) and to provide a forum to test them. Such a development Bache wants to shield from the public's "empty compliments." Bache then turns to François Arago's adventurous story that is entertaining but also serves as an illustration for his earlier arguments. "His is, in the main, the directing mind of French science," Bache writes, "though occasionally not the directing voice." Bache confirms that leaders of science care less about public or professional acclaim than for making a relevant intellectual contribution. Arago's role in French science, in other words, was similar to the one Bache assumed at the Washington meeting.

In discussing the conditions of science in the German states, Bache focuses on their universities. He keeps his comments on Germany short, as short as his comments on Belgium (which he later chose to omit altogether). He concludes this section with a discussion of the "magnetic crusade" which was initiated in Germany and coordinated by the Royal Society in England. The United States, he points out, had not joined this international project, and he advertises its political benefits. "Here is a new bond of union between nations," he argues, "a new interest to be broken up by wars."

On page fifty of his manuscript Bache finally turns to a consideration of science in his own country:

> The very fact of being one of the players and not a mere looker on, makes me distrust my own judgment as soon as ~~it~~ *we* come~~s~~ ~~at~~ *near* home and I have the difficulty before referred to of generalizing the details crowded so thick upon me.

Bache made this point even though he had initially argued that the challenge lay in understanding a foreign culture (Europe), not his own (America).

In his depiction of conditions for science in the United States, Bache distinguishes between three phases, all of which were part of a developmental trajectory towards an independent pursuit of science in the context of a sovereign nation-state. In an extension of his argument at the beginning of his talk, Bache points out that while Americans were eager to have France, England, and Germany acknowledge their scientific work, they failed to honor this work themselves. This appreciation is central for Bache: "There is no cause half so depressing to American science as the want of an American feeling in regard to it." How could this be changed? In this key passage of his speech, Bache continues to consider the U.S. in terms of a federation rather than a nation-state but he argues on behalf of developing an American nation because science was "still essentially national."

What are Americans to do? Bache rules out an active national, a federal, engagement on behalf of science. Even though individual states provided funds for scientific projects such as geological surveys, Bache considers this support ambivalent at best.

The danger in regard to attempts to aid science out of the ordinary channels is, however, not to be concealed. It is conceded that the laws of the land require special study for their comprehension + it is not to be supposed that the laws of nature can come by intuition. Hence however favorable individual members of Congress may be to science, they are exposed to error in attempts to encourage it for want of special scientific training.

Bache considers the support of science by Congress or the states a replacement for something else ("out of the ordinary channels"), and he was likely referring to the profession. Bache opposes the idea of turning the National Institute into an "expert commission" for the government:

An institution at Washington uniting the knowledge of those employed in the various departments of the government, might seem to be a safe depository for the confidence of Congress, a safe—but would it be a willing one. Time + knowledge are both commodities of too great money worth to be disposed of in return for mere honour, unless by the favoured few who have ample means to meet the claims of family, [kindred] & friendship.

Bache takes for granted that members of a commission in Washington, employees of federal organizations there, would be granted neither time nor support from their employers to live up to their added responsibilities. If such a commission were to consist of men without a public office, however, only those with adequate financial means would be able to serve. The criteria for serving on the commission would not correspond to the profession's own, i.e. experience and standing among one's colleagues on the basis of good work.

It would be no small relief to Congress now forced by Committees, or individually, to listen to multitudes of claims to rewards for inventions & discoveries, to be able to throw off so irksome, unprofitable + unpopular a load. If those who took the labouring oar in such an institution were remunerated, others would be found + easily to ~~fill~~ *discharge a* less toilsome duty.

As an alternative to a commission consisting of government employees or financially independent scientists outside of such jobs, Bache considers the idea of remunerating committee members. He suggests that this would make it easier for Congress, and that the committee would have a lot of work on their hands, more than the "less toilsome duty" they leave behind.

But here would be at once a government institute, the entering wedge it might be said of the pension system of the old world. Happily there is no need to choose between the neglect of science + such a scheme. The country has received by donation half a million for the express purpose of diffusing knowledge. The faith of the country is pledged. Congress has become a trustee of science.

Bache considers the Smithsonian bequest a solution to the peculiar difficulties of establishing a "government institute," but the way in which he introduces this solution makes clear that he would prefer a different one. In this section of his speech, Bache effectively proposed a national academy of sciences, the existence of which would reflect the political sovereign's dedication to the principles of science. Not to establish such an institute and not to create an institutional device for advising Congress, Bache calls a "neglect of science." This "neglect," Bache implies, would be a problem, not for science, but for the American state. Congress needed adequate information for putting into law enlightened policy. As long as Americans chose not to honor science, however, they would remain dependent on European science. The Smithsonian bequest was a stopgap measure because it provided the country with the means but not the political will to support science. Bache acknowledges that the bequest's acceptance represented a cultural and political shift. "The faith of the country is pledged," Bache writes. "Congress has become a trustee of science." Such a pledge, in other words, was all but certain. Bache's speech indicates that the Smithsonian bequest was significant because he wished to conceive of it as having initiated a change in the country's political attitude towards science.

Guarding the Palladium

On page eighty-three of his manuscript, Bache returns to a consideration of the state of scientific organizations in the United States:

> The effects upon American science of an association like the British association have been much talked of, if not publicly discussed among us of late years. Our Geologists have led the way, and held yearly meetings for mutual communication, beginning, with commendable caution, by assembling those connected with the state surveys.

Bache distinguishes between two areas in which discussions took place: a rather informal conversation about the consequences of the founding of the BAAS among colleagues, on the one hand, and a public discussion about scientific institutions, on the other. The BAAS obviously provided a new point of reference for these discussions, and Bache responds to the idea that such an organization should perhaps be founded in the United States as well. Existing organizations such as the American Philosophical Society or the American Academy of Arts and Sciences were unable to play a comparable national role. The issue at stake was whether American scientists were ready to consolidate their profession.

As a scientist, Bache could not have referred to geologists as "*Our* Geologists" because we have seen that he considered his audience to consist of "friends of science" as well as scientists, and because, even if he was speaking to scientists only, the possessive pronoun would not have made sense. But the use of this pronoun does make sense if Bache spoke as an American citizen, directing the audience's attention to those fellow Americans engaged in the international pursuit of geological research. While assuming this political perspective, Bache distinguishes between two overlapping groups of geologists: an "ideal group" of those who identify with the profession and another group that is actually involved in and remunerated for survey work. The association had been introduced to provide a platform for discussing practical questions related to this work, not as a vehicle for geological discussion in general. The "ideal group" of geologists must have had other channels for discussing their ideas, we may infer, such as personal correspondence or existing science journals. Bache insists that scientific organizations in the United States should not seek to represent the "ideal group," i.e. the profession at large, and instead assume a more limited practical role. Why does Bache find such caution "commendable"?

His caution falls in line with his initial hesitation to endorse the meeting's objectives. Bache sought to protect the relevance of professional discussion, particularly of questions of basic research, by way of an informal and therefore flexible "ideal" group of scientists and through organizations with a clear practical focus that would naturally limit the range of discussions and regulate membership with respect to the success of producing relevant work. Bache was interested in protecting the profession towards the inside (by providing colleagues with the guidance of practical questions) and towards the outside (by shielding the profession from criticism directed at leisurely pursuing unpractical questions of basic research). Finally, by using the definite article in "have led *the* way" (my emphasis), Bache indicates that he does not consider the development of professional institutions to be random but following a definitive pattern or logic.

The American Philosophical Society at their hundredth anniversary ~~called about them~~ invited *the [attendance] of* the men of science of the country. The National institution at Washington has ~~issued notice of an intended meeting next spring and gives invitations to attend it~~ *called about it the friends of science, + with cheering alacrity they have come at the call.* These are indications that we ~~shall soon~~ ~~decide~~ *are about to settle [definitively]* the question of the policy of these meetings.

Bache introduces two examples that show that American scientists were eager to organize an American version of the BAAS or by revitalizing an existing institution. He signals to his audience that he is aware of a restlessness to go ahead with national scientific development.

Why would "the question of the policy of these meetings" matter? Bache concedes that the question is no longer whether an American organization should be created but what the shape of such an organization should be. If questions of policy had been contentious, what could have been the sources of this conflict? Did policy questions relate to issues of membership and the preservation of standards?

It becomes men of science to ponder the matter well before deciding whether they will *or will not* take part ~~or not~~ in them, ~~or not~~. By holding back they may retard science among us, by *precipitately* moving forward they may plunge us into difficulties, out of which years may not extricate us.

Bache feels responsible for a consolidated development of American science and he issues a stern warning. Founding an American organization similar in scope to the BAAS has significant consequences not to be taken too lightly. He urges his colleagues to think hard about the feasibility of a given plan. Whatever his colleagues choose to do, their decision will have consequences for everyone ("us"), including those who did not support the plan.

Is it the diffusion of science, or the encouragement of research that American science requires?

Bache is interested in supporting the small circle of American scientists who promise to make a relevant contribution to research. A broad support of science, providing avenues into its practice and conveying knowledge, Bache considers less important than setting high standards of achievement. He had earlier pointed to the important role of schools. By stressing the role of research in this part of his speech, Bache implies that schools will not provide a relevant framework for research science any time soon but that the profession needs to insist on the ideals of research science anyway.

Is it sympathy and kindly communion of which we have most need? acquaintance with each other and our several doings? or opportunities, means and appliances for research? Is it to be directed in the route we should follow, or to have the means of journeying at all?

Bache develops his point further by equating "diffusion of knowledge" with "sympathy and kindly communion," thereby suggesting that many among the audience are less interested in scientific work than in the sociability offered by meetings.

In the last sentence of this excerpt, Bache posits two alternatives, movement and stalemate. He distinguishes between a desirable route for developing science, on the one hand, and not developing science, on the other. Bache has his own standards for developing science in America. By reiterating his idea of not creating an institutional framework for American science, he implies that he would prefer doing nothing to having a new institution veer off in the wrong direction. This again implies that Bache does not feel that America provides conditions for a reliable institutional setting.

The superintendent is aware of the consequences of creating national scientific institutions. There could be only one national institution such as a national academy or a national association. The proposed creation of a National Institute could not be ignored by anyone pursuing science in America. Bache seeks to block further development of this organization, and to support individual researchers and their projects instead. (He had spoken of "opportunities, means and appliances for research.") Bache was convinced that the country was not ready for a national scientific institution but we must also assume that he was also interested in developing the U.S. Coast Survey. This organization, after all, provided a viable contemporary version of a national scientific institution, and he was thoroughly in charge of its development. We have no reason to doubt the sincerity of his assessment that America was not ready for the National Institute, but Bache's opposition to its development also served to consolidate his own nationally prominent role in American science.

Is it to be ~~moved~~ *roused* up to greater activity? Yes—surely we need this, and as surely meeting each other will promote that activity; it may be a healthful activity, it may be febrile to be succeeded by languor and depression. Is it to improve the mode in which our scientific men bring their researches before the public? or to induce research *that we have [need]*? Do we need talkers or workers? Is it to drag modest merit from the corners in which it hides itself, and leaving from its groves and cloisters, or to repress charlatanism that we most need? Is it periodical work *under excitement* or regular steady work which we should encourage?

This paragraph falls in line with Bache's previous arguments. He contrasts the excitement attached to annual meetings with the consistency of uninterrupted research. By speaking of "febrile" and through his reference to "languor and depression," Bache pokes fun at the joys of scientific meetings. The "workers" are Bache's model, those who explore questions to which they are wholeheartedly committed. He does not flesh out the standard that prompts him implicitly to criticize at least some among the audience, but it would make little sense if Bache had in mind a picture radically different from that of the self-propelled laborious worker curiously investigating problems shown to be relevant by his or her own sincere standards. Bache's model scientist, in other words, does not need an institutional context for successfully advancing relevant questions. An institution may help but it is no prerequisite. At the same time, Bache does take for granted a community of scientists but it exists outside and independent of an institutional framework.

Bache writes: "Is it {i.e. American science} to drag modest merit from the corners in which it hides itself, and leaving from its groves and cloisters, or to repress charlatanism that we most need?" Bache regards "charlatanism" as a desire for public recognition. He thus attacks those unable to face the consequences of his implicit question about the motives for scientific work. Why would Bache mind? One can easily see that "charlatans," according to his implicit definition, impede ongoing scientific work because their arguments cannot be ignored if their proponents use public acclaim as a lever for decisions on budget and personnel. But why would Bache take for granted that a national organization and its annual meetings would support second-rate scientists who become a problem for first-rate ones? Bache assumes that his American peers coordinate their work with one another and with their European colleagues. Because no national organization existed in 1844, Bache took for granted the existence of informal networks, an "ideal" scientific profession of which he was a spokesman.

These and other questions, which each man's mind will suggest according to the wants of the circle in which he moves;—these and other questions are to be answered, in making up our minds as to the question of association or no association. We want sympathy and closer fellowship,—an association will give it,—the freer the better for it increases the number of the brotherhood,—the more varied the places of meeting the better, for opportunities are offered to more men of joining in its pleasures and profits, and of contributing *to* them.

Does Bache contradict himself here? He had mobilized arguments against founding an organization but he now discusses its advantages. Even though his own preference has become abundantly clear, Bache continues to weigh both sides and to illuminate to his audience the meaning of the decision they are about to make: "association or no association."

We do not require more knowledge of our doings among our own men of science, nor more able lecturers or talkers if you will. [So] the oral proceedings of an association can never be *in reality* of the first importance. We do not want an association like the German association.

This paragraph confirms that Bache took for granted that scientists in America could exchange ideas through established channels. Bache must have had in mind publications and letters, and this preference is suggested also by his dismissal of oral presentations (such as those at the Deutsche Gesellschaft der Naturforscher and Ärzte).

Bache then moves on to suggest what an American organization should look like:

We do require more work, and more aid for work, and an association with funds which would enable individuals to make researches without starving their families, would be of great service [in] not attempting to absorb the labors of its members through committees, always more or less anonymous, but furnishing means to individuals to occupy those fields which time and ability may enable them to cultivate. We must *then* deviate from our predecessors in this respect.

Bache further explicates his preference for supporting individual scientists and because his overall argument falls in line with his earlier comments, I will merely highlight a few details.

He takes for granted that scientists will be "amateurs" in the sense that they derive their means of support from sources outside of science and that an impulse for scientific work will not be institutional. He suggests providing gifted candidates with the means (i.e. time) to develop their work. The financial support of scientists, therefore, is important for Bache but he does not consider it an end in itself or even an aspect of the definition of either "scientist" or "profession." It is nothing more than a tool to provide curious researchers with room for focusing on the development of promising ideas. Bache's attitude and expectation towards scientists contrasts with that of a labor union: The latter would never expect its members to continue working without pay. Bache takes for granted that scientific work is *usually* pursued without remuneration, and that stipends and research jobs merely provide an opportunity to engage in science. Whereas a union member works to have an income, the scientist, in Bache's logic, needs an income in order to be able to work.

In his last sentence here, Bache views European organizations as "predecessors" to American ones. He implies again that his colleagues and his audience have assumed responsibility for creating institutions that will enable the country to become a new leader in science. This echoes Bache's initial assessment of the United States: the country's commitment to surpass Europe and to make the most of America's advantages including not being "parcelled out into independent states."

> We need more activity, but not to be stirred up every year ~~or two~~ by exciting means. The excitement of the meetings of an association is exceedingly powerful, the mind and body give way under its continuance. We cannot to advantage work by fits and starts + if our meetings are too frequent they will injure science. We need *such a* union ~~to~~ *as would* repress charlatanism,—not the form of association which would deliver us up bound hand and foot as its prey.

Bache repeats his argument that scientific meetings may not serve the purpose of science in the United States. But why is he so insistent on this point?

Bache is strongly concerned that his audience's interest in scientific work is driven by the wrong motives. He had even dismissed biannual meetings but then corrected himself. His firmness in this matter (Bache's fear of working "by fits and starts") reflects his interest in establishing sound scientific investigation as a routine, as an ongoing, self-reliant, and in this sense autonomous professional activity. Conversely, he does not consider American science to have established such a professional routine yet. How does this match our observation that Bache takes for granted the existence of a group of peers to which he belongs himself? Throughout his paper, Bache had viewed the American situation from a European perspective, measuring up his country's developments against those across the Atlantic. Publications and correspondence by mail facilitates exchange, not only among American scientists, but with colleagues in Europe as well. Bache is unwilling to give up the standard provided through such means of exchange, and to let American science develop without such checks. Giving a bad paper is easier than to have a bad article reviewed and published. Criticizing a published paper is easier than criticizing an oral presentation because a publication is accessible to anyone. An oral presentation can be criticized by an audience but it is difficult to respond to it later, and it is impossible to do so from across the Atlantic. Bache's firmness is prompted by his reluctance to lower academic standards and by his reliance on both European oversight and oversight by himself and his colleagues. An American organization and annual meetings would have provided an avenue of development away from such oversight and toward American scientific independence, but Bache was not certain that his country was ready for it.

I move over a paragraph in which Bache describes the problem of not excluding "a single man of science" while making sure not to endorse "a single quack." He then rephrases his overall assessment in the following metaphorical way:

> The great influence which an association including all the men of science of our country would have upon our local and national government and upon the public,—the peculiarly great influence which it would have in consequence of the nature of our political and social institutions is not to be ~~for a moment~~ lost sight of *for a moment.* The mighty steam engine which drives the vessel through the waves carrying it, in spite of all opposition to the desired haven, may crush the fabric upon the floating iceberg or the sea-girt rock, and even expend its resistless power in rending to pieces the structure which contains it. ~~We~~ *The men of science of our country* must study the intricate parts of the machine ~~our~~ *them*selves, and be prepared to do duty about every part of it.

Bache turns to the extra-professional, political impact of an American scientific organization. He suggests that an organization may have signifycant influence "in consequence of the nature of our political and social institutions." While European scientists may rely on the leverage provided by feudal and aristocratic traditions or institutions, in other words, American democracy provides for no such support. Bache deduces from the absence of an American aristocracy the need for professional colleagues in the United States to replace its uncompromising representation of scientific rationality.

In his somewhat cryptic steamboat metaphor, Bache likens the motivation to develop American science to an engine driving a vessel forward, a metaphor for the profession as it is being carried along by the waves of social and political developments. This engine, we may suppose, is working at full throttle, as there is much "opposition" to be overcome. Bache infers that there is a general disinclination to go where the scientists want to go, to get to their "haven." What do we make of the contradiction implied by Bache's use of "in spite of," a contradiction between the "opposition" the vessel is sought to overcome, on the one hand, and the potential unfortunate consequences, on the other?

The only way to make sense of this is to assume that Bache has merged several ideas into this sentence. He is suggesting that a strong motivation is indeed required to push forward the cause of science in the American context. In this sense, the "opposition to the desired haven" explains and justifies the steam engine's level of engagement and his audience's eagerness in the context of the National Institute's meeting and plans. This very momentum, however, may also have the consequence of destroying the framework it seeks to create and move towards. It may prompt resistance to its plans from the outside and falter on such an "iceberg," or it may be unable to control its own institutional setup and tear "to pieces the structure which contains it." This connects to the very argument Bache has been trying to make. He acknowledges his audience's eagerness to develop American scientific institutions and to direct the scientific vessel into the calm conditions of a "haven" but this very eagerness may work against the profession if such impetus is not checked by thoughtful consideration of each step's appropriateness.

In the last sentence of this quotation, Bache implies that groups other than scientists may have an interest in developing professional institutions, by stressing that scientists "themselves" needed to understand institutional prerequisites and consequences. This corresponds to the opening of his paper and Bache's distinction throughout between scientists and "friends of science." A "scientist," Bache suggests, needs to have merged his personal investigative motivation and aims into a professional whole and its universalistic goals—as a scientist, he needs to be "prepared to do duty about every part" of the profession's vessel.

Bache took out most of the following paragraph but it shows what kind of interference he had in mind, and how he conceived of the relationship between politics and science:

> ~~We must study the intricate parts of the machine ourselves~~ ~~*make and officer the vessel*~~ ~~and take care how we admit passengers. We cannot consent to have~~ *They must regulate* the fires regulated and the *control the* machinery *or they hazard the safety of the vessel* ~~controlled by those who do not belong to us, to put our safety into the hand of politicians however estimable and trustworthy in their own walks in life. They sail the ship of state and we as part of the crew follow their guidance; if they would volunteer with us let them do likewise. We are not ambitious of command, or~~

~~desirous of control, but we have a palladium to guard and we may not intrust it to others.~~

Bache's steamship metaphor had conveyed most of what he is saying here, and I will not discuss this paragraph in detail. But his reference to "a palladium" remains fascinating. Towards the end of his talk, Bache is obviously trying to encapsulate his message in a formula, and his metaphorical reference captures what he has been saying. Palladium had been discovered in the early nineteenth century. The extremely resistant and rare metal was named after the asteroid Pallas (which, in turn, was named after the epithet of goddess Athena who had slain the giant Pallas). Bache and his colleagues are religiously dedicated to guarding a precious icon, a refined and durable emblem. Science was pursued for reasons other than pecuniary ones. This complements Bache's image of a vessel plowing through the waves. The two metaphors consider science from two distinct perspectives: the realism of coordinated laborious activity aboard a steaming vessel contrasts with its dedication to an abstract and elevated cause.

American Science by an American Union

The consistency of Bache's metaphors throughout his long speech suggests that he had a firm idea of what he conceived science to be. The superintendent had a clear model against which he evaluated ideas for developing science in America. In the final sections of his ninety-page manuscript, Bache outlined his own practical recommendations for the National Institute.

~~It is plain that we want no mere gathering of men of science without organisation, no indiscriminate~~

We want in short some safe rallying points, some mode of assembling & combining the advantages of foreign associations, with features peculiarly adapting it to the condition of American science. May we not find such a mode by collecting around the societies already established among us, and having the confidence of men of science.

Will not such societies be willing to encounter the labors and difficulties of enacting the host? Will they not be willing to risk the effects of irregular action upon themselves and the reaction of stimulus? Will they not give up a part of their ease and peril somewhat for the cause of ~~science~~ American science? Let them lead us as they have heretofore done, and we will follow, but they must be jealous and stirring if they would come up to the work.

> If our local societies already distinguished by years of successful labours, fear this irregular action. If they are of opinion that more injury will be done to the regular channels through which their contributors flow by the sudden rise of the waters than will be compensated by the increased products [borne] onward by the swelling waves, we cannot ask them to take part in a movement which would be to them so destructive. This is a question for them & not for us to decide.

Bache proposes a compromise by inviting existing local societies to organize national scientific meetings. How can Bache assume that the arguments against a national organization will not also be mobilized against their local counterparts? How are regional organizations different from national ones and how are they less prone to infiltration by charlatans? Bache assumes that such organizations, unlike a new national one, are embedded in a given regional culture, in a regional community. He favors a solution growing out of established local structures that have proven sustainable ("already established among us"). Bache implies that science is best supported if such support is part of a friendly competition among regions, because a region may thereby realize and strengthen its own ambitions and identity ("but they must be jealous and stirring if they come up to the work"). There is no reason at all why this model should not work on the level of nations. The support of science would then be a matter of pride in a competition among nation-states on the basis of a universal exploration of knowledge. Bache, however, considers the state of regional scientific organizations to reflect a federal cultural structure, and he is too cautious to let go of this "natural" basis for support of science in America. His proposal implies that the United States has not yet evolved in a way that would make a competition of "American" (as opposed to regional) science with European science feasible.

> The National Institute is peculiarly situated in this respect and may well lead onward, and safely. Careful to exhibit moderation in its acts, to exhibit in ~~mode~~ its youth that defer~~r~~ence for age which so well becomes youthful vigour, it may conciliate the regard of the older societies of the country. By ranking itself as a local institution and following the modes of usefulness of local institutions it must lose the advantages of its peculiar position and resources. By occupying new ground, it will avoid conflicts with old occupants and even jealousy. It should aim at nationality, adopting such plans of advancing science as are national in their scope. Among these the connexion of all ~~modes~~ *men* of real science in every part of our union as its members and their periodical gathering together at the Capitol.

Keeping in mind the thrust of Bache's arguments above, the range of activities embraced by his recommendation that the National Institute "aim at nationality" and adopt "such plans ... as are national" must be limited indeed. Bache seeks not to disconnect himself from the meeting's momentum (the "youthful vigour" driving the organization like the figurative steam engine) but to encourage his audience to respect the given structures of American culture rather than to revolutionize it.

Bache ends his speech with the following sentence:

> The union with them of the friends of science to aid its progress. Thus united we will find that sympathy and support which we need, shoulder to shoulder we will press forward in the cause. Thus united *we*[45] will advance defend elevate American science by an American Union.

Does Bache again contradict himself? He had suggested that membership in a scientific organization be restricted to "men of real science" but now he celebrates a "union with them of the friends of science." Bache covers both bases: He had suggested that participation in a meeting organized by the National Institute, as well as membership in it, be indeed limited to "men of real science." In his final paragraph, however, he also embraces the "friends of science." Bache balances his unyielding claim for the profession's integrity with trying to steer the meeting's development. Stressing the idea of the "Union" serves his intention, for it highlights the country's diversity rather than its homogeneity, and this goes along with Bache's recommendation to rely on and strengthen the experience of local organizations.

In conclusion, Bache's speech confirms the assessment that he was "lukewarm" towards the idea of developing the National Institute, particularly as that organization seemed to upstage the meeting of the Association of American Geologists and compromise the latter's emerging role. But his speech also makes clear that his agenda was not confined to discouraging the National Institute from pursuing such plans. Bache laid out the many dimensions of his beliefs on the nature of scientific inquiry and deduced from them a tacit institutional program for developing science in America. The core of his belief ("guarding the palladium") consisted in the scientist's investigation of the "fleeting" traces of his subject matter. Bache took for granted that while there existed a scientific profession in America, it remained without an institutional equivalent. Interaction was possible through letters and journals, and these could be exchanged with European

45 This word was not inserted by Bache, if the handwriting is a clue, but by Bache's amenuensis.

peers. Scientific activity was a matter of self-cultivation, and in the United States it was connected to local societies such as the American Philosophical Society and the Franklin Institute in Philadelphia.

Bache presupposed that unlike local societies, any national society would lack the kind of cultural foundation required to keep scientific investigation on track. Bache feared that without a connection to a community (local or otherwise), scientific organizations would be unable to preserve the integrity of their discourse. What prompted him to fear the role of "charlatans" in such a development? Bache supposed that scientific investigation was relatively safe as long as it related to a curious public interested in its results. The development of the National Institute and Bache's assessment, however, suggest that in 1844, a national American public was only just emerging and that such a national counterpart for science did not yet exist. The United States was still adding territories and states. Despite its federal political framework and national election campaigns, the country, stretching over a vast territory and pursuing continental ambitions, remained decidedly regional. Bache implied that as long as there existed no common focus for political life in the Unites States (a focus that would effectively transcend and integrate these regional centers) it would remain dangerous to establish national scientific organizations because it was difficult if not impossible to check the relevance of their work against standards derived from life in a political community. Bache considered the ethic of arduous commitment to tracing the many facets of a relevant idea (scientific exploration) to be rooted, not in institutions, but in a culture that preceded and embedded organizations. His educational work in Philadelphia had shown that it was difficult enough to establish institutions within a city or a state. On the national level, Bache could see no equivalent of an urban community ready to conceive of the problems of the day as issues that concerned all citizens in the national state, and trying to find national solutions for them.

Bache's equation of America and Europe suggested that he did not conceive of his country as an integrated national state. He engaged in the discussion at the meeting of the National Institute because it had become clear that cultural institutions would be created in the U.S. and because he could not ignore the organization's initiative. But Bache feared that scientists would not be in control. While Bache was sincere in doubting the viability of a national scientific organization dominated by "friends of science," therefore, his caution was also tied to the prospect of preserving the initiative for himself. By blocking the National Institute's initiative, Bache's helped solidify his own future role in developing his profession's institutions.

In the absence of a national "sovereign" that would acknowledge science for its own sake, Bache's authority was based on the practical relevance of his scientific work. In his speech, he had argued that practical national problems justified professional organization. The Association of American Geologists was leading the way in this respect, and the U.S. Coast Survey fulfilled a similar role. Even though they were legitimized in different ways, both catered to the nation's practical needs, and the application-oriented nature of their work enabled them to turn back and to help foster a national outlook and culture as a basis for further scientific development. Such an emerging national perspective, engrained in his family tradition and represented by Bache since his West Point days, could prompt him to criticize colleagues not helping to construct a new national scientific culture. But this call for professional cohesion prefigured the future of the nation's promise since independence as conceived by Bache: to develop a national cultural life, a sovereign ready to endorse science and the arts.

Chapter 8

Bache's Program for National Consolidation III

The American Association for the Advancement of Science

While the National Institute hobbled on with the aid of a twenty-year charter, it was unable to muster significant scientific and political support after 1844.[1] The Association of American Geologists and Naturalists, meanwhile, transmogrified into the American Association for the Advancement of Science in 1848. While the American Philosophical Society and the American Academy of Arts and Sciences were regional organizations with limited membership, the AAAS was the first comprehensive national scientific organization in the United States.[2] The organization did not alleviate Alexander Dallas Bache's fears of "charlatans" entering the ranks of a national organization with few checks on membership. With the aid of close friends and colleagues such as Joseph Henry, Benjamin Peirce, and Louis Agassiz, he was able to dominate the organization during the 1850s. Bache's perception of the profession's role in America has already come into focus but what remains to be done is to test our observations against samples of his professional involvement in the AAAS and within the close group of scientific friends around him that referred to itself as the scientific "Lazzaroni."

Bache had moved to Washington at a time when national efforts to organize science were just getting underway. His concern was that national science be put on the right footing. As pointed out in the previous chapter, science had not been endorsed nationally through either the founding of the Smithsonian Institution, a "universal institution" made possible

1 Sally Kohlstedt, "A Step toward Scientific Self-Identity in the United States: The Failure of the National Institute, 1844," *Isis* 62, no. 3 (Autumn 1971): 361.

2 Sally Kohlstedt, *The Formation of the American Scientific Community: The American Association for the Advancement of Science, 1848–1860* (Urbana: Univ. of Illinois Press, 1976).

through the bequest of a foreigner, or the continued support of the U.S. Coast Survey, a large but temporary organization.[3] The AAAS, furthermore, was a professional organization chartered but not otherwise meant to signal federal acknowledgment and support of science. In the 1850s, therefore, science remained without national political endorsement. A national academy, which would have represented this acknowledgment, was not in the works even though Bache and his colleagues frequently mentioned it. In the absence of their nation's political endorsement, how did scientists feel justified in pursuing research in the United States? The most obvious answer is that they continued to rely on their sense of professional community that had preceded the founding of the AAAS. In his 1842 and 1844 speeches, after all, Bache had taken such a community for granted.

This suggests that in the absence of a national academy and the political approval it implied, Alexander Dallas Bache had personally assumed the role of such an institution prior to its foundation in 1863. The evidence so far has suggested that Bache was able to represent comprehensively the range of implications of scientific work in America, extrapolating from his understanding of the nature of scientific investigation consequences for science policy under the given political circumstances. The massive influence and the range of opportunities attached to his Coast Survey post hardly followed from his administrative responsibilities. They were an effect of the broad endorsement he was able to muster among colleagues and politicians in Washington. He was quite obviously very well connected because relatives occupied influential positions in the federal government. As we have seen, however, Bache's perspective on his work was not guided by administrative ambitions for expanding control. His understanding of the development of science in the particular American context derived, biographically, from the unique position he assumed within his family, and from his successful pursuit of his personal "mission" to help build the American nation by implementing a rational discourse best represented by science. Bache's 1844 speech provided a glimpse of his entering the national arena at a time when that arena was beginning to receive the attention of the scientific profession. His dominant role suggests that the perspective of explaining the rise of the scientific profession in the nineteenth

3 Both observations are Dupree's. The quote is from A. Hunter Dupree, *Science in the Federal Government: A History of Policies and Activities to 1940* (Baltimore: Johns Hopkins Univ. Press, 1986), 330.

century in the context of the American nation must be turned around: Bache's speeches imply that even though it lacked a professional organization until the AAAS was founded in 1848, the scientific profession was well established. To Bache, the problem was not how to implement the profession in the nation, but how to implement a nation that could provide the institutional framework for the profession. It must have been apparent to everyone that a national consolidation would take time. In its absence, it was Bache, from 1843 the head of the federal government's largest scientific enterprise, who represented the political legitimacy of science in America. In this sense he prefigured an academy he eventually helped to found.

Bache's 1851 Speech as Outgoing AAAS President

One speech that cannot be ignored in this context is Bache's August 1851 address as outgoing president of the American Association for the Advancement of Science. Anyone interested in the major institutional developments of American science in the nineteenth century will have acknowledged its existence, referred to it, or gone through it minutely.[4] In his speech, "a magisterial discourse on the state and needs of American science" that, according to Robert V. Bruce, "confirmed him as the leading spokesman and mentor of the American scientific community," Bache laid out the institutional agenda for American science as understood by himself and his closest supporters and friends such as Joseph Henry and Benjamin Peirce.[5] As A. Hunter Dupree points out, Bache in this address explicitly introduced the idea of a national academy.[6]

The purpose of my discussion of Bache's text will be more restricted than my discussion of his other speeches. I will limit my observations report to highlighting passages I believe to be important in answering the question whether Bache was the "mentor of the American scientific community," as Bruce suggests, or whether Bache assumed the perspective of developing the nation as a prerequisite for founding a national academy,

4 Ibid., 116–19; Merle M. Odgers, *Alexander Dallas Bache: Scientist and Educator* (Philadelphia: Univ. of Pennsylvania Press, 1947), 168–70. For other authors who refer to this speech, see this chapter's references in notes below.

5 Quote from Robert V. Bruce, *The Launching of Modern American Science, 1846–1876* (New York: Knopf, 1987), 264.

6 Dupree, *Science in the Federal Government*, 459.

which would match my own observations.[7] An interpretive discussion of these alternatives will also relate to the question of the financial support of research.[8] The following examination, therefore, will be much briefer than previous ones. The aim is to zoom in on just a few passages in order to answer questions such as: Where does Bache deviate from positions laid out in 1842 and 1844? What role does he envision the AAAS playing, and what does this signify for the role of science, and of the United States as a nation-state?

Bache starts out by reviewing the Association's brief history and rationale. The very fact that he feels impelled to do so suggests that these things could not be taken for granted. Rather than congratulating his colleagues on having developed the organization during a third successful year, the outgoing president declared his wish to formally submit, "in conformity with usage, to its members, a few remarks in relation to the circumstances attending its organization, and to its progress, and some considerations of the direction in which we may look for its greatest usefulness."[9] The AAAS, of course, had hardly developed such "usage." Bache's insistence on tradition confirms its fragility. Consider his second paragraph:

> The conditions of society and of science of the day seems to have called for the organization of general associations for the promotion and advancement of science in nearly every country where its cultivators are numerous, zealous, and not closely gathered in one community; the precursors of more general unions for the same good purpose. To render such meetings practicable, modern facilities of communication are indispensible; and when these shall have brought Berlin and New-York as near as were Berlin and Paris at the close of the last century, we may pass from our present organization to something characteristic of the day of railroads and the dawn of telegraphs.

---

7 Even though Dupree is aware of the difficult context of developing American science in a restrictive climate, he points out the particular ideas Bache, in his speech, expressed on the development of the scientific profession, not the development of its political context. He suggests Bache argued that "only through the professionalization of scientists and the 'minute supervision' of their efforts in specialties could real research go forward" and that "'science' meant to him [Bache] essentially those branches which the surveying and exploring enterprises of the government had stimulated." Ibid., 118.

8 "The heart of his [Bache's] proposal was appropriations from the 'public treasury.'" Ibid., 117.

9 Alexander Dallas Bache, "Address of Professor A. D. Bache, President of the American Association for the Year 1851, on Retiring from the Duties of President," American Association for the Advancement of Science, *Proceedings* 6 (1852): xli.

Bache suggests that the American Association was a useful tool but that its usefulness was bound to pass. The conditions not merely of science, but of science and society, warranted an organization such as the AAAS but improved means of communication and transportation would eventually call for a different kind of organization. Instead of annual meetings as singular events, communication would eventually facilitate closer cooperation and exchange among colleagues. Bache's larger issue is how to connect American science to its European equivalent of which he considers Berlin to be the center and Paris and New York to be contemporary and future peripheries. This confirms that Bache was concerned about attaching an American scientific discourse to its European core. The founding of the AAAS, in his view, did not signal that American science had arrived. The organization served as a temporary vehicle during a particular historical and technological phase in the country's development. Much like in 1844, Bache considers true scientific exchange to be at home on the regional and local level but he agrees that national exchange is necessary. In the United States, "cultivators" of science are "not closely gathered in one community," after all. Another, and perhaps decisive function of the AAAS, is to coordinate and to distribute information about recent researches, much of which takes place in Europe.

To recoup the AAAS's founding situation and contemporary raison d'être connects to Bache's 1844 talk on the "wants of science in the U.S." But Bache shifts the focus. Seven years earlier, he had compared the American situation to that of Europe and insisted that rather than following any given model, the design of American institutions would need to respond to specific American needs. In 1851, he relates the emergence of the AAAS to the country's historic development and situation. "As the country was explored and settled," Bache writes, "the unworked mine of natural history was laid open, and the attention of almost all the cultivators of science was turned towards the development of its riches." He adds: "Descriptive natural history is the pursuit which emphatically marks that period," a period the country was leaving behind.[10] Bache insists that the United States now partook of a "wider and deeper pursuit of natural, physical, and mathematical science," and even though his assertion was perhaps an ambitious rather than a realistic depiction of science in the United States, it connects to his earlier point. The latter fields of investiga-

10 Ibid., xliv.

tion were comparatively more prominent in Europe while the exploration of the west could be conceived as a decidedly American project. An "empiricism in forms …," according to Bache, "threatened the very life of science" in the earlier phase.[11] Why? In the absence of theory, science, in the way Bache conceived of it, could remain without traction because its accumulative and organizational bent required neither divergent views nor the premise that such conflicts could be resolved by "accredited tribunals" based on reason. Regardless of the validity of his perception, Bache's reference to tribunals is significant, not because it implied that he wanted himself and a few others to be in charge of them (even if he expected this to be the case); instead, it is the impossibility of assessing an idea and the inherent abandonment of intellectual and cultural progress which falls in line with Bache's previous comments.

After further remarks on the organization's history, Bache turns to the future:

> I would throw out for your consideration some reasons which induce me to believe that an institution of science, supplementary to existing ones, is much needed in our country, to guide public action in reference to scientific matters.[12]

Bache's proposal here was a key to the Lazzaroni-program of subsequent years and, as mentioned above, it has frequently been referred to by historians of American science. What is Bache suggesting? Does he argue that science needed a new institution to secure permanent political and financial support and to establish the American scientific profession? Bache makes clear that the institution he has in mind is not to compete with existing organizations such as the American Philosophical Society, the American Academy of Arts and Sciences, or the AAAS. It would work on a different plane without the intention to coordinate or stabilize. So how would the new organization be different? Rather than coordinate work being done by American scientists and to diffuse results of ongoing research, the new organization would be an advisor to the public and to the government. Why would such an institution be necessary? Bache assumes that the public and its political representatives were in need of guidance, that the results of scientific work and the mode of its investigation needed to be made available to the government. This idea is reminiscent, of course, of Bache's institutional efforts at the Franklin Institute in the 1830s. And it connects

11 Ibid., xliii.

12 Ibid., xlvii–xlviii. Bache's emphasis.

to what he referred to in the early sections of his speech here, his concern for connecting life in America to a universalistic rational discourse represented by science.

On the surface, Bache does not seem to develop his point further. He turns to a consideration of the roles of local societies such as the American Philosophical Society. "Not one of these associations is well endowed," he points out. "For our only endowed national institution (the Smithsonian Institution), we are indebted to the liberality of a foreigner."[13] For comparison, he turns to European models and mentions the Franklin Institute and its efforts to support research. He had initially suggested that the public was in need of guidance. Is he now suggesting something else, that scientists were in need of funds? "Among the obstacles to the progress of science with us, must be reckoned, as one of the largest" Bache points out, "the want of direct support for its cultivators as such." Is his intention that a new organization support research science and provide funds for employing scientists? Bache invites such questions and begins a new paragraph:

> It is, I believe, a common mistake, to associate the idea of academical {sic} institutions with monarchical institutions.

At first, this would seem unrelated to the issue of funding science. On the surface, Bache seems to be changing the subject. From the perspective of science and of research it hardly matters whether financial support is provided by a monarch or a parliament. But his comment falls in line with his earlier train of thought which had been guided by a concern for attaching the United States to a universalistic discourse. He is taking up common American reservations, not only towards an aristocratic form of government, but towards anything that symbolizes or represents centralized political control. Bache implies that Americans fail to see that political authority, in the United States, has a republican basis and that the national government ought to be viewed as representing its citizens rather than pursuing unrelated private interests. He assumes that identification with the national government is limited because political authority and aristocracy, republican government and monarchy, and institutions supported by a democratic sovereign and those supported by a king were conflated. As a prerequisite to his criticism, Bache shares the enlightened notion that it should be possible, in a republic, to create institutions dedicated to an

13 Ibid., xlviii.

"aristocratic" cultural development such as that of science representing ideals such as truth, adequacy, integrity, and relevance. Are these comments part of a strategy for backing up the demand for financial support of science through "an institution of science, supplementary to existing ones {Bache's emphasis}"?

Bache returns to the idea of a scientific institution sponsored by government in his subsequent paragraphs and seeks to shield it from misconceptions such as the one just discussed. He rejects the idea "of a necessary connexion between centralization and an institution," thus commenting on another popular American tenet. And with respect to matters of financial support, he writes that it could be "engaged in researches self-directed, or desired by the body, called for by Congress or by the Executive, who furnish the means for the inquiries. The detail of such an organization," he continues,

> could be marked out so as to secure efficiency without centralization, and constant labor with its appropriate results. The public treasury would be saved many times the support of such a council, by the sound advice which it would give in regard to the various projects which are constantly forced upon their notice, and in regard to which they are now compelled to decide without the knowledge which alone can ensure a wise conclusion.[14]

Bache's comments are strategic in that he seeks to counter the impression that the proposed institution was a mere luxury with little benefit to the nation's citizens. His focus and concern is the rationality of decisions on the federal level and he takes for granted that his listeners shared his concern or could be made aware of it. Bache points out that the national government cannot avoid shaping policy in areas that will affect life in America:

> Our country is making such rapid progress in material improvement, that it is impossible for either the legislative or executive departments of our Government to avoid incidentally, if not directly, being involved in the decision of such questions. … If all examination is refused, the good is confounded with the bad, and the Government may lose a most important advantage. If a decision is left to influence, or to imperfect knowledge, the worst consequences follow.[15]

While the institution proposed by Bache would need funding to do its job, Bache focuses on other issues. He doubtless represents interests specific to

14 Ibid., l.
15 Ibid., l f.

the scientific community, but his main concern here continues to be the development of the American nation-state. He takes for granted that contemporary political decisions on the federal level were a consequence of political bargaining and that a common national interest was disregarded in favor of particular groups. Bache considers this to be particularly valid for a context of "rapid progress in material improvement," a situation in which economic and industrial leverage allow for significant increases in production, infrastructural improvements, and rising profits. The rules according to which these developments take place (from industrial standards to taxes and implied ideals of economic and social justice) would need to be agreed on politically, and Bache points out that even if the federal government chose to not set standards or impose taxes so as to avoid taking a stance, it was in fact taking one. In his view, the problem consists not only in providing ideas for a rational policy but to initiate discussion about a coherent and inclusive national perspective. To begin to "examine," therefore, goes beyond the support of science for its own sake or providing financial support for colleagues. Bache envisions a United States in which federal policy decisions are aligned with universalistic standards of rationality.

But wouldn't an integrated positive national policy perspective need to be the result of a political transformation? Bache was neither a politician nor an intellectual making his point in public. As outgoing AAAS president, he represented the scientific profession. His comments are in line with his role because he considered his country's present and future from the perspective of his profession: He is concerned with the development of science in the United States. In looking at federal science policy, he finds that the nation's political development lags behind the profession's ambition for coherent development. Once the nation decides to install an institution such as the one Bache suggests, however, it would have indeed chosen to take seriously the profession's advice, for otherwise the implementation of the proposed institution would make little sense. Unlike the Smithsonian, which Congress had installed in response to a foreigner's initiative, and the AAAS, which was created by the profession itself, the decision by Congress to implement an institution such as the one Bache proposed would commit the federal government (and the nation) to science and its standards.

In subsequent passages, Bache raises issues that, on the surface, seem unrelated to his overall theme but he in fact continues to spell out the significance of a national acknowledgment of science. When he points out, for example, that a scientist will start out in his investigative work on a level established by previous research and by education, Bache hints at the importance of tradition and information relay that require a coherent educational system and sense of community to base it on. The United States, however, is not in a position to provide this setting for the pursuit of science because the country was busy developing a coherent national area and infrastructure. With respect to the role of existing organizations such as the AAAS, Bache writes:

> Separated by vast distances, scattered in larger or smaller communities, the daily avocations of men of science in the United States keep us asunder. Our small numbers at any one point produces all the bad influences of isolation. We feel cut off from the world of science, and sink discouraged on account of the isolation; or having a position in the community about us, we become content to enjoy this, and forget that we owe a duty to the world outside; that we ought to increase, as well as to diffuse; to labor, as well as to enjoy the labors of others. Our country asks for other things from us than this ….[16]

In this paragraph, Bache provides an assessment of the avant-garde role of American scientists in the context of their emerging nation. It takes an extra effort, he writes, not only to convey but to innovate and challenge, and this is because there does not yet exist an infrastructure for research as an ongoing routine. But Bache picks up his earlier topic when he suggests that American scientists have "a duty to the world outside" to develop science, to challenge ideas and to help shape them. American scientists face the twofold difficulty of having to labor against their feeling of being removed from relevant scientific discourses in Europe and against their country's tendency to dismiss their work as irrelevant. To the extent that Bache's assessment did indeed reflect his colleagues' feelings and outlook, their sense of cohesion and inner dedication to the cause of science must have been strong indeed. Unlike other countries, there existed no political and little cultural support of science. In America, associations played a particularly important role: "Organization here, for good or for evil, is the means to the end." Bache continues with a phrase that has frequently been quoted:

16 Ibid., lii.

While science is without organization, it is without power: powerless against its enemies, open or secret; powerless in the hands of false or injudicious friends.[17]

Bache's dictum has been used to highlight his group's quest for political influence. The specific textual setting of Bache's statement, however, suggests a slightly different emphasis. Though he does not mention Europe explicitly, he has been weighing in his speech the comparative organizational integrity or "power" of science across the Atlantic and in the United States. While European organizations are ends to themselves, American organizations such as the AAAS assume a function. They do not exist for their own sake but aim at developing something else. What is this "end" for which they are the "means"? Is it the scientific community's establishment, its integration and development? Bache has in mind opponents outside of the scientific community. While European organizations may rely on the political support of scientific principles, American scientists have nothing similar to rely upon but need to shoulder the responsibility for upholding "aristocratic" standards themselves. It is the profession's standing which Bache seeks to bolster. His idea of expanding the profession's power and interests, however, has little in common with those of a labor union. Bache instead suggests that scientists in American need to organize and coordinate their efforts in order to represent the principles of science, ideals of intellectual integrity, and a dedication to truth, within the larger social and political context in which they find themselves. They assume this wider responsibility because unlike European countries, no one else represents these principles in America.

17 Ibid., lii f. See, for example, Hugh R. Slotten, *Patronage, Practice, and the Culture of American Science: Alexander Dallas Bache and the U.S. Coast Survey* (Cambridge: Cambridge Univ. Press, 1994), 32. Slotten uses this phrase, among others, to verify his argument that Bache and Joseph Henry "favored a restructuring of American science that would ensure two major conditions: that the 'best' men of science would retain control and that support for original research, especially from the government, would be maximized" (ibid., 33). Sally Kohlstedt refers to this phrase in order to show that for Bache, "the intellectual agenda envisioned by the AAGN [American Association of Geologists and Naturalists] founders was important perhaps primarily as a means for the organization to gain authority and for that authority to be based on the scientists' status as experts." Kohlstadt, "Creating a Forum for Science: AAAS in the Nineteenth Century," in *The Establishment of Science in America: 150 Years of the American*, ed. Sally Gregory Kohlstedt, Michael M. Sokal, and Bruce V. Lewenstein, (New Brunswick, NJ: Rutgers Univ. Press, 1999), 13 f. The phrase is quoted by Odgers, *Alexander Dallas Bache*, 170, but he does not elaborate on it.

That Bache speaks of "for good or for evil" in this context indicates that he was ambivalent about the additional burden. What could this "evil" exist of? One quite obvious problem concerns the basis for evaluating the relevance of scientific work. In a context where scientific organizations may exist for their own sake, these standards may be developed from the relevance of such work for solving problems of explanation which arise in scientific investigation, without regard to their practical applicability. In the American context, the scientific community's wider role may create a situation in which standards of relevance are deduced from the usefulness of a perceived application. Perhaps more simply, scientists have less time to concentrate on relevant scientific work.

Bache moves on to discuss the more immediate purposes and goals of the AAAS before concluding with the following general observation. Bache states that not

> wedded to existing forms, this country is alive to everything which promises improvement; and the public mind, in this or that place, or in the whole country, made almost a physical point by the electric telegraph, runs irresistibly in one course, the results of wise or evil counsels, of knowledge or half-knowledge.[18]

As American citizens who were also scientists, there devolved on the members of the AAAS the responsibility to help work towards an institution suggested by Bache, a national academy of sciences. Bache implied that its creation by the national sovereign would indicate that the country was accepting as its standard of development "wise" instead of "evil counsels" and "knowledge" rather than "half-knowledge."

18 Bache, "Address," lii.

Chapter 9

Bache, Benjamin Peirce, and the Lazzaroni in 1854

A National Club

Alexander Dallas Bache had been a member of small collegiate groups during the 1830s, using infrequent meetings of "The Club" in Philadelphia to exchange with colleagues and friends ideas and institutional strategies. In that city, there had long been a tradition of semi-formal meetings of gatherings of science-minded men.[1] Unlike organizations such as the Franklin Institute or the American Association for the Advancement of Science, these smaller groups selected and admitted new members carefully, providing a stronger sense of cohesion and mutual acknowledgment than larger organizations.

After their return from Europe in 1838, Bache and Joseph Henry had begun to coordinate their efforts in national scientific institutions. Bache was supported by Henry in his efforts to secure the Coast Survey post in 1842. Henry became the Smithsonian Institution's first secretary (or director) in 1846 with the help of Bache, who was a Smithsonian regent. In 1847, the famous biologist Louis Agassiz arrived in the United States from Switzerland and accepted a professorship at Harvard. The following year, the American Association for the Advancement of Science was founded. As we have seen, these events were generally perceived to facilitate Ameri-

1 For a later variety, see, for example, the Vaughn Club Papers, 1838–1841, American Philosophical Society. The secret club was apparently founded in 1834, the idea having originated with John Torrey. Joseph Henry was also a member. Joseph Henry, *The Papers of Joseph Henry* (Washington: Smithsonian Institution, 1972–2008), 2:290 n. 12. In 1836, Bache complained that the club seemed dispersed. Ibid., 3:13.

can science on a new national plane.[2] During the next few years, the group around Bache, which was interested in advancing national scientific institutions and coordinating efforts to control them, grew to about seven: Charles Henry Davis, formerly a naval officer, was part of a strong group of astronomers in Cambridge where he worked for the Naval Observatory's *Ephemeris and Nautical Almanac*. Astronomer and mathematician Benjamin Apthorp Gould had declined a Göttingen professorship in order to dedicate himself to the cause of supporting science in America. Oliver Wolcott Gibbs, a chemist, joined in the 1850s. John F. Frazer was on the group's periphery, a chemist and former Central High School student and Bache's successor to the chair of natural philosophy at the University of Pennsylvania. Most "members" of this group lived in Washington or in Cambridge (except Frazer who worked in Philadelphia).

The circle around Bache was involved in most major national institutional efforts or controversies in the late 1840s and in the 1850s, including the founding of the AAAS (Louis Agassiz and Benjamin Peirce, with Henry D. Rogers, wrote the organization's constitution in 1848), the founding of the Lawrence Scientific School at Harvard University in 1847, unsuccessful attempts to found a national university, the controversy over the Dudley Observatory in Albany, and the founding of the National Academy of Sciences in 1863.[3] The annual meetings of the AAAS provided the group with the occasion to meet for an opulent dinner but it also stayed in close contact through letters and by visiting one another. Even though Bache had known Henry since the 1830s, and Benjamin Peirce since 1842, the group did not refer to itself by name until about 1852, when Peirce addressed Bache as the "president of the Florentine Academy."[4] Following Bache's 1851 call for a national academy, the idea for such an institution

2 This prompted Robert V. Bruce to have his overview of American science in the nineteenth century begin in 1846. Bruce, *The Launching of Modern American Science: 1846–1876* (New York: Knopf, 1987).

3 Lillian B. Miller, *The Lazzaroni: Science and Scientists in Mid-Nineteenth-Century America* (Washington: Published for the National Portrait Gallery, Smithsonian Institution, 1972), 7–12. For the Dudley Observatory fight, see Mary Ann James, *Elites in Conflict: The Antebellum Clash over the Dudley Observatory* (New Brunswick: Rutgers Univ. Press, 1987). See also Bruce, *Launching of Modern American Science.*

4 Ibid., 220–21. For a reference to a "Florentine Academy," see ADB to Benjamin Peirce, September 25, 1853, Benjamin Peirce Papers, Houghton Library, Harvard University (hereafter cited as "BPP").

was in the air and occasionally discussed by the group.[5] Peirce's phrase suggests that Bache and his close colleagues liked to view themselves as an informal precursor. The name the group would then come to use, the "Lazzaroni," humorously referred to idling Neapolitan beggars who took the side of the Bourbons in late eighteenth and early nineteenth century uprisings.[6] It would not be surprising had Bache suggested the name. He had visited Naples during his European tour and written about Lazzaroni in a letter to his mother.[7] The Romantic reference to a band of outsiders asking for support of idle pursuits was highly ironic, of course, and it matched the group's frequently witty prose. The group fizzled out in the late 1850s. While Bache and Peirce continued to visit the AAAS meetings, Agassiz, Frazer, and others did not. Bache and Henry grew distant due to the conflict over the Dudley Observatory. In March 1860, Bache wrote that the Lazzaroni were "defunct."[8] However, the group's last, albeit no longer unified, effort led to the founding of the National Academy of Sciences in 1863.[9]

Instead of revisiting the institutional fights in which the group around Bache was involved during the 1850s, I have selected one exchange of letters between Bache and Benjamin Peirce to characterize the group's inner dynamics, common motivation, and overall institutional goals. Among the many letters exchanged among Bache and the members of his circle, Benjamin Peirce's correspondence stands out for its unusual style. For a close analysis, I have chosen a letter that has frequently been referred to in historical accounts of the Lazzaroni and of the AAAS because of

---

5 Cochrane refers to a July 12, 1858 letter to John F. Frazer in which Louis Agassiz charted "a plan of membership and organization" for an academy. Rexmond Cochrane, *The National Academy of Sciences: The First Hundred Years, 1863–1963* (Washington: The Academy, 1978), 46. Merle Odgers points to a September 1853 letter in which Bache mentions a meeting for a "dawning project for an Academy of Sciences or a near approach to it." Merle Middleton Odgers, *Alexander Dallas Bache: Scientist and Educator, 1806–1867* (Philadelphia: Univ. of Philadelphia Press), 170.

6 Miller, *The Lazzaroni*, vii, 5.

7 ADB to Sophia Dallas Bache, April 29, 1838, box 2, vol. 2, Bache Papers, SIA.

8 ADB to John F. Frazer, March 5, 1860, Frazer Papers, American Philosophical Society.

9 Bruce, *Launching of Modern American Science*, 217–24; Sally Kohlstedt, *The Formation of the American Scientific Community: The American Association for the Advancement of Science, 1848–1860* (Urbana: Univ. of Illinois Press, 1976), 154–89. Literature on the Lazzaroni also includes Miller, *The Lazzaroni*; Mark Beach, "Was there a 'Scientific Lazzaroni'?," in *Nineteenth-Century American Science: A Reappraisal*, ed. George H. Daniels (Evanston, IL: Northwestern Univ. Press, 1972), 115–32; Nathan Reingold, *Science in Nineteenth-Century America: A Documentary History* (London: Macmillan, 1966), 127–61.

Peirce's comments on the role of the organization and of the Lazzaroni itself. This letter and Bache's response to it provide an opportunity to assess the cooperational style of leaders of science in mid-century America. It provides a glimpse of Association politics, of conflicts within the Lazzaroni group, and of Bache's role in both of them.

"The Dark Prospect Appalls Me"

From April 23 to May 3, 1854, American scientists gathered at the Smithsonian Institution in Washington D.C. for the annual meeting by the American Association for the Advancement of Science. Benjamin Peirce had been the organization's president for the preceding year. Just before he wrote the following letter to his friend Bache, therefore, Peirce had given his speech as the association's outgoing president. Peirce, whom a historian of his field has called the "father of pure mathematics in America," had been teaching at Harvard University for more than twenty years. Peirce had been noted for criticizing Urbain J. J. Le Verrier's calculations of Neptune's orbit. Peirce had argued that two orbits, not just one, could be deduced from the planet's position. At Harvard , some colleagues considered Peirce a failure at teaching. They argued that "it was useless for anyone to study mathematics without a special aptitude for them."[10] Through his efforts, mathematics became an elective at Harvard, ensuring that only motivated students would join the department.[11]

As a backdrop for the following letter, it is useful to note that the circle around Bache was then involved in a number of organizational controversies. The American Association for the Advancement of Science was only a few years old and there had emerged disagreements about the enforcement of standards. The Lazzaroni group had significant influence over the selection of papers for presentation and publication in the Association's *Proceedings* but their dominant role was increasingly criticized. Under the leadership of William Barton Rogers (founder of the Massachusetts Institute of Technology in the 1860s), their opponents demanded a more significant role for the sections representing distinct subfields, and even though Rogers, too, was unwilling to compromise on quality, he considered the

---

10 Sven R. Peterson, "Benjamin Peirce: Mathematician and Philosopher," in: *Journal of the History of Ideas* 16, no. 1 (Jan. 1955): 93.

11 On Peirce, see Edward R. Hogan, *Of the Human Heart: A Biography of Benjamin Peirce* (Bethlehem, PA: Lehigh Univ. Press, 2008).

Lazzaroni to be high-handed in their influence and decisions. In the context of this fight, and just a few months before Peirce wrote the following letter in 1854, the group around Bache had first begun to refer to itself by name and a more definite sense of group interest and cohesion was evolving. The Lazzaroni were becoming a not always coherent but nevertheless singular directing force in American science.

Immediately after the Washington AAAS meeting, Bache wrote to Peirce on May 5, 1854, inquiring why Peirce had not taken the opportunity of his visit to explain the "troubles in the American Academy, but imperfectly known to me."[12] He asked Peirce whether they were "such as to treat at all upon my decision of the question whether to publish the results of the C.S. [Che.] expedition in the Transactions? I mean have you any feeling which should guide me in the matter."[13] Bache was most likely referring to the naval astronomical expedition to Chile from 1849 to 1852 under the direction of James M. Gilliss. The expedition's goal had been to follow the recommendation of Christian Ludwig Gerling of Marburg for a new determination of the solar parallax. Gilliss, then working for the Coast Survey, was able to win support for an expedition to Chile and for making the necessary measurements, with similar observations to be made in Washington. The expedition failed in its main object (no observations had been made in the United States) but returned with abundant material for publication, astronomical and otherwise.[14] The American Academy (along with the American Philosophical Society) had supported the idea of an exploration. Bache wished to discuss with Peirce basic issues of science policy. Publication of the expedition's records would of course have corroborated the Academy's prestige. Bache's previous letters to Peirce show that the Lazza-

---

12 ADB to Benjamin Peirce, May 5, 1854, BPP.

13 I was unable to identify what "troubles" Bache was referring to.

14 It seems unlikely that Bache spoke of Commodore Matthew C. Perry's mission to Japan at about the same time because it was a navy undertaking and because it was still going on whereas Gilliss had already returned from Chile. See Matthew Calbraith Perry, *Narrative of the Expedition of an American Squadron to the China Seas and Japan* (Washington: A. O. P. Nicholson, 1856). Another exploration, the U.S. Surveying Expedition in the North Pacific Ocean was still underway in 1854. A. Hunter Dupree, *Science in the Federal Government: A History of Policies and Activities* (Baltimore: Johns Hopkins University, 1986), 96–99. For Gilliss and the Chile expedition, see also Steven J. Dick, *Sky and Ocean Joined: The U.S. Naval Observatory, 1830–2000* (Cambridge: Cambridge Univ. Press, 2002), 142 f., James Melville Gilliss, *The U.S. Naval Astronomical Expedition to the Southern Hemisphere* (Washington: A. O. P. Nicholson, 1856), and Reingold, *Science in Nineteenth-Century America*, 135–45.

roni were ready to reconsider and to shape the institutional landscape of American science, and Bache's comment here indicates that with respect to an institution dominated by Cambridge and Harvard men, he relied on, or did not want to ignore, his friend's assessment.

In his letter to Peirce, Bache also discussed the role of the American Association. He reflected that when "the smoke of the late conflict has passed off I hope to be able to consider impartially what the Assoc. has done + is doing." Bache implicitly granted that he was unable to view these matters impartially in the context of the conflict in which he was himself involved but that he would have liked to be able to consider the Association's value from a detached perspective. This confirms that Bache took a comprehensive view of the role of science and its organizations in the United States, and that the AAAS, as an institution, stood for a particular phase of development. He suggested that he wanted to talk to Peirce "so as to … conclude what it shd do for the future," and takes for granted that his group was in charge, and had the means, of influencing the Association's direction. Bache warned Peirce, who was on the Association's committee responsible for publishing the Washington proceedings,[15] of papers by Thomas Bassnett and George W. Coakley, and then continued:

> As now advised there is more hope in me than there was a year ago, + yet I see more clearly the dangers from breakers ahead. Should even a few 'leading men' (Rogers) stay away from the meetings the whole tone of things would at once alter. How far Agassiz has determined to abandon the ship + how far his example of this year will be followed I can only [yet] conjecture.
>
> The antagonism between the 'leading men' in sects A + B, was sufficiently shown by the way things worked between the two sections. On this head I have heard more since the meeting than I knew when we met.

Bache was obviously concerned about the organization's development. He insists on active leadership by men such as Peirce, himself, and others in their circle. He quotes the emerging opponent William Barton Rogers ("leading men") and implicitly uses that reference as a token of legitimacy: Rogers, too, believed that Bache and his supporters were in charge. That Bache fears the "whole tone at once [to] alter" should they stay away suggests that he considered their influence and guidance to be of great significance. Nevertheless, Bache is cautious and concerned about the prospect of the Lazzaroni acting in concert to sustain the AAAS momentum.

15 American Association for the Advancement of Science, *Proceedings* 9 (1855): xii.

Louis Agassiz had skipped the Washington meeting. Bache's concern is that influential scientists ignored the AAAS. Bache does not consider the mere existence of the AAAS as evidence for its sustainability. Towards the end of his letter to Peirce, Bache includes his opponent Rogers among the "leading men" without whom the Association cannot do. At the same time, he acknowledges Rogers' opposition and writes to Peirce that it had become much more obvious than Bache himself had expected it to be. Despite his leadership of one of the opposing "sects," Bache thus preserved an integrative (and professional) view of the Association's leadership including Rogers.[16]

Peirce responded to Bache's letter three days later:

Monday, May 8, '54

My dear, true, noble—sound friend.

I intended to have had a long and confidential talk with you about the Am. Ac. when I was at Washington—such an one as I cannot have with any body else.[17]

In addressing Bache, Peirce does not confine himself to underlining the intimacy of their friendship. He invokes specific characteristics: "true" points to openness between the two men, but also to their friendship's durability; by calling Bache "noble," Peirce suggests that he appreciates and looks up to his friend's identification with moral and intellectual ideals beyond immediate personal ambitions, and that Bache assumes a broader cultural perspective. The adjective "sound" is set off by a hyphen, and it seems to bring together the three preceding adjectives: Peirce concludes his opening by asserting the stability of their friendship. It is too early to say for what reason Peirce chose this emphasis. Is it merely a gesture, or is it connected to an issue Peirce will raise in his letter, an issue that causes Peirce to reassure himself of his friend's good feelings toward him?

Peirce refers to a meeting that did not take place but that he "intended to have had" with Bache during the recent meeting of the American Association for the Advancement of Science (AAAS). Peirce, as the organiza-

16 This assessment falls in line with observations in preceding chapters and this is why I am not extrapolating it here.

17 Benjamin Peirce to ADB, May 8, 1854, BPP. Lurie understood this letter to have been written in 1855, and the year is indeed difficult to decipher. Peirce wrote on a Monday, and this implies that he must have written in 1854 as May 8 was a Tuesday the following year. See Edward Lurie, *Louis Agassiz: A Life in Science* (Chicago: Univ. of Chicago, 1960), 182 f. nn. 27, 28. See also Hogan, *Of the Human Heart*, 121 n. 141. Formatting rules used in transcribing Peirce's letter are in chap. 3 above, 58 n. 23.

tion's outgoing president, had given the presidential address. The meeting had ended on Wednesday, May 3, and Peirce likely writes Bache after he had been back in Cambridge for a few days.[18] In his opening sentence, Peirce uses the simple past of "to intend" in combination with the past perfect infinitive "to have had." In this way, Peirce looks back on his stay in Washington and his intention to see Bache, or, rather, to have seen Bache. Strictly speaking, Peirce had no intention to speak to Bache, but to have spoken to him. He wished to be able, while he was in Washington, to look back on this conversation. This perspective focuses on the talk's completion, and this suggests that Peirce had been reluctant to call on his friend because the issue he was going to raise would create an uncomfortable situation. Why else would he have wished to have put behind him this conversation with Bache? That Peirce writes that their "talk" would have been "long and confidential" also suggests that they had to deal with a difficult matter and that it required time and effort to negotiate divergent positions. This reading corresponds to Peirce's elaborate opening and his appeal to the resolute character of Bache's friendship.

Peirce closes his sentence by stressing that there was no one else with whom he shared the same kind of intimacy. There is a tension here between Peirce's dedication to his friend, and his reluctance to meet him in Washington. The best explanation for this tension is that it is *because* of his friendship with Bache that Peirce hesitated to see him. Peirce was reluctant because he feared that their positions were sufficiently distinct to break their friendship. Whatever the issue may be (Peirce refers vaguely to the American Academy of Arts and Sciences), it does not involve matters of strategy alone, but deeply held convictions. At the same time, Peirce refers to the American Academy in a way that makes it seem as though there was a need to discuss the institution as such, its existence and role, not merely his own role within the Academy. By not meeting Bache in Washington, Peirce, perhaps consciously, retained the option of carefully editing his words in a letter rather than having to confront his friend.

To you alone, dare I open the inmost recesses of my heart upon this matter.

---

18 The meeting took place between April 23 and May 3, 1854. Benjamin Peirce, "Address of Professor Benjamin Peirce, President of the American Association for the Advancement of Science for the Year 1853, on Retiring From the Duties of President," AAAS, *Proceedings* 9 (1855): 1–17.

Peirce continues to celebrate his friendship with Bache. He limits his empathy, however, by writing that he would not open his heart to Bache on any topic, but only "upon this matter," i.e. with respect to his views on the American Academy. Peirce connects professional matters to his most private feelings. There is a balance here: Peirce's relationship to science is not strategic in the sense that it constitutes a matter distinct from private aspirations and motivations. It is intertwined with them. At the same time, Peirce does not let personal friendship direct his views on professional issues. Peirce and Bache are connected through their dedication to science; hence, there is no distinction between private matters and those pertaining to professional organizations. Peirce's letter to Bache so far suggests that the two men are not "in love" with each other, but with the American Academy's object, i.e. science. That Peirce speaks of "dare" shows that to expose himself to Bache is to allow himself to be vulnerable, and he tacitly suggests to his friend to be careful in his response.

> However bitter a cup it may be to me to have your C. S. investigations published by the Am. Ac. I shall not discuss the subject with you as a private thing, but only upon public grounds, …

Bache had inquired whether it would be advisable to publish results of investigations undertaken by the Coast Survey with the American Academy in Boston. Peirce, however, responds as though his friend had favored that idea. Why does Peirce feel so strongly about this? Is it because this would honor an institution falling short of his academic standards? There is nothing here to counter this impression, but Peirce writes that Bache's decision was a bitter cup "*to me* {my italics}." Had Peirce left out these two words, he would have taken a critical view of Bache from a generalized point of view, but including them suggests that Peirce had a personal quarrel with the Academy. Peirce still does not discuss the issue itself. His letter to Bache suggests that he was afraid of losing a friend in a conflict in which his Washington friend is not immediately involved but in which he could be an important ally.

Peirce remarks that he would "not discuss the subject with you {Bache} as a private thing, but only upon public grounds." He distinguishes between his friendship with Bache, on the one hand, and their relationship as colleagues, on the other. Peirce wants to shield his friendship with Bache by discussing the publication of papers with the American Academy as a separate matter.

Fig. 10. Benjamin Peirce

(Image Archives of the Historical Map & Chart Collection, theb3557, Courtesy of Office of Coast Survey/National Ocean Service/NOAA)

> … and you will of course take care that the arguments of a confessedly prejudiced man do not bend you from the high, generous, and national course ~~in~~ which you have hitherto [pursued] with such [mainly] beneficial results, and the [justice] of which will never make [me] love you less, or trust you less.

Peirce proposes to discuss the matter "on public grounds," but by taking a humbling position vis-à-vis his friend, he prepares the ground for presenting his private feelings.

Peirce hints at why the American Academy ranks so low in his esteem. He does so implicitly by referring to the "high, generous, and national course" which Peirce appreciates Bache to have taken. The adjective "national" is used emphatically here, not merely in a descriptive way (such as in "national government"), because it trails other such adjectives ("high" and "generous"). The use of these adjectives indicates that in Peirce's view, a policy pursued by Bache is not restricted to narrow scientific questions but represents an ambitious agenda for basic research on a national scale. These ambitions go well beyond the immediate responsibilities of Bache's post in Washington ("generous"). If Peirce's remarks have any relevance at all in the context of the American Academy, Peirce feels that the Academy lacks those kinds of qualities and that its scope is too narrow for publishing scientific findings such as those by the U.S. Coast Survey. This reading is corroborated by Peirce's somewhat restrictive assessment of his friend's work ("[mainly] beneficial"). He speaks of "justice," thereby invoking standards against which Bache's achievements can be measured, and the ideals underlying his friend's work, not their implementation, prompt Peirce to highlight his own unswerving admiration. Peirce stresses his friendship with Bache but he also suggests that his friend may not live up to their common standards of a national development of science.

That Peirce invokes national ambitions is significant because it shows that even though he taught at a prominent American university, Peirce tied the progress and future of scientific development to that of his country. From such a perspective, Bache is at the center of this development in Washington, and Peirce, in Cambridge, at its periphery.

> Strange to say, the Am. Ac. was so little in my mind while I was at Washington, that I forgot to speak about it, but let me now say to you—I have no secrets from you, and you need never hesitate to ask me about anything whatever, I will always answer you and open my heart when you knock for admission.

Peirce seems surprised at having forgotten to discuss the contentious matter with Bache in Washington. If the matter was important to him, how could this happen? Peirce uses the conjunction "but" in "but let me now say to you:" This indicates that Bache may think that because Peirce did not raise the issue, he may have secrets from his Washington friend. However, if Peirce forgot to mention the issue, how could his friend doubt their friendship? Only if Peirce had *chosen* not to raise the issue could he expect Bache to have such doubts, and this serves as additional evidence that Peirce preferred not to mention the matter in his conversations with his friend. (Peirce writes that he forgot "to speak" about the issue, thus suggesting that he met Bache but did not raise the matter. Had he used a phrase such as "to talk to you about it" it would be obvious that the two men had not met.)

While Peirce suggests that he forgot to mention the matter in Washington, he now intends to correct his oversight and to proceed to write what he would have said. And the first issue Peirce raises is that he has no secrets from his friend. If Peirce had assured Bache in Washington that he had nothing to hide, then their friendship was in doubt even then, and it follows that their disagreement about the American Academy must have had a prelude. The matter has been lingering for some time.

Peirce's metaphor at the end of this sentence is double-sided. On the surface, it goes along with his overall intention to highlight his friendship with Bache. At the same time, however, his metaphorical heart is made of solid wood. The door is shut, and instead of being asked in, Bache is to knock for admission. Because the Academy-matter had had a prelude, Bache must have been aware of it while Peirce was in Washington, and the latter now signals that it was not his fault alone that they did not speak during his visit. Bache "hesitated" as well. Because it is so subtle, Peirce may have not been aware of his somewhat defensive tone.

With regard to personal feelings in the Am. Ac. I will say at once that they did not begin with me, but with Agassiz, Gould and Gray.

At the end of the first page of his letter, the conflict at the American Academy finally bursts out into the open. The conflict does indeed not involve Bache, but others in Cambridge. Peirce has three opponents, two of them, Louis Agassiz and Benjamin Apthorp Gould, are fellow Lazzaroni. While Peirce had promised that he would discuss the American Academy only in public terms, he now does the opposite. He suggests that "personal feelings" are involved, and if his reservations about the American Academy derive from this personal conflict alone, this may contradict his earlier celebration of universalistic standards and a national development of science. By writing that he did not "begin" the fight, he implies, of course, that he was unable to keep a cool head himself. His criticism of Bache publishing Coast Survey reports with the American Academy is tied to fears of losing his last ally.

What was this fight about? Peirce does not extrapolate and assumes that Bache knows what he is referring to. Agassiz, Gould, and Gray all worked in Cambridge but during the AAAS meeting, Bache, if he had not learned about the conflict earlier, probably spoke to at least one of these Cambridge colleagues.[19]

In his next paragraph, Peirce moves on to further describe the problems in Cambridge:

> But I will not hesitate to declare it now as my settled conviction that Gray is as false as the Rogers's. His [train] is as certain to [wreck] as theirs. Upon this point, Davis differs with me as we differed for a long time, with regard to Agassiz and Desor.

Peirce announces his "settled conviction" that the botanist Asa Gray was as "false" as the four brothers Rogers he considers a negative standard, i.e. Robert, Henry, William Barton, and James Rogers (all of whom were scientists). Peirce highlights this apodictic statement by leaving a visible gap. That he is now ready to "declare" his "settled conviction" suggests that he had previously held it back. Something had taken place that prompted Peirce to lose faith in his colleagues.

Another aspect comes into view here. Peirce had not mentioned his colleague Charles Henry Davis before. Davis, born in 1807, a year younger than Bache, and the youngest of thirteen children, had been a midshipman, then lieutenant in the navy before joining the Coast Survey in 1842. His efforts on behalf of an independent American computation of ephemerides (tables of the position of celestial bodies over time) paid off in 1849 when the *Ephemeris and Nautical Almanac* was established. Davis was put in charge of its work in Cambridge, and Benjamin Peirce became a consulting astronomer. The two men were close in other ways as their wives were sisters and the two households were across from one another in Cambridge.[20] So why does Peirce refer to Davis? Wasn't his issue with Agassiz, Gould, and Gray? Because we have to assume that Bache did know something about the problems at the American Academy, it seems that a conflict between Peirce and Davis was at the heart of the matter, and that this was the reason why Peirce hesitated to mention Davis.

This conflict entailed differences of opinion with respect to Asa Gray's scientific work. Agassiz and Gould had taken Davis' side and left Peirce isolated and looking for support from Bache as the group's leader. Peirce refers to an earlier episode when his view was later accepted by Bache. Pierre Jean Édouard Desor had claimed authorship of publications by

19 I am unaware of sources that would shed further light on this conflict.

20 Hogan, *Of the Human Heart*, 118–20.

Louis Agassiz for whom he had worked, and Desor had spread rumors about Agassiz' private life. Once the matter had been resolved by private tribunal and Desor shown to be wrong, Peirce had tried to have him expelled from the American Academy. During this episode, which took place between 1848 and 1852, Davis had not always been certain that Agassiz was not to blame. He later regretted this and ended all cooperation with Desor.[21] In reference to this episode, Peirce asks Bache to view his present disagreement with Davis in the same light: Peirce's judgment had not failed him in the past and neither would it fail him now: At some point, Davis would concur.

Peirce's rashness suggests that his opinion is based on an instinctive, personal, and irrational rejection of Gray. But his use of the noun "train" adds a different motivational component because it would be an awkward metaphor for personal antipathy. It works much better as a reference to a particular scientific program, which, in Peirce's view, is on the wrong track. While Peirce's impulsiveness does come through here, it is based on a commitment to standards for scientific work. Gray's work as a botanist was devoted to collecting and organizing rather than deducing results from samples of data, and Peirce's criticism of Gray may involve such differences of scientific paradigms.[22] Peirce's bitterness, however, is not so much directed at Gray than at Davis. He assumes that there were two distinct professional circles: The brothers Rogers, Agassiz, and, in his view, Gray, are members of the profession but outside of the closer-knit group that includes himself. In Peirce's perception, Bache belongs to the latter, of course, and Davis should belong to it as well. It is because Davis is so close to his professional and private life that Peirce is profoundly shaken.

> [~~Knowing~~] Perceiving long ago the tendency of his mind ~~upon this~~ in regard to Gray, I avoided all discussion with Davis, and did not even consult him or any body else. He thinks that I am led by Gould. He does not know how exactly it is the reverse; and how ludicrously mistaken he is.

Why did Peirce avoid "all discussion" with Davis? Peirce takes for granted that the issue could not be resolved and that it affected all professional aspects of his relationship with Davis. In this matter, Peirce was unwilling to have his sentiment reviewed ("consult"). His comment makes clear that

---

21 Edward Lurie, *Louis Agassiz: A Life in Science* (Chicago: Univ. of Chicago Press, 1966), 152–61.

22 See Hogan, *Of the Human Heart*, 121, 191.

a different evaluation of Gray's work undermined his trust in Davis as a colleague.

In addition to a disagreement about Gray, Peirce goes on to introduce another aspect of his conflict with Davis: "He thinks that I am lead by Gould." Benjamin Apthorp Gould was fifteen years younger than Peirce was. Before he had gone to Germany to work with Carl Friedrich Gauss, he had studied under Peirce at Harvard.[23] How was the idea that Peirce was unable to follow his own judgment connected to the difference in evaluating Gray's work? As a botanist, Gray was interested in taxonomy and classification whereas Peirce, in a much-quoted phrase, would later call mathematics a science "that draws necessary conclusions."[24] Did Davis view Peirce's and Gould's criticism of Gray as being too radical? Because Gould had received training with the famous German mathematician Gauss after he had studied with Peirce, did his former teacher feel as though others perceived him as doubting his own ability to convey to his students the leading scientific problems of the day? Seen in this light, the source of Peirce's criticism of Gray's scientific program was also perceived as being unduly German (or European) since Gould's supposed intellectual advantage derived from his training there. Peirce had hinted at the national and cultural context of science when he admired and appealed to Bache's national perspective on the development of American science. Peirce's radicalized insistence on scientific inquiry different from that pursued by Gray goes hand in hand with his dedication to the development of science in America. According to this interpretation, Peirce claims for his own country a radical attitude that is universal but associated with Europe. According to this perception, such an outlook was not yet established in America. Peirce did not routinely mention the national context of science in letters to Bache. It seems that Peirce answered reservations about his self-sufficiency by insisting that his ideals, though they might look European, were valid not only for himself but for American science in general. This is where Bache comes into play. Peirce asks his Washington friend to support him, not against Gray whom Peirce considers Bache to shun also, but against "false friends" such as Davis who prefer to take an accommodating view.

23 For biographical information on Gould, see James, *Elites in Conflict*, 51–67.

24 Benjamin Peirce, *Linear Associative Algebra*, quoted in: Ivor Grattan-Guinness, "Benjamin Peirce's Linear Associative Algebra (1870): New Light on Its Preparation and 'Publication'," *Annals of Science* 54, no. 6 (November 1997): 602.

Fig. 11. Charles Henry Davis

(From Life of Charles Henry Davis)

I have, as I say, avoided this subject with Davis, and fearing lest it should disturb the harmony of Cambridge society, I closed my lips to every body, even to my wife.

Peirce had indeed implied earlier that he tried to avoid raising the subject. He suggests that the conflict between two prominent Harvard scholars might reverberate through the small town. The intellectual community's integration and his family ties to his neighbor Davis made it difficult for Peirce to regain his footing. That he did not confide the matter to his wife shows that he was afraid she might talk about it to someone else—to her sister Harriette, for example, who was married to Davis.

But today Davis seemed resolved to force the subject [off], and said so many things, which were bitterly severe upon my course, that I opened my lips upon Gray and we separated in anger. The breach is destined to widen. I shall leave the Almanac, I shall leave the scientific club of Cambridge, I shall leave Harvard. I shall go to N.Y.. I see my [destiny] [awash] and the dark prospect appalls me. But, nevertheless, I will not turn back.

It is easy to see that Peirce wrote in response to his confrontation with Davis. He has prepared the ground to make clear that he was not the one who started the conflict, and that he sought to avoid it. In hindsight, it is clear that Peirce was forced to write to Bache in order to forestall his friend learning about the fight from his opponent. This corresponds to Peirce's initial remark that he had no secrets from Bache.

The conflict had opened up and Peirce not only anticipates further problems, but his dissolving of all ties to Cambridge society. That Peirce considered Davis' words to have been "bitterly severe" shows his painful realization that their collegiality had been destroyed.[25] Peirce's report to Bache would have been complete without these details, but by supplying them, Peirce invites his friend to understand how deeply he was hurt, and this implies that he valued Davis as a colleague and friend. By stressing that Davis said "*so many* things {my emphasis}," Peirce suggests that he sought to hold back his comments on Gray, but that Davis, unrelentingly, kept going. He felt he had no other choice than to speak his mind. Peirce signals that he was willing to hold back his criticism as long as he was not confronted with the matter in which case he was unable to compromise even for the sake of friendship. Again, this suggests that Peirce had no personal grudges against Davis, but that he had a keen sense of personal integrity and a strong conscience that made it impossible for him to take these things lightly.

Why would the breach be bound to widen? Peirce takes a pessimistic view of his friendship with Davis. To underline this sentence is to stress that he is unable to mobilize resources for reconciliation. Since Davis remains important to Peirce, his letter and depiction of the conflict signaled to Bache that he badly needed help in resolving it.

To leave the Nautical Almanac, the Cambridge Scientific Club, and Harvard: Peirce escalates the conflict's consequences. By leaving the Almanac and the Scientific Club he abandons the projects in which the two men were cooperating. By leaving Cambridge, furthermore, Peirce acts as though he was no longer able to tolerate residing in the same town as a bygone lover, as though the possibility of running into Davis would trigger memories too painful to endure.[26] Peirce quite likely envisioned these dra-

25 Peirce and Davis would continue to cooperate (see Hogan, *Of the Human Heart*, 119–21). Nevertheless, this letter reflects Peirce's state of mind at the time he wrote to Bache.

26 The Cambridge Scientific Club had been founded in 1842 and comprised about a dozen members of the Harvard faculty such as Asa Gray (a co-founder of the club), Joseph

matic consequences because of his proximity to this brother-in-law and neighbor. He had not abandoned Cambridge after Asa Gray had moved there in 1842. It is his deep sense of betrayal and the difficulty of avoiding Davis that prompt Peirce to consider this step.

Peirce seeks to replace the close intellectual circles of Cambridge with the metropolitan anonymity of New York City. In a romantic pose, he intends to abandon his career in order to remain true to his professional convictions and beliefs. A close look at the calligraphy of the last sentence shows that Peirce initially omitted, and later inserted, the adverb "not" in "I will not turn back." A possible and even likely motivation for the initial omission is that Peirce wished he did not have to take steps as severe as the one he was about to take. He had unconsciously "turned back" before he caught his error.

> My duty to my day and generation demands the sacrifice of myself and I will make it. O, that this cup might [pass by me]! I shall have nobody to [strain] me, not even my mother and my wife. But duty, <u>as I see it</u>, points but one way, and I will follow it even to the breaking of my heart. I am very sad. However to you only am I sad. I shall be [game] to the last.
>
> Leaving feelings to their own consumption, I will now give my argument against the Am. Ac.. Whatever may have [been] my personal feelings, I should not have left it for any other reason than one which I regard to be sufficient for so serious a step.

Peirce strikes a romantic pose of self-neglect in arguing that it was his "duty" to "sacrifice" himself. He immerses himself in the cause of his generation to the point of neglecting his career. The forty-five-year-old takes an "idealistic" view of contemporary problems and his generation's mission, implying that it is more important to preserve ideals than institutions. This leads him into personal isolation, for Peirce is not discussing these matters with anyone but Bache, who is handed the responsibility for helping him ("However to you only am I sad"). His "bitter cup"-metaphor connects Bache's decision to publish with the Academy (Peirce's use of the metaphor above) and his own departure from Cambridge and a scientific life of consequence (Peirce's use of it here).

Lovering, and Louis Agassiz. Charles Henry Davis and Agassiz were leading members of the Club that met every two weeks. See, for example, Charles Sanders Peirce, *Writings of Charles S. Peirce: A Chronological Edition* (Bloomington: Indiana Univ. Press, 1982), 1:xvii; A. Hunter Dupree, *Asa Gray, 1810–1888* (Cambridge: Belknap Press of Harvard, 1959), 121 f.

While Peirce had announced at the beginning of his letter that he would not speak about the personal dimensions of his conflict with Bache, he has done just that. In the next paragraph, he again announces to discuss the "public" dimensions of the American Academy, and he again moves on to speak about his own decisions with respect to the organization. This goes along with his distinction between "personal feelings" and "reasons … sufficient for so serious a step." Peirce had apparently left or had decided to leave the Academy, and he felt the need to explain this step to Bache.[27] On the heels of his break with Davis, Peirce intends to remove himself from the academic community in the Boston area. Peirce's problem is that by abandoning professional organizations, he loses all positive influence and the social basis for his work. It will be interesting to see how Peirce justifies this step, for it will be difficult for him to escape the charge that he would be removing himself into oblivion.

I did not [fly from] an unequal contest, and avoid a battle from any dread of the consequences. On the contrary, I was well aware that I was yielding the [field] [appar]ently to the opposition. I had even gone [further], + I had assured myself that the victory was mine. If I chose to have it. But I would not have taken it for the gift.

Peirce denies that he is evading a conflict but his comments are defensive. He denies being a coward, and stresses that he had sufficient information to know that he would have won a fight. He is so thoroughly disgusted with his colleagues and the organization, however, that he "would not have taken it [the victory] for the gift." Peirce chooses to resolve the conflict with his colleagues in the Academy by ejecting himself into a position of private supremacy. Whatever the particulars of the conflict may have been, it is certain that Peirce has weakened his position by avoiding it. Had he prevailed in the conflict, he would have been in a stronger position to propose changes to the Academy. Had his colleagues then chosen to ignore his suggestions, he would have been able to portray his departure as hurting the institution rather than himself.

Peirce's emotional disposition now seems clear. I will not discuss the rest of his letter in detail and will instead look for clues to fill in the picture. Peirce goes on to write:

27 I was unable to find evidence that would suggest that Peirce did indeed give up his membership at this time.

I turned from it with contempt as soon as I had clearly ascertained that scientific influence did not and would not predominate in it. The [Am. Acad.] is a creature of patronage, and it fancies at the foot of wealth. I despise it with all my heart, and I regard it as our duty to put it aside and fill its place with a society worthy of the country and of science. Naturally it is a mere private association and has no claim to be [viewed] a government institution; and upon this account alone, it should not be the organ of government publications. You must not be angry with me, if I say this aloud. [In my] opinion, the interests of science and the proper respect of the Coast Survey for its own researches, demand that its results should not [flow] through any private channel into the ocean of knowledge. Let it publish its own researches for itself, [and] let there be some secretary, established by government, [to] which such researches may be officially communicated. Why should not the Coast Survey publish the [—] ~~all if [——]~~ as such as the observatory, the almanac and the exploring expedition? Am I wrong?

Peirce argues that the American Academy was influenced by those who supported it financially, but he does not conclude that this influence ought to be limited. He suggests that another institution take on the responsibility of an idealized American Academy, an organization to be put under government auspices such as the one Bache had proposed in his 1851 speech. This goes along with the earlier appeal to Bache's "high, generous, and national course." Peirce's suggestion for replacing the American Academy, however, shows that he has given little thought to the details of such a plan. He speaks of a "society," then a "secretary," and finally suggests that the Coast Survey publish its own results. The heart of the matter is that Peirce seeks legitimacy for his radical criticism that ejected him from his circle of friends and colleagues in Cambridge, and that he is appealing for help. His position remains weak because he seeks such support at a time when he has been put on the defensive by Davis' forcing the conflict. As we have seen, Bache had not planned to publish the Coast Survey papers with the American Academy at all, and had merely asked his friend's opinion, but Peirce was using this as an opportunity for appealing to his friend on rational grounds.

But it remains significant that Peirce's reintegration efforts would lead him to suggest the creation of a government-sponsored institution. Peirce could have argued that another private organization be founded that would then adhere to his more rigorous standards. Instead, Peirce turns to the issue of private versus public legitimacy, suggesting that it is a matter of honor to have Coast Survey results published by a public "secretary," a government-sponsored clearing house for scientific results. His proposition makes sense only if such an institution provided a professional framework different from the Academy's. How could this be? Would it not be just as likely that politicians in charge of providing the funds for the organization (or the "secretary") would seek to influence it in ways similar to the patrons of the Academy? It is important to point out in this context that Peirce had not criticized the Academy's sponsors for influencing the society's business. He had disparaged his colleagues for playing up to them: The Academy, he wrote, "fancies at the foot of wealth," and he dismissed his colleagues for seeking attention from prominent extra-professional members of the society. This contrasts with Peirce's characterization of scientific results flowing "into the ocean of knowledge." His metaphor suggests that Peirce's scientific radicalism was tied to a secularized view of its rewards.

Subjectively, Peirce had thus maneuvered himself into a fragile and isolated position but this was a result of his uncompromising conscience. As he fears "losing" Bache, he envisions, vaguely, a professional institution that would adhere to his radical standards. In his own view, what sets himself apart from his despised colleagues are different expectations for scientific work. His motivation for suggesting a national clearing house for science is twofold: To legitimate his departure from the profession by proposing to the one friend left to him a new organization; and to find an organizational equivalent for his selfless truth-seeking attitude. A national institution was a natural solution, for it would not compete with the American Academy of Arts and Sciences. Because it represents the nation and because funds would be drawn, not from private individuals and their particular eccentricities, but from the political sovereign as represented by the federal government, it would surpass the latter's limited private and regional legitimacy. Finances, however, were unimportant to Peirce. What mattered was a new personal legitimacy for radical and uncompromising standards.

As to the asses of science, Agassiz is evidently mortified that he did not go, and is seeking in all possible and impossible pretences some good apology for his absence. He promised to go next time, and I think that he will go, although he is not to be depended upon. But then he is true to science, and we must pardon his human frailty. I love him dearly but I have long ceased to rely upon his support in any emergency. How different from you, my certain support, my guardian angel! We must go on, and I think that the gain of [—] is worth everything to [us.] We must go on. If we stop we shall dry up into a useless mass of attic salt. The devil will be there at any rate and the saints must not stay away.[28]

Peirce moves away from his initial topic as he summarily refers to the "asses of science," the members of the American Association for the Advancement of Science. He invokes his own term as president of that organization which he had concluded at the Washington meeting. The term "asses" was occasionally used in Lazzaroni correspondence.[29] In a play on the organization's acronym, it invoked, among other things, the uninspired and sedate tempo of mules. By using this term, Peirce asserts a derogatory Lazzaroni perspective. He thus includes Louis Agassiz with everyone else in the organization, as an outsider to the small group around Bache.

Peirce had used such language before but it is significant that he does so at this particular place in his letter: By mentioning Agassiz' absence from the Washington meeting he highlights his own participation. By assuming the view of the former AAAS president, Peirce stresses his dedication to developing that organization. This contrasts with his reaction to the trouble in the American Academy that he is ready to leave behind. He acknowledges that because of his "human frailty," Agassiz' absence should be pardoned. This implies that to participate in the AAAS takes strength and that Peirce is willing to take this effort. An obvious way in which this organization differs from the one in Boston, a "mere private organization," is its national scope. Having been president of the AAAS, Peirce compensates his frustration with his close colleagues in Cambridge with a renewal of his dedication to a national development of science. By announcing to his friend his departure from Cambridge, Peirce had taken exception to the idea that scientists should stand together. He now urges Bache to stand with him in dedicated efforts on behalf of science.

Peirce did not contradict himself if we assume that he was strongly identified with a national perspective for American science. His visceral dedication to such a trajectory allowed him to view his Cambridge environment with skepticism, detachment, and sometimes even with aversion; and it would allow him to situate his "real" identity on a new national plane represented by his cooperation with his friend Bache in Washington D.C. Peirce had recently been president of the profession's only national organization in America. Perhaps his return to Cambridge after handing over

28 The gap between "will" and "be" left by Peirce.

29 Among them, the Lazzaroni frequently used such shorthand. On variations of "asses," see Kohlstedt, *Formation of the American Scientific Community*, 169.

the baton to his successor went along with the unwelcome realization that once again he could not take his national prominence for granted. He had just announced that he was ready to abandon his career. As though he was once again invoking a presidential role, he now called on his friend Bache for continued dedication to their common goals: "We must go on." But Peirce's repetition of the phrase indicates just how weak he continues to be.

Peirce's reference to saints and the devil summarizes the point he is trying to make in this paragraph. He discusses the AAAS and its support by Agassiz, Bache, and himself: three out of five former presidents of the organization. (The other two had been Joseph Henry and William C. Redfield.) Peirce's standard expectation is that the organization is bent on doing much evil, and that the country's prominent scientists must live up to their responsibility of keeping up the standards. By calling Bache, Agassiz, and himself "saints," he implies that their work is beyond earthly criticism, that they have proven their exceptional moral integrity and guiding abilities, and that there exists a community of believers for which this carries special significance. The comparison, of course, is self-congratulatory. In the context of his letter, we can see that Peirce allowed himself to indulge in self-appreciation because he felt misunderstood by his Cambridge colleagues. He focuses on Agassiz because unlike Henry and Redfield, the biologist is both a Cambridge colleague and a former AAAS president. In the wake of his own AAAS presidency, however, Peirce cannot but acknowledge his predecessor's relevance.

> Bassnett's theory of storms shall not get into the proceedings until it is substantiated by verified prediction, and I will make the computations myself in view to be seen that there is no mistake. As to his hypothesis, it seems to me to be a harmless pigment on pigment.
>
> Coakley cannot be published, if he gave a correct account of himself. But perhaps when he was looking in the glass he was brighted by his own ears.
>
> In general as to the Association, it does much good. The positive is remembered and its negative is forgotten; but then we need the other society for science, and this for sociability.

Peirce comments on papers presented by two colleagues at the recent AAAS meeting, papers Bache had mentioned in his letter to Peirce. He criticizes Thomas Bassnett for not testing the results of his mathematical computations on winds.[30] Peirce's dismissal of George W. Coakley is even harsher, for he jokes that Coakley may have had correct results but that he might have presented his results under the influence of alcohol.[31]

Speaking of his two colleagues leads Peirce to his earlier suggestion for a scientific organization sponsored by the national government. He becomes aware that his suggestion may be countered by reference to the AAAS, a nationwide organization for science in America. He takes a skeptical view of this organization, conceding that "it does much good." The organization's value, Peirce considers not with regard to its benefit to the profession at large but to people like Bache and himself ("we need"). The "other" society is the one Peirce had in mind when he suggested the creation of a new national organization. This goes along with his use of the conjunction "but." Peirce points out that the AAAS was not useful for "science." Following his criticism of Bassnett's and Coakley's papers, Peirce has in mind an organization with stricter academic standards. In his view, these standards did not have an institutional home in 1854. While he is now engaging in a humbler consideration of the AAAS, he insists on his idea that the country needed an organization of a different type.

> The present society is a very good punch bowl in which the wise men may sail with the slower for sociability in fair weather. But when the storms arise, we must have a frigate at hand in which the wise men may rest in safety while the prunes go to the bottom.[32]

Peirce suggests that unlike many of their colleagues in the AAAS, men such as Bache and himself possess intellectual experience (they are "wise"). It is important to Peirce that they are "saved" in situations in which the AAAS provides little protection. What is the nature of "storms" that may threaten the organization? Peirce has come to this metaphor by way of criticizing the quality of others' work. His conflict with Davis involved different standards of evaluation. In Peirce's view, the problem was that many colleagues did not live up to the standards he is unwilling to give up

30 Thomas Bassnett (from Ottawa, Illinois) had presented a paper on "A Theory of Storms." This paper was not published in the Association's *Proceedings* as "no copy of [it] has been furnished." AAAS, *Proceedings* 9 (1855): 226.

31 George W. Coakley (Professor at St. James College, Maryland) had presented a paper on "Moon and Star Culminations." Ibid., 58–61.

32 Lurie quotes this sentence in this way: "a frigate in which the wise men may seek safety." He understands "frigate" as a reference to the Lazzaroni and the institutions controlled by them. He dates this letter as 1855. Lurie, *Louis Agassiz*, 183.

or water down. They are comparatively new to science and lack the experience of "wise men" such as Bache and himself. Peirce uses a definite article in "the storms." He does not have in mind unforeseeable conflicts but is aware of the kinds of problems that may arise. If the AAAS, as a "punch bowl," includes experienced scientists as well as less experienced ones, the most obvious sense for Peirce's storm metaphor is that he cannot take for granted that the society is securely wedded to rigid scientific standards. "Wise men" and their ideals may be put on the defensive. This implies that Peirce felt responsible for upholding scientific ideals in the face of an opposition by those unable to live up to them. But his "frigate" does not serve as a weapon against his lesser colleagues. It serves as a protection and as a retreat for science as he merely seeks to "rest" in it. The "prunes" do not "go to the bottom" because they have no immediate defense against an attack by the frigate; the picture of them sinking to the bottom nicely illustrates their becoming irrelevant over time. Those aboard the frigate have no responsibility to aid their colleagues. Their safety is assured by their ability to get away in their agile vessel, leaving their shipwrecked colleagues behind. The "wise men," in other words, have a responsibility, not for saving their peers, but for preserving their own ability to advance science.

> I presume that our geological friends have resolved to abandon us, and that they do not find it comfortable sailing in company with Physics. The earthen vessel is afraid of the iron retort.

Peirce extends his sailing metaphor to comment on the relationship between disciplines within the AAAS. Geologists considered leaving the organization, and Peirce interprets this not as an attempt to benefit from a more specialized platform for discussion, but as a sign of shying away from the thorough scientific standards of physics.[33] Vis-à-vis Bache, of course, this comment serves as an assertion of their common rigidity. At the same time, however, Peirce considers the AAAS as a means for implementing such standards, and he tacitly blames geologists for not following through with them. One could argue, if the geologists abandoned the AAAS, why would physicists no longer be able to attack them? The two "vessels" may no longer have travelled alongside one another but they would not be out of each other's reach. Peirce takes for granted that the AAAS provided a framework within which it was difficult to ignore the validity and strength

33 Kohlstedt, *Formation of the American Scientific Community*, 176–78.

of arguments. In this sense, it served as a cultivating force across disciplinary boundaries within the scientific profession.

Following his reassurance of friendship with Bache, Peirce had started this letter by attacking immediate colleagues such as Davis and Gould. He now extends his criticism to all scientists who seemed to slow the progress of science because of their homeliness ("earthen vessel"), unwillingness to let ideas go, and an inability to face tough realities.

Does Henry need [air]? The more I think of it and talk with others, the more am I satisfied that the library and museum should be wholly abandoned. I mean just what I say, I do not see how the Smithsonian is justified in diverting any of its funds to such local purposes. It cannot do better than to keep a supervision over all the libraries of the world America included, and see that ~~that~~ the proper books are every where brought here to the right men and now and then assist in purchasing them. The same is true in regard to museums, its high and peculiar [premise] is to ask where are the [small] naturalists and what can be done to bring them to their proper materials or the materials to them? The hand of the Smithsonian should be free to act everywhere, stimulating and restraining true science against demagogism and treachery. It seems to me that it ought not even to be [mason-ruled] by such a mass of brick and stone housing about its neck, as that abortion of a building. However you are wise and Henry is wise, and we trust you without asking ~~what~~ too many questions, or overloading you with impertinent advice.

God bless you
my sincere[34] friend
Benjamin Peirce

B.A.D.

I shall at once prepare my report. [—] is almost done and can be drawn from the field at any time. Will you send a [—]?[—]?

The Smithsonian Institution was created by Congress in 1846 after James Smithson, an English chemist in France, had died and left a bequest to the United States to establish in Washington an institution to advance knowledge. The act was a result of several competing interests. In line with the transnational scope of Smithson's bequest, it created an institution dedicated to research but it also included a library and a museum, and this implied a much more national and local role. From the institution's inception, the research side had had the upper hand because of the regent's selection of Joseph Henry as Secretary. The conflict with the "library

34 In the original, this word has five lines under it, not just two, but these lines could not be reproduced here.

wing," which had remained unresolved in the founding process, came to a head in 1854 when Assistant Secretary Charles C. Jewett convinced regents to undo the earlier compromise and to turn the Smithsonian into a library. Henry later dismissed Jewett, was sustained by the regents, and in the 1870s sent to the Library of Congress the volumes Jewett had amassed at the Smithsonian. When Peirce wrote his letter, however, the library conflict was out in the open and remained unresolved. Bache had been one of the Smithsonian's initial regents and instrumental in installing Henry there. Peirce's friend in Washington, therefore, understood the institution's raison d'être and the unfolding course of events very well.[35]

In his view of the Smithsonian's role, Peirce follows Henry's initial 1847 view. Henry had insisted that the Smithsonian was a universal institution and that it should not, therefore, spend money on books and objects which were useful only for those who were in (or could easily travel to) Washington.[36] This was also the position taken by Henry and his supporters (including Bache) in 1854. By suggesting that the museum and the library be given up because it primarily served "local purposes," therefore, Peirce merely repeated the long-standing views of his Washington friends.

In his concluding sentence, Peirce assumes a pose of self-restraint that only serves to highlight his transgression into his friend's business. He ends his letter in a tacitly self-assertive tone by granting himself the privilege of evaluating Bache's and Henry's professional standing ("wise"). Peirce's greeting ("my sincere friend") suggests that he had not entirely written off the issue. Finally, to address his letter to "B.A.D." (instead of "A.D.B.") Peirce was playfully reminding Bache of his disloyalty by publishing with the American Academy.[37]

35 These episodes are relayed in Dupree, *Science in the Federal Government*, 66–90. For correspondence and a more recent discussion of these events, see Henry, *Papers of Joseph Henry*, vol. 9.

36 Dupree, *Science in the Federal Government*, 81.

37 Peirce had played with Bache's initials before. See his letter to Bache, October 28, 1853 (RH 1980, box 16, Rhees Collection, HL) where he had addressed his letter to "a.d.b. not bad." Bache himself had once signed his letter as "A.B.D." and he had inserted a "C" in its "proper" alphabetical place. See his letter to Peirce, [ca. Nov. 1853], letterbook, BPP. In Peirce's letter to Bache discussed above, Peirce added a postscript that is hardly legible, and it appears that he addresses practical Coast Survey matters there.

"A Victory for the Evil One"

A detailed explication of the premises of Benjamin Peirce's letter not only throws light on his personal crisis but also on where his views on science in America coincided with Bache's. Peirce felt that his stern and demanding assessment of other scientists had isolated him from his more accommodating Cambridge friends such as Charles Henry Davis. This opened up the imaginary abyss of moving to New York City, and much more significantly for our purposes, this crisis reminded Peirce of the Lazzaroni's common goal to establish a new and expanded legitimacy for a radical understanding of scientific norms. According to this idea, the American Academy would be outflanked by a national institution of the kind Bache had envisioned in his 1851 speech and which Peirce's friend, his close ally in Washington D.C., seemed to personify.[38] Peirce perceived Bache not only as a friend and consolidator, but as the protagonist of a future national relevance for the rigid and uncompromising standards that would allow America to stand up to European science. This confirms what we know about Bache: Peirce shared with his Washington friend the very same motivational goal, a deeply-held belief that the American "sovereign" would eventually have to acknowledge and protect their radical scientific standards. For most of his career, Bache had been the protagonist of a national implementation of science. Peirce's letter suggests that after Bache had moved to Washington, he grew into a role in which his closest friends viewed him, the Lazzaroni "chief," as an unofficial "president" of an invisible national academy. The superintendent of America's largest scientific enterprise had come to personify an academy nine years before it came into being.

From the letter which Bache had written to Peirce after the Washington meeting, the letter to which Peirce was responding, we may assume that the superintendent took issue with Peirce's exclusive attitude and the motivation for insisting on their ambitious goals. It will be interesting to see in detail just how Bache responded to his friend's letter. Bache responded as follows:

---

38 The consistency of his view may be inferred from the fact that Peirce, in his speech as outgoing AAAS president just a few days earlier, had reiterated Bache's call for the founding of a national academy. "Of all the virtues, patriotism is the least selfish, and that which is most kindred to the grand sentiments of the heroic soul," Peirce wrote in the published version of the speech. Peirce, "Address," 16 f.

Washington, May 11/54

My very dear friend,

I cannot refrain tho' I know that there will be so many interruptions that I shall be incoherent from acknowledging for the warmth of your outpourings of brotherly affection, which are so very welcome, + find such entire reciprocation.[39]

Bache's salutation is somewhat less playful than in other letters to Peirce where he occasionally used the endearment "Functionary." Bache anticipates frequent interruptions. He is probably at his office. Despite upcoming interruptions, he proceeds to write, which indicates that he wishes to lose no time, that his response is urgent, and that there is much to say. Bache signals that Peirce can count on his friendship, and that his trust is not misplaced. By speaking of "outpourings," Bache picks up on the self-abandonment that resulted from Peirce's disheveled emotional state.

Would that I were by you to talk over what has excited you to the very depths of your heart. [Such then certain] I feel that I could pour some balm, tending at least to allay a part of your suffering.

Bache further highlights that he should be able to help Peirce. He writes that if he were closer, there would be an opportunity to "talk over" the issues and this implies that Bache did not consider the issue to be his publishing Coast Survey papers with the American Academy. He considered the problem to be Peirce's emotional disposition for only then would talking things over make any sense. This corresponds to his suggestion that he felt certain he could "pour some [balm]:" Peirce had stressed how important Bache was to him and his friend can be assured that he may be able to reach him. Bache's empathy becomes apparent from his underlining "I feel." Only a portion of Peirce's "suffering," however, might be alleviated by his friend's counsel. Bache takes for granted that in the end, Peirce will have to come out of this himself.

True affection is such a balm, + you have allowed one to penetrate so deep below the surface that I would feel sure of reaching the spot. But a letter—is just worth next to nothing. Still better than nothing.

Bache continues to engage in preparatory matter by reasserting his friendship with Peirce and his "True affection" for him. He can only infer, of course, from Peirce's letter and earlier behavior, that his friend will accept a "balm." But his insisting on their friendship is a sincere token of it. Bache

39 ADB to Benjamin Peirce, May 11, 1854, Bache Letters, BPP.

seems to pick up on Peirce's "heart-metaphor" ("knock for admission") in his first sentence and he claims almost therapeutic abilities by suggesting that he could reach "the spot," i.e. the emotional knot in which Peirce had become entangled. A "spot," of course, is a small area visibly different from its surrounding, and the word is often used to refer to a taint in character. Bache ignores Peirce's version of the story as well as Peirce's invitation to join him against Davis and, from the beginning of his letter, Bache directs his attention at a problem that is of Peirce's own making, a problem that is "below the surface" and psychical.

These unhappy affairs at the Am. Acad.! What a victory for the evil one to separate even for a moment + by error so thin a screen such friends. The measures of Satan are as perfectly carried out by such a separation, as if he had carried the measures themselves.

In a concession to Peirce, Bache now picks up on the institutional dimension of his friend's conflict with Davis and with others in Cambridge. Following his initial and spontaneous focus on Peirce's inner disposition, Bache signals that he does not ignore the conflict's institutional dimensions. To the extent that Peirce considered the conflict to emanate from differences of assessment of scientific work, he perhaps perceived this as a further tightening of the screw. Bache focuses on Peirce even though he is aware of these "unhappy affairs."

In his second sentence, Bache drives home his point. In his letter, Peirce had likened Bache, Henry, and himself to "saints" who were facing the devil (at the American Association for the Advancement of Science). Bache turns his friend's metaphor around and suggests that by letting himself be divided from his friend Davis, Peirce was the one who had turned himself into a tool of Satan. Bache goes even further by suggesting that this separation was a result of an "error," hence within Peirce's competence, entirely avoidable, and even now easy to correct ("so thin a screen").

I conjure you my dear, very dear friend, do not let the <u>breach widen</u>, be the cause what it may, between you + Davis.

Bache's plea follows naturally from what he has been saying. He does not urge his friend to bridge the gap that has opened up (this may have been too much to ask) but to view Davis as a friend rather than an emerging enemy. Peirce had insisted on, but not explained, his differences with Davis in evaluating Asa Gray's work. Bache again dismisses them and insists on the value and relevance of Peirce's friendship with Davis. Does this imply that Bache cared little for what his friends had in common scientifically? An answer to this question is contingent on a definition of "science." In the given context, Bache took for granted that even where scientific institutions were involved, the basis for cooperation was friendship and even love. One could argue that Bache took a strategic view of Peirce's conflict with Davis, and that he sought to reintegrate Peirce because he needed his support within the Lazzaroni group and within the profession, but this would ignore the risk Bache runs in confronting his friend with uncomfortable realities. We cannot assume that Peirce took Bache's letter lightly (in which his Washington friend, after all, suggested that Peirce was a tool of Satan).

> If ever a man was devoted to another Davis is to you. What is a difference in opinion of the uttermost kind between two such friends. If I were only able to tell you how I know Davis' mind + heart in this matter. You are one of the idols which he has set up in the inner shrine. A few hasty words can not [—] separate you two. However, [—ous], it makes a chill go down my head to my feet at the very idea.

Bache moves on to provide Peirce with positive arguments for reconsidering his position, conveying to him Davis' perspective. Bache portrays their friendship in the strongest terms, as the model of all friendships akin to marriage, and he explicitly belittles their disagreement as a matter of "opinion." But was not Peirce's letter evidence that his friendship with Davis was over? By invoking the strength of their friendship, Bache reasserts it. In doing so, he has no evidence other than believing in it—believing in Peirce's ability to turn around and to take a humbler view of Davis. What stands out is Bache's charismatic perspective that, in the given context, serves as a realistic and therapeutic corrective.

In his third sentence here, Bache turns himself into a spokesman for Davis. He claims to know him very well "in this matter." Did Davis speak to Bache about the conflict? This would be feasible because the American Academy matter had been simmering for some time and because Davis had attended the Washington meeting.[40] Bache notes this relationship by speaking of Davis' "devotion." In the context of science, such adoration makes him a student. Peirce is not the only idol but one of several that Davis, according to Bache, has set up in "the inner shrine." The phrase

40 Davis had presented several papers. AAAS, *Proceedings* 9 (1855): 76, 73, 237.

plays on Peirce's metaphor. Whereas Bache needs to knock for admission to his friend's heart, Peirce is an idol in Davis' inner shrine.

On the whole, Bache takes a conciliatory and sympathetic attitude towards his friend. He speaks of a "few hasty words" and grants that Peirce may have dealt with Davis in the heat of the moment. Bache's response, however, should not be misunderstood as having been "common" or "routine." In some ways, it would have been easier for Bache to support Peirce's criticism or to delay his response to give Peirce a chance to cool down. This would have been particularly true if Bache had known Peirce's temper to be easily irritated. In this case, it would have perhaps made sense to wait and not to respond for a day or two. But Bache takes Peirce seriously, and this implies that he must have been aware that by criticizing his friend and by suggesting that he was not an ally in the sense Peirce wished him to be, he was putting his own friendship with Peirce at risk.

> I see it all before me. Your silence misconstrued + chasing him. Your burning words when the [fire] did break out, enveloping him in flames.

Bache tries to reconstruct how the crisis evolved. Peirce had written that he had withheld a long time, and that when Davis raised the issue, it all came out. Bache develops his interpretation on the basis of a friendship that has evolved from a student-teacher-relationship. Because Peirce's opinion mattered so much to Davis, he was unable to deal with Peirce's silence on the subject of Gray's qualities while suspecting that his former mentor differed with him. In Bache's perception, Davis was "chased" by what he perceived to be Peirce's held-back criticism. What does Bache mean, however, when he speaks of Peirce's silence having been "misconstrued" by Davis? Was Davis not right in suspecting that Peirce held back criticism of himself since Peirce later said "burning words"? In Bache's perception, Davis must have misconstrued his former mentor's silence not with respect to what the latter had to say to him but with respect to the fact that he remained silent. In Bache's reconstruction of the conflict, Davis took this silence as a sign of mistrust, a lack of confidence in his ability to deal with this criticism, evidence for his not being accepted as a full-fledged colleague. Bache does not say so explicitly but in his perception, the conflict is of Peirce's making because of his ongoing responsibility as Davis' former mentor.

In the above section of his letter, Bache stressed the value of collegiate friendship and urged Peirce to take a more responsible attitude. He now criticizes Peirce more overtly for not speaking to Davis ("your silence misconstrued").

The almanac cannot do without you, nor can Harvard. The [curse] will pass from you. Hope says it may already have passed, God grant that it be so. Hope says that Davis came back after his irritation had subsided, + you were both sorry. A stranger cannot come between two such men, even a friend could not come between you + Davis.

Bache's reference to the passing of the "curse" is clear and unmistakable evidence that he considered Peirce to be the troublemaker. Bache wishes the crisis had already passed, and he envisions Davis getting back in touch with Peirce, not the other way around. At the end of this section, Bache implicitly rejects Peirce's offer to team up with him in the Academy fight by suggesting that Davis was much closer to Peirce than he was himself ("even a friend").

How will what I have written seem to you [dissolved] of all soul as it is by the want of presence! I cannot trust myself now to say more. Write to me as soon as you can. Tell me that this breach is healed. If Gray were a dozen Rogers [he] should not [win] it.

Bache's initial comment makes clear that he does not expect Peirce to take his letter lightly and that he was indeed taking a risk. He suggests that if he could be certain that this would not jeopardize their friendship, there would have been other unpleasant observations with which he could have confronted Peirce. Bache's witty reference to the brothers Rogers as a standard of uncongeniality introduces a humorous tone with which to loosen up Peirce. While Peirce had assumed Gray would win if he were not criticized, however, Bache suggests that this would be the result of Peirce being estranged from Davis, and that Gray does not deserve Peirce's harshness because of his relative collegiate proximity when compared to the brothers Rogers.

About Gould I will not speak now, but I have much to say + <u>to you</u>. He was in wretched spirits when here + talked a great deal to me. For the sake of the science which you love—American Science bear + forbear. Oh for a long talk face to face instead of this dull letter so imperfectly conveying my meaning + not at all my feelings. What will you say to it!

Ever truly yours A. D. B.

Peirce had mentioned Gould in his letter but he had not discussed their disagreement. Even though Bache had written that he could not trust himself to write more we may assume that his letter would otherwise have included criticism of Peirce's behavior towards Benjamin Apthorp Gould. Peirce had focused on his problems with Davis but we can now see that his issues with Gould were severe as well. Peirce's difficulties affected all of his professional relationships, particularly those with his former students. After his two-year term as president of the AAAS ended, Peirce had apparently entered a crisis of confidence. Bache picks up on this underlying issue particularly well. His criticism and support of his friend is realistic and emphatic. That he had much to say to Peirce about Gould ("to you") makes it very clear that Bache considered Peirce responsible for his student's "wretched spirits." Bache's comment also suggests that he learned a good deal about Peirce's state of mind during the AAAS meeting the previous week and that Peirce's fears in this respect were not unfounded.

In the third sentence quoted here, Bache directly confronts Peirce with the contentious issue of his personal irritation by urging him to "bear + forbear." He does so in connection with Peirce's suggestion that his unwillingness to compromise was a "duty" to his time and generation. Bache does not hesitate to deflate Peirce's claim for national and international significance and ambition. Bache turns his friend's reference around by suggesting that a more tolerant view of his colleagues would be more adequate and helpful for advancing American science. Bache urges his Cambridge colleague and former AAAS president to adopt a more conciliatory perspective for the very ideals he had proclaimed.

Looking back on this letter, we can see that it was composed in a single paragraph that contained the coherent idea of urging Peirce to reconsider his position. Bache's concluding sentence ("What will you say to it!") again suggests that he was not at all sure whether he would be successful in doing so. He ends his question with an exclamation mark, calling on his friend to acknowledge his criticism.

President of an Invisible National Academy

Peirce took his time responding to Bache's letter as he was apparently trying to cope with his difficulties in Cambridge. A week after he had sent his first response, Bache wrote that he was "bitterly disappointed in not hearing from you to-day which is the last limit I had fixed for possible contingencies."[41] He continued his letter the next day in an ambivalent, disappointed tone.

> May 19. […] Thought you knew me so well that you would answer me at once to relieve me from the excitement. I must confess that this morning I felt better as the matter is less ardent, + not so close, + after a few more mornings […] be used to the skinning.
>
> I was so occupied with your feelings in my last but I omitted to say your argument in regard to the Amer. Acad. was conclusive, especially as the appropriate bill of this year now contains an item which has been refused me for several terms [—] for publishing our results.

He thus conceded to Peirce the argument that Coast Survey papers not be published with the American Academy. The whole matter did not result in Benjamin Peirce's self-ejection from Cambridge. He remained at Harvard until his death in 1880; he became Bache's successor as superintendent of the Coast Survey in 1867; and he continued to work with Davis.[42] Furthermore, as will be seen in the next chapter, he cooperated with Davis and Bache in founding the National Academy of Sciences.

Peirce's letter has frequently been quoted to describe the Lazzaroni's dismissive perspective on the AAAS, but the specific setting of Peirce's comments is usually ignored.[43] A consideration of the entire letter is necessary, however, in order to contextualize Peirce's statements. Peirce felt betrayed by his colleagues, especially Charles Henry Davis, and prefers abandoning the profession rather than endure the prospect of isolation within it. Once Davis forced the conflict, Peirce is put on the defensive and he writes to Bache in fear that his antagonist beat him to it. Peirce, perhaps unconsciously, assumes an aggressive stance towards Bache by suggesting that the latter had decided, and not merely considered, publishing Coast Survey results with the American Academy in Boston. But Peirce really is looking for help from his last remaining ally, and between the lines, he asks Bache for assistance in resolving the conflict. Peirce's problem is that he finds himself isolated in Cambridge because he forces himself to adhere to rigorous inner standards while his friends take a more accommodating view of colleagues such as Asa Gray. Peirce was unwilling to cooperate unless he was fully convinced of the quality of someone's work.

41 ADB to Benjamin Peirce, May 18, 1854, BPP.

42 Hogan, *Of the Human Heart*, 121.

43 Kohlstedt, *Formation of the American Scientific Community*, 176; Lurie, *Louis Agassiz*, 183.

For our purposes, it is of course significant that Peirce raised the issue of a national orientation of American science in this particular context. He criticized the American Academy because it "fancies at the foot of wealth." Instead of proposing, however, that another private and more radical institution replace or compete with it, Peirce turns to Bache as the proponent of a national institution of science. In line with the perspective of a Romantic "Young America" movement consciously seeking to shape national as opposed to regional life, Peirce assumes that he shares with his Washington friend a dedication to developing science on the federal level. Peirce implicitly relates to everything Bache stood for, of course, as well as the latter's 1851 idea for a national institution, an idea that Peirce had reiterated in his own speech as outgoing AAAS president. As suggested in the previous chapters, the founding of the American Association had provided a national platform for science in the United States but its formation did not signal that science and its principles were accepted by the political sovereign. What Peirce's letter shows is that as long as such a state of affairs persisted, the national arena could serve as a projection screen for scientists like Peirce who keenly felt the responsibility of living up to European intellectual standards and for representing and implementing them in their own country. The American Academy and the American Philosophical Society had existed alongside one another for some time, but everyone knew that there could only be one national academy. Such an institution promised to provide a unique legitimacy superior to that of the Boston organization Peirce despised.[44]

In his response to Peirce's letter, Bache did not celebrate the idea of a national organization. He did not, at first, react to the idea that Coast Survey results not be published with the American Academy in order to strengthen the former instead of the latter. By focusing on Peirce's difficulties in an almost therapeutic way, Bache took seriously his friend's crisis, and he criticized him for ignoring the principle that collegiality also entails charity, and that the leaders of the American scientific profession, despite their personal differences, must continue to cooperate. Peirce's letter

44 This interpretation of Peirce's motives falls in line with other letters I consulted. For example, Peirce wrote to Joseph Henry in 1849 that he and Bache "deserve to receive mural crowns from your scientific fellow-citizens for the good, which you have done" in Washington, i.e. a sign of honor given the soldier who first climbed the wall of an enemy town (such as Scipio at Intercatia). Benjamin Peirce to Joseph Henry, April 14, 1849, Henry, *Papers of Joseph Henry*, 7:506 f.

shows that within the Lazzaroni group, Bache was the only one who could convincingly represent national ambitions. Nevertheless, he rejected Peirce's offer because his friend was obviously not in a sound condition and used a national perspective as a lifeboat for escaping his Cambridge troubles. Bache's leadership within his circle of friend, in other words, was founded on principles of personal integrity. Such a comprehensive and integrative view was not unique to Alexander Dallas Bache, but considering the Lazzaroni "chief's" standing within the profession and his broad and far-sighted perspective for antebellum American science, his role may perhaps be described as having been that of president of an invisible academy.

Chapter 10

The 1863 Founding of the National Academy of Sciences

The Timing

By signing the law creating the National Academy of Sciences in the late hours of March 3, 1863, President Abraham Lincoln seemed to fulfill the long-standing ambitions of the small group of scientists around Alexander Dallas Bache. In his speech as outgoing president of the American Association for the Advancement of Science (AAAS) in 1851, Bache had concluded his broad consideration of the prospects of science in America by calling for a national institution such as an academy established by Congress to advise the government on scientific and technological matters.[1] His views developed into a broad agenda for organizing the scientific profession in America. Bache and his circle had helped found or develop professional and scientific organizations and institutions such as the U.S. Coast Survey, the American Association for the Advancement of Science, and the Smithsonian Institution in the 1840s and 1850s. But it was not until 1863 that they chose to act on the academy idea.

This timing calls for an explanation. The Civil War, and 1862 and 1863 in particular, was a time of severe national hardship. Would it not have been sensible to wait and to push for an academy when the nation could calmly consider this proposal, a time when attention and resources could more easily be directed towards the new institution? For several years, the Lazzaroni had discussed the idea but they had not pushed it. Why did Bache and other Lazzaroni choose to pursue it now? How could Bache and his colleagues expect that, in the middle of a civil war, politicians would agree to support the idea?

---

1 See chap. 8. A previous version of chap. 10 was published in *Sozialer Sinn* 8, no. 2 (2007), 333–56.

Historians have not usually put this question at the center of their analysis but some preliminary answers may be inferred from their account of events leading up to the academy bill. Robert V. Bruce suggests that the war provided an opportunity for developing older plans. Secession, he writes, "turned Bache into an uncompromising Unionist," but Bruce does not go on to use this change or intensification of perspective as an explanation for his wartime efforts. He suggests that Bache and his circle merely considered the war in terms of political opportunities for realizing long-standing academy plans. In the absence of Southern legislators, the Thirty-seventh Congress in July 1862 passed the Morrill Act providing for land grant colleges, and "the time now seemed ripe to realize the Academy dream."[2] Accordingly, Bache and his colleagues went to work on an academy bill later that year. In his essay on the Permanent Commission of the Navy Department, founded at about the same time as the National Academy, Nathan Reingold implicitly compares both to scientific institutions in later wars and concludes that they accomplished little.[3] He explains that the Permanent Commission was created in response to the navy's need to evaluate proposals by engineers and inventors. He further limits the National Academy's historical role by highlighting that the idea for it was an afterthought to the Commission plan, that the war situation was perceived as an opportunity for implementing long-standing ideas, and that the Academy's advisory role remained limited. Reingold does not ignore a patriotic impetus in Bache and his colleagues, but this motivational aspect is last on his list. A. Hunter Dupree's review of the role of science in the federal government contains a more explicit interpretation of the political context. Dupree considers Bache to be prompted by feelings of national pride that preceded the war. The "impulse of the group" around Bache, he writes, "was heightened by the war only in that it provided the occasion for their basically nationalistic enterprise." Their aim "was to provide a young America, which they defiantly expected to survive its great trial, with a

2 Robert V. Bruce, *The Launching of Modern American Science, 1846–1874* (New York: Knopf, 1987), 274, 302. See A. Hunter Dupree, "The Founding of the National Academy of Sciences—A Reinterpretation," American Philosophical Society, *Proceedings* 101, no. 5 (1957): 434–40.

3 Nathan Reingold, "Science in the Civil War. The Permanent Commission of the Navy Department," *Isis*, no. 3 (1958): 307–18; Nathan Reingold, *Science in Nineteenth-Century America: A Documentary History* (London: Macmillan, 1966), 200–03.

worthy counterpart of the Royal Society and the French Academy."[4] His notion that the Lazzaroni were prompted by "nationalistic" fervor highlights a motivational dimension that has received less attention by more recent authors. In this chapter, I will try to deduce an explanation for the timing of the National Academy's founding from the motivation that prompted Alexander Dallas Bache to support it in 1863.

Without Bache's initial endorsement, planning, and support, the idea for an academy would probably not have moved forward. The academy idea could and would be developed without the explicit initial support of Joseph Henry but it could probably not have been done without Bache because of his standing in the profession.[5] It is therefore of particular relevance to try to explain Bache's motives for supporting and developing the idea in 1862 and 1863. I will focus on Bache's response to the war and on his correspondence with political scientist Francis Lieber to develop the idea that while plans for an academy had been circulating for a long time, the decision to move ahead with it was less a response to the profession's need for such an institution than the profession's response to the political situation during the war. In this sense, my interpretation follows Dupree but considers the Academy's founding to be an ambitious answer to the perception that the nation was threatened by a loss of aspiration and civil integrity. I will argue that Bache, by pushing forward plans for an academy, had reluctantly decided to bolster and support the nation to whose cultural development he had dedicated himself. In this way, the Academy's founding reflects the particularities of its American setting.

The Bache-Lieber Correspondence

My focus here is on Bache's correspondence with Francis Lieber, a German émigré who, in 1856, had moved to New York City from South Caro-

4 A. Hunter Dupree, *Science in the Federal Government: A History of Policies and Activities to 1940* (Baltimore: Johns Hopkins Univ. Press, 1986), 138.

5 On Henry's role in the founding of the National Academy, see ibid., 135–41; Joseph Henry, *The Papers of Joseph Henry* (Washington: Smithsonian Institution, 1972–2008), 10:xxxii–xxxvi; Leonard Carmichael, "Joseph Henry and the National Academy of Sciences," *Proceedings of the National Academy of Sciences of the United States of America* 48, no. 1 (July 15, 1967): 1–10. For Bache in this context, see also Frank B. Jewett, "Alexander Dallas Bache. A Founder, First President and Benefactor of the National Academy of Sciences," American Philosophical Society, *Proceedings* 84, no. 2 (1941): 181–86.

lina for a chair in political science. In 1860, Bache and Lieber had known each other for thirty years (they had perhaps met in Philadelphia) but their acquaintance had never grown into a friendship.[6] Lieber's work was mostly unrelated to Bache's, and the Prussian was outside of Bache's circle of professional friends and political contacts, but in 1860, they began writing each other on a regular basis. Their correspondence sheds light on how Bache perceived the war and on what prompted him to support implementing the academy idea.

Lieber was born in Berlin in 1798, which made him eight years Bache's senior. As a student at the University of Berlin, he had become involved in the *Burschenschaften*, youth associations advocating liberalism and egalitarianism in a united Germany after 1815. In 1821, his romantic political ideas led Lieber to travel to Greece where he intended to support the Moreans in their struggle against the Turks, a conflict Lieber and other volunteers perceived to be a romantic Greek fight for independence and political freedom. The project failed. Lieber returned home broke and disillusioned with Greece. But because of his political interests and activities, Prussia now barred the returning Lieber from continuing his university studies. In the winter of 1825–1826, Lieber left for England and, a year later, had moved on to Boston with letters of recommendation and a newly-wed wife ready to follow him from London as soon as he was in a position to support her. Lieber was ambitious and culturally versed, and in the United States, he quickly established ties among a political and intellectual leadership. But he lacked a steady and secure occupation, and a reliable income. After several attempts at business, publishing, and translation, he was elected to a chair at the University of South Carolina in 1835. Lieber was much more at home among intellectual circles in New England or New York. He felt stifled by the cultural and political climate in South Carolina and frequently spent the summers in the North to keep in touch with friends and fellow intellectuals there.[7]

In 1860, Lieber initiated the correspondence by asking the superintendent to support explorer Isaac Hayes. Five years earlier, the twenty-five year old Hayes had returned from an Arctic expedition led by Elisha Kent.

---

6 Lieber had of course written the proposal for Girard College which prompted the board of trustees to send to Europe for two years the new president of the institution—Bache rather than Lieber. See chap. 5 above.

7 Frank Freidel, *Francis Lieber, Nineteenth-Century Liberal* (Baton Rouge: Louisiana State Univ. Press, 1947).

He had served as a medical doctor on board the exploring vessel and was now keen on finding and investigating an "open polar sea."[8] In his reply to Lieber, Bache sweepingly endorsed the manifold scientific benefits he could see Hayes' new expedition having. "The unanimity with which men of science have spoken on this subject," he dictated his amanuensis, "authorizes the strongest appeal in behalf of Dr Hayes and his plans."[9] By September, Bache had assumed a more personal tone toward Lieber as he was responding to another request to support a young man. It was Oscar on whose behalf Lieber had now written, his oldest son and a promising geologist. When his parents and two brothers had finally left South Carolina for New York City four years earlier, Oscar had stayed and adopted a thoroughly Southern perspective on culture and politics. The rift between North and South widened at a quicker pace after 1856, and so did the one between Oscar and his father. As war was looming, Francis Lieber (now a professor of political science at Columbia University) found his son stubbornly holding on to the wrong side.

The Coast Survey must have struck Lieber as a potential "reintegrator" for his geographer son. From June to August 1860, that organization had sponsored a trip by Oscar to Labrador, and Bache now wrote to Lieber from his camp near Wachusett Mountain in Massachusetts that Oscar had granted him permission to read a paper Oscar had written there.[10] Bache was confirming this decision to Oscar's father, asking the latter to send the manuscript, and this implies that Francis Lieber could well have been the one who had suggested that Bache look at it, perhaps for publication. Oscar's affiliation with the Coast Survey, which Oscar's father sought to support and strengthen, allowed Lieber to entertain the vague hope of being able to reconnect Oscar to politics, culture, and his family in the North. Bache could invite hopes of becoming a "reintegrator" because he was not merely a scientist but represented the national dimension of Oscar's profession. But reintegration failed. Soon Oscar decided to join the Confederate army. His two younger brothers, Hamilton and Norman, were fighting for the Union, the former in the West along the Mississippi, the

8 John English, *Dictionary of Canadian Biography Online* (Toronto: Univ. of Toronto, 2000), s.v. "Hayes, Isaac Israel," http://www.biographi.ca.

9 ADB to Francis Lieber, March 14, 1860, Papers of Francis Lieber, Huntington Library (hereafter cited as "Lieber Papers, HL"). That Lieber had sent a request may be inferred from Bache's answer.

10 ADB to Francis Lieber, September 18, 1860, Lieber Papers, HL.

latter on the Peninsula south of Washington.[11] In the spring of 1861, Oscar and Norman were in the same theater of war and their units were likely to engage in battle.

Fig. 12. Francis Lieber

(Library of Congress Prints and Photographs Division. Brady-Handy Photograph Collection, LC-BH82- 4591 C)

With the exception of a letter by Lieber to Bache in December 1861, the correspondence between the two did not pick up again until March 1862 when it evolved into an almost intimate exchange.[12] Bache felt exhausted from the burdens of war and work, and instead of turning to one of his

11 Francis Heitman, *Historical Register and Dictionary of the United States Army, From Its Organization, September 29, 1789, to March 2, 1903* (Washington D.C.: Government Printing Office, 1903), 632.

12 Francis Lieber to ADB, December 14, 1861, contemporary copy enclosed in Francis Lieber to William Graham Sumner, December 14, 1861, quoted in: Freidel, *Francis Lieber*, 322 n. 12. The letter to which Lieber responded on March 22 has not been preserved, and neither has an earlier letter by Bache which reached Lieber the same day.

longtime friends and scientific peers, he reconnected with Lieber who was quick to reciprocate. He invited Bache to New York ("{if} you must leave Wash. for the restoration of health") and the following day, without waiting for Bache's reply, he wrote again. Lieber asked whether the superintendent could hold a map of Eastern Virginia for his youngest son Norman who was encamped near Washington.[13] There must have been other ways of getting the map to Norman but Lieber preferred using Bache. Perhaps Norman's father felt the need to know that his son would be in touch with a friendly spirit closer to him. He might have sensed that Bache, who did not have children, would not mind taking on the role of a surrogate parent closer to the front.

Lieber ended this letter by reporting that he was

> meditating the invention of a nail and handbrush with which my monobrachyte can wash his sole hand .... Will you not propose to Secr. {of War Edwin M.} Stanton to buy the patent of {sic} me, for the use of the army? How easily we sport with that, which but yester-day filled our eyes![14]

It was Lieber's son Hamilton who had lost an arm at Fort Donelson on the Mississippi River, and who was returning to his parents in New York City.[15] Lieber was writing this letter across the whole width of the sheet instead of folding the letter in the middle, allowing for an uninterrupted flow of the pen. To Bache he wrote openly about the burden of having three sons in the war, two of them still fighting on opposing sides in the same campaign, and one severely handicapped. If one considers Oscar's decision to join the Confederate army a loss to his father, then only Norman remained unharmed. It was Norman, too, for whom Lieber wanted Bache to hold the map of Eastern Virginia. With Oscar in the South and Hamilton returning home severely injured, Lieber was doing what he could for Norman by sending him the council of a map and the care of Bache (a producer of maps and expert in matters of orientation). By suggesting, with irony, that Bache forward his invention to the secretary of war, he signaled that he was aware that his experience was all too common. It also provided him with an opportunity to vent his pain in a letter to a friend with whom

13 Francis Lieber to ADB, March 23, 1862, box 30, Rhees Collection, HL.

14 Lieber's "monobrachyte" (one-armed) is an irregular and jocular combination of Greek words, "mónos" (alone, only) and "brachíon" (arm; Latin "bracchium"). I am indebted to Lorenz Rumpf for pointing this out.

15 Freidel, *Francis Lieber*, 324 f.

he shared the premise that they were to provide unflinching support for the Union.

But Lieber's letters soon assumed a harsher tone. On May 1, he wrote to Bache that there must be "No Armistice, for the sake of all that is sacred, sensible or worthy!" He added that "Blow upon Blow, ought to be our motto and only motto for the next 12 months."[16] "Yes Blow upon Blow Hard, Harder, Hardest" he asserted five days later.

> I think of the blacksmith, what was his name? who nailed his apron to a staff and became the founder of a Persian Dynasty. We might adopt that popular American symbol of an arm with a hammer; over {it} Blow upon Blow, under it Hard~~er~~, Harder, Hardest.
>
> You see I had your note of May 5. It makes me feel glad, for somehow, I had imagined you will, and as Patrick our sub-janitor said, when I told him that my son had lost his arm: 'That won't do at all, Sir, at all, at all', shaking his head and repeating: 'It won't do.'[17]

Lieber's uncompromising motto ("Blow upon Blow Hard, Harder, Hardest") had a homoerotic undercurrent. A new level of understanding had opened up that legitimated a rougher, less circumspect, and in this sense adolescent behavior. Lieber's reference was to Kaveh, the Ironsmith, who was canonized by the tenth century Persian poet Abolqasem Ferdowsi and was remembered to have fought a despotic ruler who had killed seventeen of his sons and fed their brains to serpents. Kaveh led a revolt against the palace and reinstated the rightful ruler Feraydun.[18] Lieber imagines himself to be the Kaveh of his own time, incensed by Oscar's treason, the loss of Hamilton's arm, and Norman's continued peril. The arm assumes additional meaning through the blacksmith and his prominent use of it. With Kaveh, a reference presupposing a significant educational depth, Lieber transposes his son's lost arm into the driving symbol for eliminating the enemy and, with reference to Kaveh, for establishing a new, an American, dynasty. The simple repetition and augmentation of the underlined motto adds to its violence.

Lieber ended his letter with this story:

---

16 Francis Lieber to ADB, May 1, 1862, box 31, Rhees Collection, HL.

17 Francis Lieber to ADB, May 6, 1862, box 31, Rhees Collection, HL.

18 Ahmad Karimi-Hakkak, "Revolutionary Posturing: Iranian Writers and the Iranian Revolution of 1979," *International Journal of Middle East Studies* 23, no. 4 (November 1991): 525 ff.

> Pay me back in kind for the following, which I have from a letter of a superior officer: The rebel general [...] at Pea Ridge [...] was shot by a German, who two days after became quite melancholy. He was comforted by everyone, and told that a soldier had not to grieve for ~~the~~ killing an enemy, general or not. Oh, said the man in the drollest German dialect, it is not that which makes me so sad, but the general had patent leather boots, and I was such an ass as to miss the only opportunity I shall ever have in my life of getting a pair of patent leather boots. It is that which I cannot comfort myself about.

While the German's comrades thought he felt sad at having killed a general, he really only cared about the general's boots. Comic relief was provided by the realization that a Confederate general's death was nothing to feel melancholic about. Where the German soldier's counterpart expects deference, even toward the enemy, the unreconstructed German immigrant comes to realize that the war opened up a rare opportunity for personal enrichment. Perhaps this perception of the war (not as a collective enterprise, but within more familiar coordinates of personal ambition and interests) was reassuring.

In his other letters, Lieber reiterated these themes. On May 12, for example, he wrote to Bache in celebration of recent Union victories and he included a drawing of a raised arm holding a hammer. As a reference to his motto, he decorated the logo by adding the letters "BUB" at the top and "I-I-I" at the bottom. "Why I write?" he asked. "For no earthly purpose but the cause is that I must talk to some one about our boys' successes."[19]

Lieber implied that there were not many people with whom he could talk about these matters, even among his family, perhaps out of deference toward his wife and his son Hamilton. They may not have appreciated the good news in the way Lieber expected Bache to, in an almost juvenile, boyish way. In their semi-paternal perception, the Union soldiers were Bache's and Lieber's "boys." But Lieber's letter suggests that he and his friend felt that to openly display their feelings and celebrate their "sons'" successes would have been inadequate.

19 Francis Lieber to ADB, May 12, 1862, box 31, Rhees Collection, HL.

Fig. 13. Alexander Dallas Bache, ca. 1863

(Courtesy of The Library Company of Philadelphia, (1)5750.F.24b)

"Ignorant of Scriptural Injunctions"

Bache must have shared some of these views, but what exactly prompted him to engage in this kind of correspondence? What made Bache overlook or tolerate Lieber's rougher side, i.e. those dimensions of his personality that caused fellow Lazzarone Oliver Wolcott Gibbs to complain that for a "man of talent" Lieber was "a perfect hog"?[20]

Few letters remain to shed light on Bache's attitude. There is a September 25, 1860 letter written in triangulation camp and an earlier one dated August 20, 1861, in which Bache made suggestions for the selection of army officers. It appears that some time in mid-May 1862, Bache visited Lieber at his home in New York. In a letter dated May 15, 1862, Bache

20 Oliver Wolcott Gibbs to ADB, March 25, 1860, box 21, Rhees Collection, HL.

responded to Lieber's May 6 story of the German Union soldier, and this letter provides important clues for Bache's perception of the war.

This is the background to Bache's letter: A few days before he sat down to write, the Chief of the Topographical Engineers with the Army of the Rappahannock, Lt. Col. John Navarre Macomb, had conveyed to Bache a story of a Union soldier mistaking a Coast Survey triangulation point for a boundary marker. Coast Survey officers had been at work in mapping the Northern shore of the Rappahannock River during General George B. McClellan's Peninsula Campaign in May 1862.[21] Having had difficulties in locating triangulation point "Scott," they learned how the problem had arisen: "{It} seems that when our forces first advanced upon the vicinity," Macomb wrote,

> one of our own men of Genl. Augur's command discovered the stone marking the triangulation point 'Scott' and seeing the initials upon it U. S. C. S at once judged it to be a monument to mark a point of the boundary between the United States and the Confederate States as claimed by the rebels! and in his patriotic ire he plucked it up and brought it into camp as a trophy; hence our difficulty in finding the exact point in the ploughed field once occupied in the progress of your great work; then pursued to advance the interests of commerce but since proved so eminently useful in War.[22]

It had not been a Confederate soldier who had encumbered the mapping of the Northern Rappahannock shore, but a Union soldier excited to have retrieved a war trophy! Bache now saw an opportunity to reciprocate for the patent leather story. I will take this opportunity to not only discuss Bache's particular version of the story, but to consider how the letter was phrased and what this implies for Bache's attitude toward Lieber, the nation, the war, and the role of an academy.

Bache's letter begins as follows:

Pr.
Washington

21 J. N. Macomb to ADB, May 2, 1862, roll 247, RG 23, National Archives. See also Albert E. Theberge, "The Coast Survey, 1807–1867," n.d., http://www.lib.noaa.gov/noaainfo/heritage/coastsurveyvol1/CONTENTS.html. For Macomb's record, see Heitman, *Historical Register*, 680.

22 J. N. Macomb to ADB, May 1862 [no exact date legible; according to the letter's location on the microfilm roll, it was written after May 7, thus reaching Bache no sooner than May 8 or 9], roll 247, RG 23, National Archives.

My dear Sir, May 15[23]

The letter has been written on small-sized stationary designed for private use. Compared to other letters from the period, Bache's handwriting is surprisingly easy to read. He marked his letter private, thus indicating that his letter was not to be shared by Lieber. "My dear Sir" is not formal but also not personal.

I think I have a match for the [patent] leather story.

Bache refers to Lieber's May 6 letter and the story of the German soldier who was "melancholic" about having let pass an opportunity to take the boots from a Confederate general who had just been killed. A "story," of course, contains narration and a dramatic plot rather than just information. It has entertaining and aesthetic qualities. When told among friends or colleagues, the presentation of a story presupposes common interests or perspectives. By declaring that he has a "match" for Lieber's anecdote, Bache turns the latter's invitation for an exchange of stories into a friendly storytelling competition.

The army of the Rappahannock desiring to base their reconnaissance upon the signals of the Coast Survey of the River, …

Bache sets the stage. The Army of the Rappahannock under Irvin McDowell in 1862 was part of the larger Union attempt to take Richmond, a main strategic goal during the first years of the war. As mentioned above, Coast Survey staff was involved in providing geographic intelligence and in charting maps for use by the armies. In this case, the terrain bordering on the Rappahannock River in Virginia was charted with the help of markers left behind by the Coast Survey in its work there since 1851. Bache's staff had taken care to mark points of triangulation. They now tried to identify them as a basis for their new mapping work.[24]

… two of our officers were sent to them + under charge of Col. McComb of the Topl. Engrs. proceeded to look up the stations used in the triangulations.

23 ADB to Francis Lieber, May 15, 1862, box 1, Lieber Papers, HL.

24 "Care was taken in all cases," Bache had written, "to mark the points used in a permanent manner for future reference." U.S. Coast Survey, *Annual Report of the Superintendent of the Coast Survey Showing the Progress of that Work during the Year Ending November, 1851* ([Washington, D.C.?]: Robert Armstrong, 1852), 50.

The Coast Survey was asked to aid in reconnaissance work by identifying the locations used for triangulation. The Coast Survey could not directly be ordered to conduct such work, we must infer, because it had remained a civilian organization. Its officers were assigned ranks but, when working for the army, they were under the command of the army's Topographical Engineers.

> Station "Scott" could not be found.

The anecdotal character of the story is now obvious. His colleagues could not identify the station shown on their Coast Survey map. Instead of referring to "one of the stations," Bache uses the station's name. It is unlikely that Lieber knew it but the precise reference adds literary depth and brevity. It does not matter where the station was. It matters, instead, that a particular station could not be identified. By keeping his paragraphs short, Bache consciously designs his story by building up narrative tension.

> A soldier in General Augur's advance, had found a stone in a ploughed field, marked U.S. C.S + took it to be the boundary mark by secession between the U.S. + the so called Confederate States!

Brigadier General Christopher Colon Augur's troops were operating close to the enemy lines.[25] As outlined above, the situation referred to by Bache was that a soldier found a marker indicating the position of station "Scott" in the Coast Survey's earlier triangulation, but mistook it for a marker by the seceding states.

If it were simply the soldier's ignorance he was trying to get at, Bache could have poked fun at the soldier in a different way. He could have arranged his sentence to conclude with the soldier's lack of common sense rather than his mistaking the triangulation marker for a boundary sign, even if the story would have been much flatter that way. The issue here is not the soldier, but the absurdity of his supposition: By seceding, southern states claimed that they were no longer part of the United States. They would therefore not call themselves "United States Confederate States" nor would they mark their territory by including, on a boundary sign, both the name of their own "nation" and that of the country from which they had seceded. The soldier completely misunderstood the marker, but the issue was not limited to ignorance or unfamiliarity with marking conven-

25 On Augur's military service, see Heitman, *Historical Register*, 175.

tions. The soldier's particular mistake must have struck Bache as ironic for another reason.

The exclamation mark indicates that this paragraph delivers Bache's punch line. One way in which it qualifies as such is to suppose that Bache considered his Coast Survey to represent the exact opposite of what the soldier took the marker to stand for. We have seen that Bache's career had been premised on the idea of developing the nation's culture and its ability to partake in a universal rational discourse that science represented particularly well. During the preceding twenty years, he had developed the Coast Survey into America's largest scientific institution. The organization stood for common national goals, for a development of American culture through an improved understanding of the environment in which the country was situating itself. The Coast Survey's size and dimension (when compared to other federal institutions at the time) indicates that this was not just any project, but central to the country's self-conception. For Bache, there was reason to feel that the Coast Survey represented larger cultural ambitions that naturally transcended the interests of any particular American region and its political interests.

So why did Bache feel that this anecdote was a match for Lieber's patent leather story? Both the patent leather boots and the Coast Survey marker are trophies, but they represent different assessments of the war. While Lieber takes a reckless Northern perspective by suggesting that Confederate boots were booty, Bache insists on a comprehensive, an American, perspective that includes both the North and the South. His criticism is directed at the North for abandoning a national perspective that includes the secessionists, a perspective to which he had dedicated his life. While Lieber finds it ironic that the German soldier would not conceive of Confederate soldiers as enemies to be looted, Bache finds it ironic that anyone could ignore the national dimension of American culture of which the U.S. Coast Survey was the spearhead. Even more pertinent: He finds it amusing that someone would even consider its work to be part of secessionist efforts to break the country apart! While the boots, in Lieber's story, are a token of Southern secession, therefore, the Coast Survey mark, in Bache's perception, was a token of the American nation, of both the North and the South. In keeping with this perspective, the superintendent ridicules the North rather than the secessionists. Bache pokes fun at the soldier's ignorance of the Coast Survey and its work, of the national dimension and ambition that the organization represents in America.

This reading conforms to a small but meaningful detail in Bache's letter: The superintendent takes particular care in drawing the letters "U.S. C.S", much like Lieber had repeatedly been drawing, in his correspondence, mottos such as "Blow upon Blow" and "Hard, Harder, Hardest." (These mottoes had then evolved into a sketch of an arm, above it the letters "BuB" and under it "I-I-I."[26]) Bache's abbreviation represented a motto also, but it referred to the Confederate and to the Union states, in the soldier's perception, and to an overarching national organization, in his own.

Full of ire + unmindful of scriptural injunctions, + he took up the stone, + brought it as a trophy into the Union camp!

Bache continues to poke fun at the Union soldier. The superintendent could have assumed a different perspective. He could have appreciated the soldier's intent. Bache also does not care about the counterproductive results of the soldier's removal of the Coast Survey marker. He is not concerned with the practical dimensions of the war effort but with the ridiculous implications of the soldier's efforts. Bache finds it amusing that the soldier, eager to support the Northern cause, conceives of a token of American nationality as a token of fraternal hostility! Bache mocks the soldier's incapacity to see the marker's significance and his implicit acknowledgment of the Secessionists as "enemies," and this reading is further solidified by the superintendent's reference to "scriptural injunctions." Bache is known for his witty style. But even if we acknowledge that Bache frequently engaged in wordplay, this particular choice of words calls for an explanation. It falls in line with the observation that by referring to the Coast Survey acronym on the marker as "scriptural injunctions," Bache tacitly bestows on the organization a semi-religious authority.

By making fun of the soldier's neglect of the cultural authority represented by the Coast Survey, Bache's perspective transcends concrete political organizations. One could argue that the Coast Survey was a federal institution and that the marker represented that state's authority, so that Bache was ultimately annoyed by the soldier's disregard of American institutions. But the focus here is different. The soldier angrily removed the marker because he thought it would be in the interest of the United States. It is the implicit ignorance of cultural meanings and their authority in general that Bache finds so bitterly comical. In his view, the soldier's limited

26 See, for example, Francis Lieber to ADB, May 12, 1862, box 31, Rhees Collection, HL.

cultural horizon nullified any benefit derived from his eagerness to help win the war.

This suggests that for Bache, the war was a disaster because he could not understand how anyone in the South or North could be willing to waste time and resources in a conflict bound to put on hold the country's ambitious and competitive race in all areas of cultural investigation and development. At the same time, however, Bache's war engagement shows that he knew that the only way to get the country back on track was to help end the Civil War, and the only way to do this was to help bring about an early Union victory. The implication here is that Bache had a strong and deep-rooted conviction of the country's developmental purpose and trajectory, a perspective that was ahead of its time in the sense that it took for granted a unified and coherent body politic to carry it out. From this avant-garde position, Bache turned back reluctantly to help his countrymen follow his trail. The war was a deep disappointment of his cultural ambition and leadership. It brought into full view the absence of a cultural and political coherence and peace for which Bache and his colleagues had worked.

> Perhaps under the direction of the Aruspex it may have been cut with razors into little trophies, as the Merrimac is split into splinters!

Of Etruscan origin, Aruspices (or Haruspix, pl. Haruspices) in the Roman Empire were "entrail observers […] whose art consisted primarily in deducing the will of the gods from the appearance presented by the entrails of the sacrificed animal." They also interpreted a range of natural phenomena. This class of diviners remained outside of established religion in Roman antiquity.[27] How does Bache come to make this connection? The singular in "the Aruspex" indicates that Bache referred to a particular individual. He is playing with words: The soldier was part of General Augur's advance, and Bache transposes the general's name (taken as a noun) into an ancient context. Bache was taking a literary and playful view, engaging with Lieber in their friendly story-telling contest.

The gist of Bache's remark here, of course, is that of generalizing the soldier's ignorance. The Coast Survey marker could be carved up and become a trophy only if the soldier's attitude was widely shared among his comrades. Against the reality of the Civil War and the Union war effort that he implicitly characterizes, Bache assumes an ambitious and self-

27 *Encyclopedia Britannica, 2005 Ultimate Reference Suite DVD*, s.v. "The Etruscans."

assured view of the country. He looks at both the Union army and the secessionists as tokens of a pagan culture akin to Etruscans in the Roman Empire. One could have argued that if even the Union army must rely on soldiers mistaking the Coast Survey sign as a border marker, the prospects for American culture were bleak indeed. But Bache upholds a different view of American culture, a view that allows him to assume this critical perspective. Bache's ability to successfully support the Union war effort indicates, of course, that he was not delusional.[28] He was a visionary: Against the reality of his time, he anticipated a sophisticated and ambitious American culture transcending political fault lines.

There is further implicit evidence that Bache assumed such a broadly ambitious view in that he connects the Coast Survey marker with Confederate engineering success. He embraces the intellectual achievements and distances himself from the common (Northern) American citizen's spontaneous, uninhibited, and uncultivated response to war. The *U.S.S. Merrimac* (or Merrimack) was burned and left behind, at the beginning of the war, by the United States navy in Norfolk Navy Yard, then raised and rebuilt as an ironclad ram by the Confederate navy. The Merrimac caused significant damage in a sortie on March 8, 1862 and the next day fought the U.S.S. Monitor, an ironclad vessel recently introduced by the Union navy, to a draw. The Merrimac's success (i.e. its ability to stand up to the Monitor) caused alarm in the North. Because the vessel was about to lose its naval base on the James River, it was then destroyed and abandoned by the Confederate navy on May 11, two weeks prior to Bache's letter. Bache compared the vessel's fate to that of the Coast Survey marker even though the Merrimac was a Confederate (and not a Union) engineering success. He imagines both being carved up and taken apart as trophies. Bache was concerned with research and engineering as part of a cultural development in the entire country and across political divisions. According to this perception, the war was an impediment and little more than a cause for regression. Bache cannot identify with strongly partisan pro-Union feelings because his allegiance is with the United States. Quite naturally for him, this includes the South.

28 Bache supported the union war effort in various ways: His letter shows how the Coast Survey sought to provide the army with accurate maps. Bache was vice-president of the Sanitary Commission, a precursor to the American Red Cross. He was also involved in designing and implementing defense works for the city of Philadelphia. Slotten, *Patronage, Practice, and the Culture of American Science*, 174.

> Is it not rather heathenish to carry Aruspices with a x'n army?

Bache likens the role of the Union army to that of the crusading Christian ("x'n") armies that brought destruction to the East while pursuing questionable religious goals. The Union army's soldiers are eager to destroy the enemy, indignant at Southern secession, but their engagement results in death and devastation. To Bache, the irony of the situation is represented by the image of Union soldiers carrying pieces of a marker by their very own U.S. Coast Survey. What is more, they assume these pieces to promise good luck and speedy victory because they are viewed as tokens of the enemy. To call such behavior "heathenish" and to contrast it with the Christian ideals of their cause implies that these soldiers have forgotten their own moral standards. In Bache's perception, the pieces are in truth part of these soldiers' own culture and its achievements—that of the United States of America and her Coast Survey. The Union, in other words, in its eager efforts to subdue the South, was in danger of abandoning its belief in a coherent American culture.

What is remarkable here is that Bache did not hold the South alone responsible for the way things were going, but the entire country. In assuming this perspective, Bache was not merely a spectator and much more than a civil servant. Bache's natural self-confidence and self-reliant moral and cultural standards prompted him to take an integrated and ambitious national perspective that was ahead of (and impatient with) contemporary political reality.

> Thanks for the patriotic song. I remember well how it affected me in the dark days! When treason spread over the land!

Bache refers to Lieber's "A Song to Our Country and Her Flag" written in 1861.[29] Bache acknowledges the song somewhat guardedly. He mentions that it affected him but leaves open what he thought of it. The "dark days," we must infer, are those of the previous year, when the South chose to secede. Even though Bache intuitively took a holistic view and considered the entire country (and not just one region) to be responsible for failing to keep the peace, he also participated in contemporary Unionist views of Southern secession.

---

29 See Lieber's letter to Bache, May 14, 1862, box 31, Rhees Collection, HL. *A song on our country and her flag by Francis Lieber. Written in 1861, after the raising of the flag on Columbia College, New York* (New York: Baker & Godwin, printers, [1861]), http://hdl.loc.gov/loc.rbc/rbpe.12303400.

Bache moves on to a different topic, indicating a change of subjects by inserting a centered line:

———

What do you think of this. I have a letter from a [secess] lady appealing to me to save her property. Her husband is a double distilled traitor + her appeal is founded on the fact which doubly condemns him, that he was once in the employ of the Coast Survey! Heaven save the mark!!

B
u *h. H.* H.
B

Yours A.D.B.

Given the woman's predicament, it was not a bad idea to contact Alexander Dallas Bache. Her property was about to be confiscated and she was looking for ways to save it.[30] Superintendent Bache had excellent connections in the federal government and would probably feel obliged to aid a former employee of the Coast Survey. But Bache did not intend to comply with her request. He calls the appealer's husband a "traitor" for supporting the South; perhaps his former employee even occupied a position of leadership ("double distilled").

For our purposes, however, the decisive matter is that Bache then goes on to suggest that the husband was to be denounced *because* he had formerly worked for the U.S. Coast Survey. Why would this "doubly condemn" him? Bache assumes that those who have worked for the Coast Survey have an even stronger obligation to uphold the country's integrity than those who have not. This implies that the Coast Survey was not merely a service agency for the federal government but an organization that represented and instilled in its staff a particular sense of national allegiance and civil responsibility. In this perspective, the Coast Survey prefigured the United States as a nation-state because it heralded a future, national perspective of the country, and this was why, in Bache's view, the former Coast Survey employee's behavior was particularly disappointing and treacherous.

Looking back on Bache's letter, we can now see that the superintendent presented two stories that both point in the same direction. Bache had started out by taking a condescending view of the Northern response to secession, and in the second part of his letter, he matched this with a story

30 Probably with reference to the Confiscation Act of 1861, United States, *Statutes at Large, Treaties, and Proclamations of the United States of America*, vol. 12 (Boston, 1863), p. 319.

characterizing Southern villainy. The two stories are connected by the underlying assumption that the U.S. Coast Survey stood for the nation's integrated future. Hence, it makes perfect sense that Bache summarizes and connects the two stories with the exclamation "Heaven save the mark!" The Coast Survey sign that the Union soldier had mistaken for a Confederate boundary mark symbolizes to Bache, not the progress of science or the particular success of his organization's work, but the United States of America as such.

Because Bache was the leader of an organization that he took to be the advance embodiment of the nation, furthermore, we now have additional and very strong evidence that he intuitively considered himself to be among the nation's cultural leaders, if not *the* cultural leader. This was not an explicit claim or assertion but an unconscious subjective standard and reality. Within the logic of this personal mythology that structured his perception of everyday life, Bache lived up to the ambitions of his great-grandfather Benjamin Franklin.

By adding "B u B" and "*h. H.* H." at the end of his letter, Bache subscribes to Lieber's mottoes "Blow upon Blow" and "Hard, Harder, the Hardest." Bache had not used Lieber's symbols in previous letters. It seems as though writing about the ignorant soldier and his former employee's pleading wife had loosened him up for a more adolescent (and somewhat regressive) posture. That these two stories would have this effect highlights the relevance of their implications. In his correspondence with Lieber, Bache perceived an opportunity for opening up in a way he could not with his professional peers. Here he had the room to more directly voice his convictions about national culture and development, the backdrop and setting for the development of science in America. While his colleagues were not ignorant of politics, of course, we must assume, on the basis of this letter, that Bache appreciated the opportunity to converse with someone who shared the totality of his perspective, the national dimension of things, and romantic ideals of driving culture forward. Apparently, he felt as though Lieber, perhaps better than others in Bache's, could grasp his marker-story's bitter irony and the broad cultural and historical perspective that prompted Bache to react to it.[31]

---

31 For the case of Lieber, see Merle Curti, "Francis Lieber and Nationalism," *The Huntington Library Quarterly* 4, no. 3 (April 1941): 263–92; Adam I. P. Smith, *No Party Now: Politics in the Civil War North* (New York: Oxford University Press, 2006), particularly 25–48, 67–91.

Bache ended his letter to Lieber with the following postscript:

> I do not think that Genl. T. has any poetry but if you choose I will try him "with the author's compliments." That may touch.
>
> He has not noticed my note in behalf of Hamilton. The least a big man can do for a little one is to answer.

Lieber had asked Bache to convey a note concerning a nomination of his son Hamilton through Adjutant General Lorenzo Thomas to Secretary of War Edwin M. Stanton.[32] It is in his postscript that Bache turns to practical matters. As we have seen, he had reserved the main section of his letter for his two stories. This priority given to an aesthetic and impractical mode is reflected in his concluding comment.

Bache picks up Lieber's comments about military leaders who were ignoring his request.[33] He considers Adjutant General Thomas to be of limited use in conveying Lieber's case to Stanton. By "poetry," Bache refers to the ability to make a good case but he also contrasts Thomas' motivation with Lieber's and his own. Whereas the General merely fulfills his duty, Bache and his friend are moved by the war's meaning and consequences: Bache, we have to assume on the basis of this letter, with respect to the war's impact on an emerging American culture, and Lieber with respect to his sons. In this sense, both men experience the war romantically, and their problem in the given situation is how to "touch," i.e. how to convey effectively the relevance of their experience to those in charge of conducting the war. To Bache, a large workload is no excuse. He feels that the political and military leadership, or at least Thomas and Stanton, did not share his own emotional perspective.

In his last paragraph, Bache refers to a note concerning Hamilton that he had written himself. He contrasts the general's role with his own, further highlighting the distinction between a romantic and a less engaged perception of the war. By suggesting that the general was a "big man," Bache also contrasts aristocratic sophistication with the general's worldly

32 Adjutant General Lorenzo Thomas had promised to see Hamilton's nomination through. Lieber had asked Bache to convey a letter to Secretary of War Edwin M. Stanton through Thomas and it is likely that this is "Genl. T." Francis Lieber to ADB, May 9, 1862, box 31, Rhees Collection, HL. Concerning Thomas' service, see Heitman, *Historical Register*, 38, 954.

33 With respect to Adjutant General Thomas, Lieber had written: "I know they have to attend to high and great business." Francis Lieber to ADB, May 9, 1862, box 31, Rhees Collection, HL.

relevance. In the given context, science and the arts (the "poetry" against which Bache measures the general) are in the same camp as Hamilton, the romantic soldier. Bache sets these dimensions apart from federal officialdom. None of them can be measured in terms of usefulness. Despite their intellectual superiority, they are at a practical disadvantage.

This is a crucial point, for it puts the foregoing into perspective. Bache was perfectly aware that science and the arts remained without immediate political influence or consequence. Lieber's and his was an "unpractical" business. Despite Bache's efforts such as his work on the Coast Survey, it was of little relevance when compared to that of the military. And yet Bache, the scientist and science administrator, considered himself to be the visionary of a coherent American culture, a visionary so captivated by the idea that (at least in his correspondence with Francis Lieber) he disdained those who were destroying it. This perspective made him both a cultural romantic and an organizational realist.

More on the Bache-Lieber Correspondence

How does Bache's letter connect to Lieber's patent leather story to which he was responding? The German-born soldier who was sorry not to have taken the dead general's boots was introduced by Lieber as a token of undue respect for the Confederate dead. Bache stuck to the idea of reporting a situation that stood for something larger, but instead of satirizing the South, he pointed his finger at the entire country. An assessment of perhaps unintentional elements in his letter (similar to Sigmund Freud's *Abhub der Erscheinungswelt*) showed that Bache viewed the crisis from a vantage point of a coherent American culture and its future, a development retarded and threatened by the war. In Lieber's patent leather story, the soldier had intuitively shared this view because he had not taken the general's boots. This was in tune with Bache's perspective that must have prompted the superintendent to perceive of the deceased Confederate general as an American citizen, not an unworthy enemy, and to assume that to take the boots was to loot a corpse, not trophy hunting. By leaving the corpse alone the soldier had initially reacted with decency, and his comrades had done the same thing after his return to camp. They assumed that the German soldier had a troubled conscience, which presupposes that it could appear to be wrong to kill a Confederate soldier. Bache's attitude corresponded to this spontaneous reaction. But Lieber's story, which played on the idea that

Confederates were traitors and enemies and that their boots could therefore be taken, indicates that the public's perception of the war was changing. Bache shared this perspective as he was unwilling to help the wife of a "double-distilled traitor," but he chose to begin his letter to Lieber with a story in which the joke was on the Union troops, not on Confederate traitors. Bache was well aware of the public discourse. Intuitively, however, he retained a comprehensive American perspective for he could otherwise not have told his lost-marker story in the way he did. Even if Bache speaks of "traitors," therefore, his perspective remained more integrative than Lieber's.

Why, then, did Bache engage with Lieber at all? Given his outlook and refinement, how could he underwrite Lieber's cruder mottoes? Their discrepancy is put into perspective by the above observation, i.e. that Bache, too, was part of the prominent political discourse, and that he shared Lieber's perspective of the "traitors." His letter shows, however, that he must have felt uncomfortable with this because it went against his innermost beliefs and everything he stood for. In the story of the lost marker and the stupidity of the Union soldier, Bache, after some reflection, saw an opportunity to put those impressions to paper. But he participated because he understood that the only way to put the country back on track was to support the Union and to help lead it to victory. In this respect, he was on the same page as Lieber, the immigrant, even if the latter assumed a more aggressive stance.

But this only shows that the two men concurred in a particular way. Many others took similar political views, and the fact that Lieber was staunchly pro-Union does not explain why Bache chose to develop and maintain this friendship in the context of the Civil War. Something else must have attracted Bache. When compared to Bache's other contemporary letters, it is the regressive, semi-adolescent nature of their camaraderie that stands out from his correspondence with the former fighter for the "freedom of Greece," a camaraderie that went along with a shared sense of responsibility for younger men including Lieber's sons. Perhaps the most obvious and significant parallel among the two was the tormenting schism the war had brought to each of them in different ways. Lieber had three sons, two of them fighting on opposite sides in the Peninsula campaign; Bache had no children, but the country to whose cultural advancement he had dedicated his life was being set back and destroyed in a fratricidal conflict. This situation was perhaps more dramatic for Lieber, if any such impossible comparison should be attempted at all. But Lieber was an immigrant and did not look back on a family history deeply rooted in the country's foundations. Lieber did not need to live up to a biographical standard of advancing the nation intellectually and culturally, even if he was in fact trying to do so. Such a standard, however, was relevant for Bache whose career had been shaped under very different circumstances. It is impossible to compare Lieber's pain at seeing his sons fight each other and Bache's torment at seeing the country tear itself apart. But Bache's letter suggests that given his identification with the country's future, he must have suffered severely from seeing its existence jeopardized. It is against the background of Bache's attitude towards the war, an attitude implicit in his letter to Lieber, that the impetus for the founding of the National Academy of Sciences begins to come into focus.

The Founding of the National Academy of Sciences

Why did Bache and his closest colleagues pursue their long-standing academy idea during the Civil War? The hypothesis that the creation of the academy had something to do with the war is corroborated by the timing: The winter of 1862 was a period of severe trial for the North and brought about a "crisis of confidence" there, and this is when the academy idea was endorsed by Alexander Dallas Bache.

It was a time of crisis for the North because of setbacks in the field. After disappointing initial defeats in a war that many expected to be brief, George B. McClellan was appointed commander of the Army of the Potomac in 1861 to reinstall discipline and panache among the troops. But McClellan remained hesitant to move from training his armies to engaging them. His caution was a response to exaggerated and misleading intelligence about the number of Confederate soldiers facing him. Instead of looking for action, he continued to drill his troops within eyesight of the Coast Survey offices on New Jersey Avenue in Washington. McClellan finally did move in May 1862, but the Peninsula Campaign (an attempt to approach Richmond from the east) ended in defeat. The retreat by Union troops and the Seven Days Battle (June 25–July 1, 1862) marked its end.[34] Other disasters followed. After the defeat of the Second Battle of Bull Run

34 For a general overview, see James M. McPherson, *Battle Cry of Freedom: The Civil War Era* (New York: Oxford Univ. Press, 1988).

(August 29–30, 1862), the battle of Antietam (in September) can be considered a turning point even if McClellan's troops were overturned. President Abraham Lincoln used it as an opportunity to announce that from January 1863, all slaves in the Confederate states would be "forever free." His Preliminary Emancipation Proclamation was conservative in that it was a military measure but it turned the war for the Union into a war for the abolition of slavery. Britain now abandoned the idea of recognizing the proslavery South. The military situation in the East remained difficult, however, for General Ambrose Burnside (who succeeded McClellan in November) failed to bring home a decisive victory at Fredericksburg on December 13, 1862. Union troops had suffered 12,600 casualties, the Confederate armies fewer than 5,000, and no headway had been made. While Antietam could be interpreted as a victory, Fredericksburg was certainly a setback. In December 1862, despite a slightly more promising situation in the West, the Union war effort was in crisis.[35] Public morale was at perhaps its lowest since the inception of war.

There is little immediate evidence to show that Bache and his colleagues, by beginning to discuss the academy idea at about this time, responded to the war situation. One may argue that this timing is best explained by Charles Henry Davis' return to Washington in November 1862 because he rejuvenated the idea. After commanding gunboat operations on the Mississippi, he was appointed head of the Bureau of Navigation (which included its Naval Observatory, a post he had long coveted).[36] In line with Bache's views of the situation in the North, Davis, in letters to his wife, had complained about how the war was perceived there.[37] But what had changed to make Bache and his colleagues feel entitled to pursue Davis' idea at the particular moment they chose to do so, in late 1862 and early 1863? On the basis of the above interpretation of Bache's letter, we have a motive: In the context of sagging public spirits, Bache and his colleagues chose to bolster the threatened nation by going ahead with their old acad-

35 The phrase is from James M. McPherson, *Ordeal by Fire*, vol. 2, *The Civil War* (New York: Knopf, 1982), 304.

36 Rexmond Cochrane, *The National Academy of Sciences: The First Hundred Years, 1863–1963* (Washington: The Academy, 1978), 50. Bruce, *Launching of Modern American Science*, 302 f. Bruce leaves open whether Davis or Bache first came up with the idea of pursuing the plan.

37 See, for example, Davis to his wife, April 12, 1863, quoted in: Charles H. Davis Jr., *Life of Charles Henry Davis, Rear Admiral* (Boston and New York: Houghton Mifflin, 1899), 292 f.

emy plan. An academy was certainly a fulfillment of long-standing aspirations but Bache's letter to Lieber has shown that an academy would not be an institution implemented by the nation to certify and stabilize the science profession's unique capabilities. In Bache's perception, the United States was too weak to bestow such honors. It was the scientific profession, represented by himself and his colleagues, which felt a need to reinforce the nation. Bache intended the Academy to become a symbol of national optimism and consolidation. It seems natural that Bache and his colleagues, acknowledged leaders of their country's scientific community, would envisage a national academy. But when the National Academy of Sciences was founded in 1863, it was intended as a tool to support and to help consolidate the nation. As the nation seemed on the brink, the scientists, in a gesture of romantic nationalism, came to the fore and sought symbolically to assume cultural leadership. In this sense, the Academy stood for the profession's support of the American nation-state rather than for the nation's support of science as a profession.

This interpretation implies that the academy was considered to be of sufficient meaning and cultural weight to represent national leadership, and it implied a particular attitude towards politics. Politics and the military, and the politicized army in particular, had been unable to stem the tide and to undo secession. Bache and his comrades, of course, did not seek to take on the problem of how to achieve victory in the field. A national academy would not raise any troops. Instead, an academy was to represent America's scientific ambitions and its future relevance among nations. If the nation committed itself to such an ambitious project of developing science, it followed that there was something to fight for. The founding was romantic in the sense that it sought to put down a "mark," a cultural rallying point and a natural extension of the role Bache considered the U.S. Coast Survey to have. This is why the founding of a national academy could be perceived to help alleviate the loss of motivational momentum in the North and among all those ready to commit themselves to the Union and the nation.

If such leadership was required, it follows that the political elite could not be relied upon to provide it. (Bache's comments on the Union soldier serve to illustrate this point.) The founding of a national academy was a matter to be acted on by Congress, and political elites would have to be involved. Because Bache and his colleagues had doubts about the nation's self-assuredness that the founding of an academy was to help alleviate, they were quite naturally determined that the academy would not become a new target. The discussion about the federal government's role in the context of James Smithson's bequest in the late 1840s had prompted some to voice concerns about the government's legitimacy to implement national institutions. If the founding of an academy was intended to lessen doubts about the nation's existence and future, why let such doubts endanger the project? The idea would have backfired and corroborated the problem it sought to address. By this logic, it made little sense to seek a public discussion of this matter. The group chose not to rely on the government's good will except where it was absolutely necessary. Once Davis had first brought up the old academy idea in late 1862, close colleagues were informed and invited.[38] Bache mentioned a possible Lazzaroni dinner in a letter to Benjamin Peirce on December 9, 1862.[39] In early January, Peirce received word that Bache and Davis had been "talking Academy of Sciences." Bache added: "How wonderful it is that just when countries are in the midst of most troublous times they get up the greatest things. Excited minds view de haut en bas. Mum is the word."[40] The strategy was to keep quiet about the matter. In this quote, Bache refers to the nation as the agent for this historic project, and on the surface, this contradicts our interpretation and Bache's appeal for secrecy. One could argue that if the nation was indeed the agent, why avoid public discussion? Bache's comment makes sense if we assume that he and his immediate colleagues *were* the future idea of the nation. It encapsulates the motivational constellation suggested by the timing of the founding process.

---

38 For accounts of the founding of the Academy, see Cochrane, *The National Academy of Sciences*, 43–78, and Bruce, *Launching of Modern American Science*, 301–05. On the basis of my interpretation of Bache's letter, I am suggesting a slightly different motivation for the founding the Academy than Bruce, at least with respect to Alexander Dallas Bache. With reference to the exclusion of important scientists from the list of the initial fifty members (see below), Bruce speaks of a "shame of its {the Academy's} founding" (ibid., 305). My interpretation also differs from Miller's suggestion that the Lazzaroni sought to maintain "the highest scientific standards for patriotic purposes." While I find this assessment to be correct, it does not include other important aspects of their motivation that I am discussing here. Lillian B. Miller, *The Lazzaroni: Science and Scientists in Mid-Nineteenth-Century America* (Washington: Published for the National Portrait Gallery, Smithsonian Institution, 1972), 5.

39 ADB to Benjamin Peirce, December 9, 1862, BPP.

40 ADB to Benjamin Peirce, January 7, 1863, microfilm N.R. II, roll G, Reingold Papers, SIA (copy of microfilm 63-3028, BPP).

On February 6, 1863, Harvard biologist Louis Agassiz wrote to Bache that he had been thinking "over the improvements needed in our university and also our old Academy of Sciences plans." He added: "I here once had an opportunity of speaking of it to {Senator Henry} Wilson {of Massachusetts} and I am satisfied the Cobbler of Nattick {i.e. Wilson} will carry the measure sooner & with more effect than the Graduate of Harvard."[41] At this point, an academy was on the group's agenda and Agassiz, instead of opening up public debate, was looking for a political instrument. It was the scientists who had taken the initiative, not senator Henry Wilson who was considered to ensure safe passage in Congress. Agassiz inferred that a senator without an educated background would more harmoniously resonate with anti-elitist or populist undercurrents in American politics. Agassiz was committed. On February 11, he wrote to Bache that he would soon travel to Washington because (with Senator Wilson's help) he was to assume the post of a Regent of the Smithsonian Institution. The Swiss-born Agassiz wrote that he had gone through the necessary steps of acquiring American citizenship in order to prevent others from doubting his loyalty.[42]

The idea for an academy was loosely associated with plans to create a "permanent commission" with the Navy Department, a committee of scientists and engineers who would evaluate inventions proposed to the navy by civilians. Both ideas were pursued and the permanent commission was instituted with little fanfare, but, in January and February 1863, the group around Bache differed as to the achievability of the academy plan. Joseph Henry later wrote that a meeting took place at the end of January with himself, Bache, and Davis present. Both a permanent commission and a national academy were discussed. Henry had several objections concerning a national academy. He did not think that it could pass with a free discussion in the House, that it would cause jealousy among scientists if it were passed, that it would be difficult to raise the funds necessary to run the academy, and that it would be in danger of becoming the victim of partisan politics. In January and February 1863, he apparently did not ex-

41 Louis Agassiz to ADB, February 6, 1863, box 34, Rhees Collection, HL. The "Graduate of Harvard" was perhaps Charles Sumner, a fellow member of the Saturday Club in Cambridge.

42 Louis Agassiz to ADB, February 11, 1863, Rhees Collection, HL. There are two letters from Agassiz to Bache that day, and Agassiz' American citizenship is mentioned in the second one.

pect Bache to pursue the idea any further and was later surprised to learn that such a bill had passed Congress and was signed into law.[43]

Fig. 15. Albert Herter's 1924 painting depicts President Abraham Lincoln signing the charter of the National Academy of Sciences on March 3, 1863. Left to right: Benjamin Peirce, Alexander Dallas Bache, Joseph Henry, Louis Agassiz, President Abraham Lincoln, Senator Henry Wilson, Admiral Charles Henry Davis, and Benjamin Apthorp Gould.

(Courtesy of the National Academy of Sciences)

On February 19, Agassiz, Peirce, Benjamin Apthorp Gould, and Senator Wilson met at Bache's house in Washington to discuss the plan's details.[44] Henry was probably not invited because his doubts seemed strong enough to jeopardize the project. In Congress, Wilson had indicated his intention to introduce a bill on the subject on February 20, and he did so the next day, Saturday, February 21. When the dinner took place that evening (with Henry present but still unaware), the law had not yet passed. It would do so on March 3, the last day the Thirty-seventh Congress was in session. It was then passed by the House without discussion and signed into law by Lincoln before midnight on the same day.

The law creating the Academy consisted of six lines of text. It also included a list of fifty persons who were to be its members. This closed the

43 Joseph Henry to Louis Agassiz, August 13, 1864, Henry, *Papers of Joseph Henry*, 10:395. See also works referred to in nn. 4, 6 above.

44 The following account follows Dupree, *Science in the Federal Government*, 138 ff.

option of making decisions of membership before a current member had passed away. During the February 19 meeting, there had been two opposing views concerning the selection of academy members. Davis had argued that twenty members should be chosen initially and written into the law, and that these members then proceed to select additional members from their respective fields.[45] In the end, the organization could only fill vacancies as they came up, not add additional members to its roster. What had made Bache, Davis, Gould, and Peirce feel as though they were legitimized to decide who was to be invited to be a member? What would they have lost by pursuing earlier ideas of selecting a handful first, and to let that group later decide whom to invite?

Once the decision had been made to push for an academy, Bache and his group had already assumed the authority of including at least themselves. Bache was set on creating a symbol of national cultural unity and purpose. To follow Davis' proposal would have been a sign of deference toward professional peers. This may have been considered a benefit, but it also would have begged the question of why the plan for an Academy had not been openly discussed within the profession. The AAAS, for example, may have provided a platform for such a discussion. This would undoubtedly have been noticed by the wider public, however, something Bache and his colleagues were seeking to avoid. Even in case they had chosen a small number of scientists who were to select additional members once the Academy was organized, the symbolic effect would have been weakened. The Academy, instead of representing the profession's united support of the nation, would have been a token of the profession's internal quarrels. Only Bache, because of his standing within the profession, could argue against Davis' proposal and for the more comprehensive and courageous plan. Gould was too young; Davis did not have the professional cache; and Peirce, though a former AAAS president, lacked Bache's broad view from the nation's capital and its institutions. So the bill passed as drafted by the four scientists and new members could be co-opted only when a member had passed away.

45 This was to avoid "the odium of exclusion." See Charles Henry Davis to his wife, March 7, 1863, Davis, *Life of Charles Henry Davis*, 292. Agassiz had apparently had a similar idea in his 1858 letter to Frazer. See also Cochrane, *The National Academy of Sciences*, 46; Bruce, *Launching of Modern American Science*, 301.

Chapter 11

Conclusion

A New Paradigm for Writing the History of Nineteenth-Century American Science as a Profession

"What was the significance of the founding of the National Academy of Sciences?" Nathan Reingold asked in 1966, and answered that "the most important aspect of the founding is that nothing happened." Skepticism about the rationale for founding the National Academy has dominated in historical writing on the subject and this comprises the institution's role for the postwar period.[1] Reingold's widely read documentary history of American science in the nineteenth century includes letters by Louis Agassiz, Alexander Dallas Bache, and other Academy instigators, but Reingold limits his selection to letters written *after* the Academy had been signed into law in March 1863.[2] His intention is to provide readers with an impression of the contemporary (and critical) reception of the new institution within the scientific profession, and this perspective has dominated critical historical writing since.

Events leading up to Congressional action in March 1863 were not unknown to Reingold, of course, but historians have viewed efforts by Bache and his cohort prior to the Civil War mainly in the context of the scientific profession's institutional development.[3] The founding of the National Academy was understood to be in line with that group's ambitions for

1 Nathan Reingold, *Science in Nineteenth-Century America: A Documentary History* (London: Macmillan, 1966), 202.

2 The first letter included by Reingold on this subject is Agassiz' March 6, 1863 epistle to Bache, confirming the Academy's existence. Louis Agassiz to ADB, March 6, 1863 [original in the Rhees Collection, HL], quoted in: ibid., 203.

3 In other sections of his book, Reingold includes documents to illustrate the outlook of the Lazzaroni. Ibid., 127–61.

leadership and expansion of influence.[4] Reingold's perception of such "dreams of power" was part of a broader ambition, now somewhat muted, of explaining the role of professions in American history with reference to older theories of the profession.[5] In reference to such documents as Alexander Dallas Bache's 1851 speech as outgoing president of the American Association for the Advancement of Science (AAAS), Bache and his peers were considered to seek for science "support without strings" and a permanent place for science in the United States. Robert V. Bruce, for example, considered crude and high-handed the selection of only fifty members by the Academy's instigators, inviting minor figures, and ignoring accomplished scientists.[6]

Explanations of the emergence of science as a distinct profession in the United States did not ignore the political and social context in which American scientific institutions were being developed. As we have seen, it was apparent to Bache and to historians who have read his speeches that the conditions for science in America deviated significantly from those in European countries. Bruce has pointed out that while scientists strove to implement national organizations, they knew that "the logic of centralized authority would grate against the ideology of individual freedom."[7] A. Hunter Dupree, in his early study of *Science in the Federal Government* emphasized the particular conditions provided by the American state.[8] But outside of observations concerning the lack of a coherent infrastructure, the vagaries of party politics, or the shifting opportunities of war, the political foundation of the American national state was usually taken for granted. In

4 Bruce suggests that the Lazzaroni fought for "paternalistic authoritarianism." Robert V. Bruce, *The Launching of Modern American Science, 1846–1876* (New York: Knopf, 1987), 263.

5 My brief sketch of different types of theories of the professions is in chap. 1 above.

6 "Support without Strings" is the title of a chapter in which Bruce describes the Lazzaroni rationale for leadership. He also discusses the intra-professional conflict about the National Academy. Bruce, *Launching of Modern American Science*, 225–39 and 303–05, respectively. See also Marc Rothenberg's assessment in Joseph Henry, *The Papers of Joseph Henry* (Washington: Smithsonian Institution, 1972–2008), vol. 10. Rothenberg focuses on Joseph Henry's reaction. See also Rexmond Cochrane, *The National Academy of Sciences: The First Hundred Years, 1863–1963* (Washington: The Academy, 1978), 58–63. As pointed out in the preceding chapter, Henry had expected an academy to create "'a great deal of unpleasant feeling'" (ibid., 59).

7 Bruce, *Launching of Modern American Science*, 251.

8 A. Hunter Dupree, *Science in the Federal Government: A History of Policies and Activities to 1940* (Baltimore: Johns Hopkins Univ. Press, 1986).

line with discussions in other subfields, the United States was considered as providing settled political circumstances within which opposing factions were fighting out ideological battles but which provided a relevant and stable framework for professional aspirations.[9] But in the eyes of leading contemporary scientists, this was not the case, and historians of American science in the nineteenth century have not formulated a new and necessary paradigm in which the role of the scientific profession is considered to be an aspect of the slow consolidation of American nationhood.

This study of the motivation and role of Bache has shown that the American nation-state could not be taken for granted to provide "support without strings." This central figure in the history of his profession considered science to be crucial for his country's future participating in a universal cultural discourse. The authoritative implementation of this discourse, its acknowledgment by an American nation, was essential for successfully competing with Europe in order to realize American ambitions for world leadership. Bache took the existence of an evolving scientific profession for granted, invoked its standards, and sought to aid its institutional development. But this was part of his much broader concern in line with his

9 There exists important literature on an evolving American nationhood, particularly with respect to national symbols and, in the context of the early history of the United States, relating to questions of federal power that were on the forefront of contemporary debate. See David Waldstreicher, *In the Midst of Perpetual Fetes: The Making of American Nationalism, 1776–1820* (Chapel Hill: Univ. of North Carolina Press, 1997); Peter Onuf, "Nations, Revolutions, and the End of History," *Revolutionary Currents: Nation-Building in the Transatlantic World*, Michael A. Morrison, Melinda S. Zook, eds. (Lanham, Md.: Rowman & Littlefield, 2004), 173–88. Concerning the absence of a national perspective among American "elites" before 1917, see Axel Jansen, *Individuelle Bewährung im Krieg: Amerikaner in Europa, 1914–1917* (Frankfurt and New York: Campus, 2003). Historians have sometimes viewed "nation" and "nationalism" to be the same thing, ethically dismissing the former along with the latter. When Sally Kohlstedt observes that nationalism "was less strident but no less real in science than in other cultural or social aspects of nineteenth-century America," she refers to a jingoistic, or at least a self-serving, variant of nationalism rather than taking into view the problem of nation-building. Sally Kohlstedt, *The Formation of the American Scientific Community: The American Association for the Advancement of Science, 1848–1860* (Urbana: Univ. of Illinois Press, 1976), 176. Important work has been done on the relationship between science and nation-building in Europe (even though "science" is not always discussed in terms of a profession). See Ralph Jessen and Jakob Vogel, eds., *Wissenschaft und Nation in der europäischen Geschichte* (Frankfurt and New York: Campus, 2002); Andreas Franzmann, "Die Krise Frankreichs von 1870 und ihre Ausdeutung durch den Wissenschaftler Louis Pasteur. Eine Deutungsmusteranalyse," in *Wissen in der Krise*, ed. Carsten Kretschmann and Henning Pahl (Berlin: Akademie-Verlag, 2004), 117–56.

family's background for helping build the American nation by providing the country with a foundation for cultural life through public institutions. While historians have not ignored the particular situation of science in nineteenth-century America, therefore, they have insufficiently emphasized its consequences for anyone who shared Bache's ambitions. All of the Lazzaroni were solidly grounded in their respective research fields but Bache was their acknowledged leader because he understood and personified the necessity of strengthening the American "sovereign" as a condition for developing American culture and its institutions.

The history of American science, therefore, cannot be understood without a thorough consideration of its peculiar national setting. The American "sovereign," the aristocracy-invoking term Bache sometimes used for the republic, had not yet developed a sense of common national action outside of expanding individual opportunities. A "crucial formative influence in the American experience," Peter Parish has observed with respect to the 1780s and 1790s, "was the establishment of a political framework, with the potential to become a national framework, even before a vigorous national self-consciousness had developed."[10] The political struggles over states versus federal rights in the context of the slavery question signify that an integrative national political perspective remained distant. In antebellum America, many contentious issues such as those of slavery involved divergent views on national policy primarily as they pertained to conditions for territories to become states. While this led to political fights over national policy and brought about a war, a positive policy endorsing common national goals with binding consequences for all states was a different matter entirely. In the absence of a coherent political perspective to integrate the nation and to provide a basis for national cultural institutions, Bache and his peers struggled with the difficulty of finding ways to live up to the expectation that the United States would some day succeed Europe in all areas of cultural achievement. How could they develop these ambitions in the absence of a national "sovereign" to which they could dedicate their work? The solution was for Bache to establish that cultural authority himself.

After his appointment to the U.S. Coast Survey and in line with his familial expectation and responsibility, Bache became a unique cultural leader of the emerging United States. His Coast Survey, in all facets of its wide

10 Peter J. Parish, *The North and the Nation in the Era of the Civil War*, ed. Adam I. P. Smith and Susan-Mary Grant (New York: Fordham Univ. Press, 2003), 58.

array of cultural work, represented in the nation's capital a project aloof from the political battles of antebellum America, a vision of activities uniting instead of separating the nation, a national academy *avant la lettre.* Bache and his comrades felt that they were the heralds of a future national culture dedicated to radical standards ideally represented by science and the arts. This went along with a sometimes condescending view of politics. They "seem to feel a desire to be backed," he once wrote to Benjamin Peirce about American politicians, "which perhaps belongs to their life generally."[11]

This perspective culminated in the founding of the National Academy of Sciences in 1863. Perhaps its rationale corresponded reversely to the founding process of the Academy's sister institutions in Paris and London. French and British scientists pursuit the creation of these institutions with the emerging role of their profession in mind. But the National Academy of Sciences, according to its instigator Alexander Dallas Bache, became the scientific profession's tool for helping solidify an emerging nation and its state.[12]

This perspective on Alexander Dallas Bache, the history of scientific institutions in antebellum America, and of the founding of the National Academy integrates the history of American science as a profession with that of the emergence of the American nation-state. Bache and his comrades viewed the universal discourse of science and the arts to provide the setting for the American nation and its politics. This shows that an explanation of intentions and strategies of protagonists of American science during this period must no longer remain disconnected from the political setting in which they evolved.[13]

---

11 ADB to Benjamin Peirce, January 18, 1849, BPP.

12 For the motivational background to the founding of the Paris Academy of Sciences, see Roger Hahn, *The Anatomy of a Scientific Institution: The Paris Academy of Sciences, 1666–1803* (Berkeley and Los Angeles: Univ. of California Press, 1971), 4–34. For the Royal Society, see Peter Münte, *Die Autonomisierung der Erfahrungswissenschaften im Kontext frühneuzeitlicher Herrschaft: Fallrekonstruktive Analysen zur Gründung der Royal Society*, 2 vols. (Frankfurt: Humanities Online, 2004).

13 Only because this study is not primarily intended to contribute to a sociological discourse on the role of the professions, I refrain here from explicating the connections between my findings and theories of the professions in general, and the evidence this study provides for buffering the revised theory of the professions. The explication of such connections would have to begin by emphasizing Alexander Dallas Bache's expectation that the scientific profession in America was to try to assume the responsibility of a "vicarious crisis management" on behalf of the national political community. Unlike

Coordinates of Alexander Dallas Bache's Career

This perspective emerges from the series of investigations undertaken in the preceding chapters. They helped resolve and explain what appeared to be inconsistencies or ambivalences in Alexander Dallas Bache's career.

Bache's early career was an extension of a highly successful family history intertwined with American independence and state-building. Sophia Dallas' marriage to a descendent of Benjamin Franklin, the quintessential symbol of American independence and national cultural ambition, had confirmed these standards. Her oldest son's name, Alexander Dallas Bache, signaled his parents' (and particularly Sophia's) aspirations for a distinct career of national leadership. This enormous biographical responsibility and expectation Bache translated into a successful career that began at the United States Military Academy at West Point and in the Army Corps of Engineers. The choice of the army for a career followed from his family's expansive and coordinated efforts to help erect the American state. The decisive first phase of Bache's biographical development then took place in the metropolitan context of Philadelphia, a city facing severe economic challenges initially balanced by its role as the nation's intellectual and, for a time, political hub. Bache established during this early phase a pattern of institutional development that he later transferred to the national stage. Even though he did not ignore science and remained committed to it, Bache did not focus on the exploratory advancement of knowledge.

His main interest concerned the implementation of a universalistic culture in America for which the transnational discourse of science was a particularly apt representation. As a social model, Bache referred to his West Point experience and to that institution's unique commitment to the national ethos of developing the continent's vast and expanding territory. In his papers on meteor showers, Bache envisioned a body of observers dedicated to rigorous standards of observation for measuring a phenomenon unique to the American continent and relevant for a scientific discourse of universal significance. In his steam boiler research and in his reports on weights and measures, Bache demanded a national implementation of regulations and standards based on scientific experiments. Even

France and England, however, the professions in the United States, in his view, struggled with the problem of having to help implement a community for which science could then become such a "crisis manager."

more importantly, as the head of the Franklin Institute's committee on steam boiler explosions, Bache claimed a cultural authority *independent* of national political power. He sought to use the committee's intellectual cache to bolster, rather than draw on, the authority of the federal state.

Bache's educational efforts in the late 1830s and early 1840s added a new dimension to his continued efforts to establish in America the foundation for participating in a universalistic discourse of science and the arts then dominated by Europe. Bache naturally identified with science and the scientific community. As president of Girard College and as principal of the Central High School, he sought to professionalize education, to connect teachers and their students to the "intellectual movements of the day" by providing possibilities and incentives for exploration. Girard College was a private institution with national aspirations but Bache's heart was with the Central High School as the apex of a public school system. The reasons for his ejection from both institutions confirm that he stood for a pattern of republican leadership that was only then emerging in the United States. He did not share a Federalist perspective on an emerging American mass democracy and considered Greek and Latin irrelevant for solving the country's problems; but he also rejected the Jacksonian Democracy's frequently populist criticism of elites and their cultural interests and symbols. Bache stood for a third position in line with his family's mission and his West Point experience: Creating avenues for intellectual achievement in areas that were practically significant for the political community. Science supplied an intellectual content and framework whose universalism corresponded to those of America's republican ideals.[14]

Bache's selection for the post of superintendent of the U.S. Coast Survey in 1842 coincided with the scientific profession's growing awareness of the potentials of national scientific institutions. In speeches held in 1842,

14 These findings deviate from Hugh R. Slotten's assessment. Slotten suggests that Bache participated in an antebellum reform culture in which innovation as well as "[p]lanning, order, and control were important ideals." This study confirms that innovation was important for Bache. But his institutional efforts in both education and science suggest that Bache, instead of "imposing" moral discipline, sought to evoke it. The difference is crucial. Rather than cementing an empty elitism by reducing intellectual curiosity to a claim, Bache's leadership in nineteenth-century professionalization aimed at advancing the disciplined pursuit of such curiosity, and, by implication, that of personal, cultural, and political autonomy. Slotten, *Patronage, Practice, and the Culture of American Science* (Cambridge: Cambridge Univ. Press, 1994), quotation 27. See also Hugh R. Slotten, "The Dilemmas of Science in the Unites States. Alexander Dallas Bache and the U.S. Coast Survey," *Isis*, no. 84 (1993): 32, 43.

1844, and 1851, Bache reflected and explained his earlier work, and he drew conclusions for the national phase of developing scientific institutions. He viewed things from the perspective of a metropolitan cultural setting but he programmatically looked to the nation's development as key to everything else. The scientific profession he considered to be guarding a "palladium," making it impossible to compromise on standards of intellectual integrity. In the absence of an American aristocracy, he believed that the profession was to represent the standards associated with it. In Bache's perception, the scientific profession assumed a role that was functionally similar to that of the Central High School for Philadelphia's public school system. It was to provide the "crown" of intellectual achievement and an incentive for intellectual aspiration and development for a country rapidly expanding into new unconsolidated territories. In 1843, however, the nation's scientists cooperated through journals and letters and there existed no national scientific organization.

As the new superintendent of the U.S. Coast Survey, Bache took his Philadelphia experience to Washington. In the context of the debate on the role of the National Institute in 1844, it became increasingly apparent to Bache and his colleagues such as Joseph Henry that a national academy of sciences would eventually come into existence. They had discussed the institutional development of American science for some time and the matter was now resolved in two steps. Bache favored the organization of a national platform for the scientific profession first, the AAAS founded in 1848. Neither this association nor the Smithsonian Institution created in 1846, however, symbolized the nation's acknowledgment of science. Bache and his close friends continued to discuss the second step of founding a national academy privately but they shied away from pushing it.

Bache had provided the rationale for this hesitation in his 1844 speech on the "wants of science in the United States." In line with his earlier work for the Franklin Institute and his positions in the Philadelphia school fights, he insisted that scientific investigation be relevant for the political community. Bache welcomed the founding of a national organization by geologists because it was to facilitate practical survey work rather than science. The founding of the AAAS then created a general platform for American science but Bache and his circle continued to have reservations because, as Bache had insinuated in 1844, it lacked the corrective of a national political community and public, a general appreciation of aristocratic standards by the democratic "sovereign."

Contemporary colleagues as well as historians have criticized Bache and his "clique" for their high-handedness in the 1850s.[15] Bache's family background, career pattern, and ambitions help explain the Lazzaroni's reluctance to cede control where they had it. In his 1842 speech, Bache had broadly contextualized the American national project, aligning it on a trajectory that extended from historic instances of great cultural achievements under difficult settlement conditions. Such references were part of a more general contemporary discourse, of course, but this does not cancel their pertinence for Bache's subjective standards of achievement. Bache considered his generation to be laying the tracks for his nation's culture. Inside of the political framework provided by the U.S. Constitution, he viewed his cohort to be responsible for creating the country's intellectual and institutional infrastructure. Nothing less than the future success of America was at stake, measured against the standard of European culture, which the United States was supposed to surpass. Only then would America confirm its particular mission to turn the universality of its political design into an advantage in all areas of life. This resulted in a tremendous responsibility for putting the country's culture on the right track. In science and education, consolidation could *only* be achieved by creating *public* institutions. Bache had learned as principal of Central High School that many Americans preferred exclusive private schools to public institutions, but he insisted on the relevance of education and of research even for those American citizens who were not immediately engaged in it. Bache, in other words, considered both education and science to be *professions* in the sense that they were engaged in activities *on behalf of every* U.S. citizen. At the same time, however, he was well aware that few of his contemporaries assumed such a national perspective. As a result, Bache chose to be careful in developing national scientific institutions if, in his assessment, they had outpaced an evolving American nation that he considered to be an important counterpart to the professions. He oscillated between endorsing national institutional initiatives such as the founding of the AAAS, on the one hand, and skepticism towards their viability without Lazzaroni command, on the other.

The *Lebensgefühl* was that of facing unprecedented national opportunities and dramatic personal responsibilities. While colleagues such as Ben-

15 See n. 6 above.

jamin Peirce shared this view, Bache's background made him particularly responsive to it.[16]

In keeping with the ideas Bache had laid out in his 1844 speech, the Lazzaroni discussed but did not pursue a national academy in the 1850s because they felt that the nation was not ready for it.[17] Neither the Smithsonian (an international entity) nor the AAAS (a professional platform) had filled the spot. Bache hesitated. He had developed the academy idea in his 1851 speech but refrained from pushing it. In the meantime, the nation needed technical advice on matters related to such things as exploring and settling the continent. In lieu of an academy, the federal government used the Coast Survey as a "general scientific agency."[18] Because of the Survey's unique significance during this period, Bache became, metaphorically speaking, the president of an invisible national academy. No one else had a similar national standing nor was anyone else in charge of an organization as significantly "national" as the U.S. Coast Survey.

Bache's fear that the war was a threat to everything he stood for provided the rationale for founding the National Academy in 1863. During the 1850s, he could still hope for a consolidation of the American nation and a changing view on its dedication to science and the arts. At a time of severe national trial, however, Bache turned the old academy idea into a tool for propping up the nation. Instead of becoming a symbol for the nation's proud support of science and its universalistic principles (parallel to the universality of human rights implied in the Emancipation Proclamation), the Academy was to be a token of the scientific profession's support of the nation. James M. McPherson has suggested that the war "fused the several states bound loosely in a federal *Union* under a weak central government into a new *Nation* forged by the fires of a war in which more Americans lost their lives than in all of the country's other wars combined."[19] The

16 All of this naturally corresponds to Bache's institutional radicalism based on the idea of informal professional cooperation as the nonnegotiable basis for all administrative concerns. Bache once wrote that if he had to choose between "a good head without any organization of the institution, or an incompetent head with a beautiful plan," he would prefer the former (ADB to [Deloutte?], August 5, 1850, Gratz MSS, Historical Society of Pennsylvania).

17 As pointed out in the previous chapter, Joseph Henry continued to argue in 1863 that Congress and the public would reject the proposal for an academy.

18 Dupree, *Science in the Federal Government*, 104.

19 James M. McPherson, *Battle Cry of Freedom: The Civil War Era* (New York: Oxford Univ. Press, 1988), viii. McPherson's concept of "nation" refers to post-war efforts to integrate the former opponents by referring to the common war experience. The "heroism"

Academy's founding rationale suggests that such a "nation" was only beginning to emerge during the Civil War. The reality of the war attested to the fact that the states remained disunited. The Constitution had provided a political framework since 1789; after 1865, it was evident that a state's membership in the United States was irreversible. The immense and unprecedented human toll demanded by the Civil War, as well as setbacks in the field and diplomatic considerations, made a shift of war aims essential in 1862. The Emancipation Proclamation established the United States as a country that accepted and fought for certain human rights. In January 1863, this opened up a new perspective on federal institutions as an activist Congress began to consider providing institutional means for other areas of national life. The Academy's instigators joined in the transformation of a war fought for the preservation of the Union into a war fought for the abolition of slavery.[20] Bache's motives for founding the National Academy suggest, however, that the nation's consolidation remained incomplete. He intended the institution to be a "mark," a symbolic representation of the viability of an enlightened American nation. Rather than providing evidence for the nation's acceptance of its standard and logic, therefore, the Academy was implemented as a symbolic reminder to the nation to live up to its cultural promise.[21]

---

of the war could become a banner for symbolizing "nationhood" but this merely transposed the neglect of common national goals during the war. See Cecilia Elizabeth O'Leary, *To Die for: The Paradox of American Patriotism* (Princeton: Princeton Univ. Press, 1999); and Jansen, *Individuelle Bewährung im Krieg*, 251–82.

20 This shift was independent from views on slavery. Benjamin Peirce, at the beginning of the war, had taken a proslavery stance but he was nevertheless committed to the union. In April 1860, he proposed in a letter to Bache that the South be undone by turning universities there into army camps. The underlying logic of this idea corresponded to Bache's perception of a wartime role of a national academy though Peirce considered the problem from the point of view of universities. Benjamin Peirce to ADB, April 30, 1861, box 27, Rhees Collection, HL.

21 There is no room here to extrapolate on obvious connections between the findings of this study and other political and cultural developments. One such connection I would not want to omit, however, and that is how Charles Sanders Peirce's philosophical work and his concept of the "community of inquiry" related to the experience of his father Benjamin Peirce and others in his circle, including Alexander Dallas Bache. This point has been made by Thomas L. Haskell in *Objectivity Is Not Neutrality: Explanatory Schemes in History* (Baltimore: Johns Hopkins Univ. Press, 2000), 103 f., and in *The Emergence of Professional Social Science: The American Social Science Association and the Nineteenth-Century Crisis of Authority* (Urbana: Univ. of Illinois Press, 1977).

Shortly after the new Academy began to conduct business and after it had elected him its first president, a stroke forced Bache to give up all work. His wife Ency took him on a journey to Europe to recover and he again visited some of the places and colleagues he had seen in 1837 and 1838. Alexander Dallas Bache died in Newport, Rhode Island, in 1867 and bequeathed, upon the death of his wife, the income of his estate of $40,000 to the National Academy of Sciences.[22]

22 Cochrane, *The National Academy of Sciences*, 98 f. A year after Ency Bache's death in 1870, the Academy received about $40,000.

Figures

Selected Bibliography

1. Manuscripts and Archival Material

American Philosophical Society
Alexander Dallas Bache Papers
John Fries Frazer Papers
John Kintzing Kane Papers
John Lawrence LeConte Papers
J. Peter Lesley Papers
Robert M. Patterson Papers
John Warner Papers
Quetelet Correspondence

Girard College
Various documents pertaining to Alexander Dallas Bache's presidency

Historical Society of Pennsylvania
George Mifflin Dallas Papers, Phi(1490)
Gratz Collection
    Bache Correspondence
    "Philadelphia Public Schools"

Houghton Library, Harvard University
Benjamin Peirce Papers [cited as "BPP"]

Huntington Library
Papers of William Jones Rhees ["Rhees Collection, HL"]
    Alexander Dallas Bache Papers
    National Institute for the Promotion of Science
    Smithsonian Institution
Papers of Francis Lieber ["Lieber Papers, HL"]

Library of Congress
Alexander Dallas Bache Papers

Lewis Reeves Gibbes Papers
William Chauncy Langdon Papers
Robert Wilson Sufeldt Papers

Massachusetts Institute of Technology (MIT), Institute Archives & Special Collections
William Barton Rogers Papers (MC 1)
Rogers Family Papers (MC 2)

National Archives
RG23, Correspondence of A. D. Bache, Superintendent of the Coast and Geodetic Survey, 1843–1865
RG45, Permanent Commission on Science and Art

Princeton University Library, Department of Rare Books and Special Collections
Bache and Wistar Family Correspondence

Smithsonian Institution Archives
Record Unit 7053, Alexander Dallas Bache Papers ["Bache papers, SIA"]
Record Unit 7470, Nathan Reingold Papers ["Reingold Papers, SIA"]
Includes microfilm copies of documents in other depositories: Bache Correspondence on the Dudley Observatory from the Bache Correspondence in the Benjamin Peirce Papers, Houghton Library, Harvard University; Davidson Papers, Bancroft Library, University of California, Berkeley; Elias Loomis family papers, Manuscripts and Archives, Yale University.
Record Unit 7050, George Brown Goode Collection
Record Unit 7177, George P. Merrill Collection

2. Printed Primary Sources

American Association for the Advancement of Education. *Proceedings.*
American Association for the Advancement of Science. *Proceedings.*
American Journal of Science and Arts.
American Philosophical Society. *Proceedings.*
Bache, Alexander Dallas. "A National University. Remarks at the Opening of the Fifth Session of the American Association for the Advancement of Education." *Proceedings of the Session of the American Association for the Advancement of Education Held at the City of New York, August 28th, 29th, 30th, and 31st, A.D. 1855* (1856): 19–21.
— *Address delivered at the close of the twelfth exhibition of American manufactures: held by the Franklin Institute of the state of Pennsylvania, for the promotion of the mechanic arts, October, 1842.* [Philadelphia]: [The Institute], n.d.

— "Address delivered at the close of the Twelfth Exhibition of American Manufactures, held by the Franklin Institute of the State of Pennsylvania, for the Promotion of the Mechanic Arts, October 1842." *Journal of the Franklin Institute* (1842): 379–94.
— "Address of Professor A. D. Bache, President of the American Association for the Year 1851, on Retiring from the Duties of President." American Association for the Advancement of Science, *Proceedings* 6 (1852): xli–lx.
— "Alarm to be applied to the interior flues of steam boilers." *Journal of the Franklin Institute* 14, no. 4 (October 1832): 217–23.
— *Eulogy on Hon. James Alfred Pearce, of Maryland, United States Senator: One of the Regents of the Smithsonian Institution.* [Washington D.C.]: Smithsonian Institution, 1863.
— "Experimental illustrations of the radiating and absorbing powers of surfaces for heat, of the effects of transparent screens, of the conducting power of solids, &c." *Journal of the Franklin Institute* 19, no. 5 (May 1835): 303–09.
— "Experiments on the efficacy of Perkins' steam boilers, or circulators." *Journal of the Franklin Institute* 19, no. 6 (June 1835): 379–86.
— *Lecture on the Gulf Stream.* The Association, 1860.
— "Meteoric observations made on and about the 13th of November, 1834." *American Journal of Science and Arts* 27, no. 2 (January 1835): 335–38.
— "Note relating to the hardening of lime under water, by the action of carbonate of potassa, &c., and to the hardening of carbonate of lime in the air, by potassa and soda." *Journal of the Franklin Institute* 19, no. 1 (January 1835): 6–8.
— *Observations at the Magnetic and Meteorological Observatory, at the Girard College.* Washington: Gales and Seaton, printers, 1847.
— "Observations on the disturbance in the direction of the horizontal needle during the occurence of the Aurora of July 10th, 1833." *Journal of the Franklin Institute* 17, no. 1 (January 1834): 1–9.
— "Proposed forms of diagrams for exhibiting to the eye the results of a register of the direction of the wind." *Journal of the Franklin Institute* 22, no. 1 (July 1836): 22–27.
— "Replies to a circular in relation of an unusual meteoric display on the 13th Nov. 1834, addressed by the Secretary of War to the Military posts of the United States, with other facts relating to the same question." *American Journal of Science and Arts* 28 (1835): 305–09.
— *Report of Experiments on the Navigation of the Chesapeake and Delaware Canal.* Philadelphia: J. Harding, 1834. First published in the *Journal of the Franklin Institute.*
— "Report of the Committee on Weights and Measures. Appendix to the Report of the Committee of the Franklin Institute on Weights and Measures. Abstract of the reports on Weights and Measures which have been submitted to the Congress of the United States, or to the Legislature of Pennsylvania." *Journal of the Franklin Institute* 8, no. 4 (April 1834): 232–47.

— *Report of the Secretary of the Treasury on the Construction and Distribution.* Washington: A. O. P. Nicholson, 1857.

— *Report on Education in Europe: To the Trustees of the Girard College for Orphans.* Philadelphia: Printed by Lydia R. Bailey, 1839.

— *Report on the Organization of a High School for Girls, and Seminary for Female Teachers.* Philadelphia: J. Crissy, printer, 1840.

— "Safety apparatus for steam boats, being a combination of the fusible metal disk with the common safety valve." *Journal of the Franklin Institute* 11, no. 4 (April 1831): 217–21.

Bache, Richard. *Oration, delivered at Spring Garden, July 5, 1813: to a very numerous and respectable company of Democratic Republicans, of the city and county of Philadelphia.* Philadelphia: Printed by John Binns, 1813.

Barnard, Henry. "Report on a System of Common Schools, For Cities and Large Villages." *The District* 3, no. 2 (August 1, 1842): 17–31.

Barr, Ernest Scott. *An Index to Biographical Fragments in Unspecialized Scientific Journals.* University, AL: Univ. of Alabama Press, 1973.

Biddle, Nicholas. "The Address." *The North American Magazine*, August 1833.

Brewster, David. *A Treatise on Optics.* 1st ed. Philadelphia: Carey, Lea, & Blanchard, 1833. Published by Alexander Dallas Bache.

Espy, James P. "Remarks on Professor Olmsted's Theory of the Meteoric Phenomenon of November 12th, 1833, denominating Shooting Stars, with some Queries towards forming a just Theory." *Journal of the Franklin Institute* 15, no. 1 (January 1835): 9–19.

Franklin Institute. "Annual Meeting of the Franklin Institute." *Journal of the Franklin Institute* 11, no. 2 (February 1833): 85–86.

— "Annual Report of the Board of Managers." *Journal of the Franklin Institute* 11, no. 2 (February 1833): 87–91.

— "Annual Report of the Board of Managers." *Journal of the Franklin Institute* 13, no. 4 (April 1834): 226–31.

— "Annual Report of the Committee on Arts and Sciences." *Journal of the Franklin Institute* 15, no. 3 (March 1835).

— "Forty-first Quarterly Report of the Board of Managers of the Franklin Institute." *Journal of the Franklin Institute* 13, no. 6 (June 1832): 371–73.

— "General Report on the Explosions of Steam-Boilers (1836)." In *Early Research at the Franklin Institute: The Investigation into the Causes of Steam-Boiler Explosions, 1830–1837*, edited by Bruce Sinclair. Philadelphia: Franklin Institute of the State of Pennsylvania, 1966.

— "Quarterly Report of the Board of Managers." *Journal of the Franklin Institute* 12, no. 3 (September 1833): 158–60.

— "Report in Relation to Weights and Measures in the Commonwealth of Pennsylvania." *Journal of the Franklin Institute* 14, no. 1 (July 1834): 6–14.

— "Report of the Board of Managers." *Journal of the Franklin Institute* 9, no. 2 (February 1832).

— "Report of the Joint Special Committee of the Expediencey of Opening the Girard College." Philadelphia: L. R. Bailey, 1842.
— "Report of the Special Committee Appointed by the Common Council on a Communication from the Board of Trustees of the Girard College." Philadelphia: s.n., 1840.
— "Thirty-fourth Quarterly Report of the Board of Managers of the Franklin Institute." *Journal of the Franklin Institute* 10, no. 6 (December 1832).
— "Thirty-seventh Quarterly Report of the Board of Managers of the Franklin Institute." *Journal of the Franklin Institute* 15, no. 5 (May 1833): 303.
Girard College, Thomas Ustick Walter, Nicholas Biddle, and Joseph Ripley Chandler. *A Description of the Girard College for Orphans.* Philadelphia: Crissy and Markley, 1848.
Girard College Magnetic and Meteorological Observatory, Alexander Dallas Bache, and United States Army Corps of Engineers. *Observations at the Magnetic and Meteorological Observatory, at the Girard College, Philadelphia, Made Under the Direction of A. D. Bache, LL. D. and With Funds Supplied by the Members of the American Philosophical Society, and by the Topographical Bureau of the United States.* Washington: Gales and Seaton, printers, 1847.
Goodwin, Daniel R. "Obituary Notice of Samuel Vaughan Merrick, Esq." *Proceedings of the American Philosophical Society* 11, no. 81 (January 1869): 584–597.
Hargreaves, Mary W. M., and James F. Hopkins, eds. *The Papers of Henry Clay.* Vol. 6. Lexington, KY: Univ. Press of Kentucky, 1992.
Henry, Joseph. "Biographical Memoir of Alexander Dallas Bache." National Academy of Sciences, *Biographical Memoirs,* 1 (1877).
— *The Papers of Joseph Henry.* 10 vols. Washington: Smithsonian Institution Press, New York, 1972–2008. Edited by Nathan Reingold, Marc Rothenberg, et al.
Ingersoll, Joseph Reed. *Third Circular of the Committee, Relating to the National Institute.* s.n, 1844.
Jefferson, Thomas. *Writings.* New York: Literary Classics of the United States, 1984.
McEwen, T. "Report giving an account of the system of weights and measures of France." *Journal of the Franklin Institute* 17, no. 3 (March 1834): 160–71.
National Institute for the Promotion of Science. *Bulletin of the National Institute for the Promotion of Science.* Washington: Gales and Seaton, printers, 1841.
Olmsted, Denison. "Facts respecting the Meteoric Phenomena of November 13, 1834." *American Journal of Science and Arts* 29, no. 1 (January 1835): 168–70.
— "Facts respecting the Meteoric Phenomena of November 13th, 1834." *Journal of the Franklin Institute* 16, no. 6 (December 1835): 367–69.
— "Observations on the Meteors of November 13th, 1833." *American Journal of Science and Arts* 25, no. 2 (January 2, 1834): 363–411.
Peirce, Benjamin. "Address of Professor Benjamin Peirce, President of the American Association for the Advancement of Science for the Year 1853, on Retiring

From the Duties of President." American Association for the Advancement of Science. *Proceedings* 9 (1855), 1–17.

Peirce, Charles Sanders. *Writings of Charles S. Peirce: A Chronological Edition*. Bloomington: Indiana Univ. Press, 1982.

Perry, Matthew Calbraith. *Narrative of the Expedition of an American Squadron to the China Seas*. Trubner & Co, 1856.

Rogers, William Barton. *Life and Letters of William Barton Rogers*. 2 vols. Boston: Houghton Mifflin, 1896.

Seager II, Robert, ed. *The Papers of Henry Clay*. Lexington, Kentucky: Univ. Press of Kentucky, 1982.

U.S. Coast Survey. *Annual Report of the Superintendent of the Coast Survey Showing the Progress of that Work during the Year Ending November, 1851*. 35th Cong., 2d Sess., Exec. Doc. 6 (Ser. 980), 1852.

University of Pennsylvania. "Catalogue of the Officers and Students of the University of Pennsylvania." http://www.archives.upenn.edu/histy/features/1800s/1830s/catalogue.html.

Walker, Sears. "Appendix to the report of the committee of the Franklin Institute on weights and measures." *Journal of the Franklin Institute* 17, no. 2 (February 1834): 94–109.

Wilson, Edwin B. "National Academy of Sciences: Minutes of the Meeting for Organization, April, 1863." *Proceedings of the National Academy of Sciences of the United States of America* 36, no. 4 (April 15, 1950): 277–92.

3. Books and Articles

Abbott, Andrew. *The System of Professions: An Essay on the Divisons of Expert Labor*. Chicago: Univ. of Chicago Press, 1988.

Adams, Henry. *History of the United States of America During the Administration of Thomas Jefferson*. New York: Literary Classics of the United States, 1986.

Agnoli, Paolo, and Giulio D'Agostini. "Why does the meter beat the second?" *physics/0412078* (December 14, 2004). http://arxiv.org/abs/physics/0412078.

Aldrich, Mark. "Earnings of American Civil Engineers, 1820–1859." *Journal of Economic History* 31, no. 2 (June 1971).

Allen, Henry Butler. "The Franklin Institute of the State of Pennsylvania." *Transactions of the American Philosophical Society* 43, no. 1. New Series (1953): 275–279.

Ambrose, Stephen E. *Duty, Honor, Country: A History of West Point*. Baltimore: Johns Hopkins Univ. Press, 1966.

Arey, Henry W. *The Girard College and Its Founder: Containing the Biography of Mr. Girard*. Philadelphia: Printed by C. Sherman & Son, 1869.

Baltzell, Edward Digby. *Philadelphia Gentlemen: The Making of a National Upper Class.* Glencoe, IL: Free Press, 1958.

Baritz, Loren. "The Idea of the West," *American Historical Review* 66, no. 3 (April 1961): 618–40.

Bates, Ralph Samuel. *Scientific Societies in the United States.* Chapman & Hall, 1945.

Beach, Mark. "Was there a 'Scientific Lazzaroni'?" In *Nineteenth-Century American Science: A Reappraisal,* edited by George H. Daniels, 115–32. Evanston, IL: Northwestern Univ. Press, 1972.

Belohlavek, John M. *George Mifflin Dallas: Jacksonian Patrician.* University Park and London: Pennsylvania State Univ. Press, 1977.

Ben-David, Joseph. *The Scientist's Role in Society; a Comparative Study.* Englewood Cliffs, N.J: Prentice-Hall, 1971.

Bledstein, Burton J. *The Culture of Professionalism: The Middle Class and the Development of Higher Education in America.* New York and London: W. W. Norton, 1976.

Blissett, Marlan. *Politics in Science.* Boston: Little, Brown, 1972.

Brent, Joseph. *Charles Sanders Peirce: A Life.* Bloomington: Indiana Univ. Press, 1998.

Brown, Chandos Michael. *Benjamin Silliman: A Life in the Young Republic.* Princeton: Princeton Univ. Press, 1989.

Brown, John K. *The Baldwin Locomotive Works, 1831–1915: A Study in American Industrial Practice.* Baltimore: Johns Hopkins Univ. Press, 1995.

Bruce, Robert V. *The Launching of Modern American Science, 1846–1876.* New York: Knopf, 1987.

Burke, John G. "Bursting Boilers and the Federal Power." *Technology and Culture* 7, no. 1 (Winter 1966): 1–23.

Burnham, John C. *Science in America: Historical Selections.* New York: Holt, Rinehart and Winston, 1971.

Cajori, Florian. *The Chequered Career of Ferdinand Rudolph Hassler.* New York: Arno Press, 1980. First published in Boston by Christopher Publishing House, 1929.

Carmichael, Leonard. "Joseph Henry and the National Academy of Sciences." *Proceedings of the National Academy of Sciences of the United States of America* 48, no. 1 (July 15, 1967): 1–10.

Carter, Edward C. *One Grand Pursuit: A Brief History of the American Philosophical Society's First 250 Years, 1743–1993.* Philadelphia: American Philosophical Society, 1993.

Chapin, David. *Exploring Other Worlds: Margaret Fox, Elisha Kent Kane, and the Antebellum Culture of Curiosity.* Amherst: Univ. of Massachusetts Press, 2004.

Chartier, Roger, and Pietro Corsi, eds. *Sciences et langues en Europe.* Paris: Centre Alexandre Koyré, 1996.

Cheyney, Edward P. "The Connection of Alexander Dallas Bache with the University of Pennsylvania." *Proceedings of the American Philosophical Society* 84, no. 2 (1941): 151–160.

— *History of the University of Pennsylvania, 1740–1940.* New York: Arno Press, 1977. First published in Philadelphia: Univ. of Pennsylvania Press, 1940.

Cochrane, Rexmond. *The National Academy of Sciences: The First Hundred Years, 1863–1963*. Washington: The Academy, 1978.

Cohen, I. Bernhard. *Science and the Founding Fathers: Science in the Political Thought of Thomas Jefferson, Benjamin Franklin, John Adams, and James Madison*. New York: Norton, 1995.

— "Science in America: The 19th Century." In *Paths of American Thought*, edited by Arthur M. Schlesinger Jr. and Morton White. Boston: Houghton Mifflin, 1963.

Commemoration of the Life and Work of Alexander Dallas Bache and Symposium on Geomagnetism. Vol. 84. *Proceedings of the American Philosophical Society*. Philadelphia: American Philosophical Society, 1941.

Coulson, Thomas. "Alexander Dallas Bache, 1806–1867." *Journal of the Franklin Institute* 263, no. 6 (June 1957).

— "The first hundred years of research at the Franklin Institute." *Journal of the Franklin Institute* 256, no. 1 (July 1953): 1–25.

Cowan, Ruth Schwartz. *A Social History of American Technology*. New York: Oxford Univ. Press, 1997.

Curti, Merle. "Francis Lieber and Nationalism." *The Huntington Library Quarterly* 4, no. 3 (April 1941): 263–92.

Cushing, Thomas. *History of the Counties of Gloucester, Salem, and Cumberland, New Jersey: With Biographical Sketches of Their Prominent Citizens*. Woodbury, N.J: Gloucester County Historical Society, 1974.

Dallas, George Mifflin. *Life and Writings of Alexander James Dallas*. Philadelphia: Lippincott, 1871.

Daniels, George H. *American Science in the Age of Jackson*. New York: Columbia Univ. Press, 1968.

— "The Process of Professionalization in American Science: The Emergent Period, 1820–1860." *Isis* 58, no. 2 (Summer 1967): 150–166.

Davis, Charles Henry. *Life of Charles Henry Davis, Rear Admiral, 1807–1877*. Boston: Houghton Mifflin, 1899.

Dawson, Andrew. *Lives of the Philadelphia Engineers: Capital, Class, and Revolution, 1830–1890*. Aldershot, Hants, England and Burlington, VT: Ashgate Publishing, 2004.

Dick, Steven J. *Sky and Ocean Joined: The U.S. Naval Observatory, 1830–2000*. Cambridge: Cambridge Univ. Press, 2002.

Dupree, A. Hunter. *Asa Gray, 1810–1888*. Cambridge: Belknap Press of Harvard, 1959.

— "The Founding of the National Academy of Sciences—A Reinterpretation." *Proceedings of the American Philosophical Society* 44 (1957): 434–40.

— *Science in the Federal Government: A History of Policies and Activities to 1940*. Baltimore: Johns Hopkins Univ. Press, 1986.

— "Science Policy in the United States: The Legacy of John Quincy Adams." *Minerva* 28, no. 3 (September 1990): 259–71.

Dupuy, Richard Ernest. *Sylvanus Thayer, Father of Technology in the United States.* West Point, NY: Association of Graduates, United States Military Academy, 1958.

Edmonds, Franklin Spencer. *History of the Central High School of Philadelphia.* Philadelphia: Lippincott, 1902.

Elliott, Clark A., and Margaret W. Rossiter. *Science at Harvard University: Historical Perspectives.* Bethelehem, PA: Lehigh Univ. Press, 1992.

Emerson, Edward Waldo. *The Early Years of the Saturday Club: 1855–1870.* Boston: Houghton Mifflin, 1918.

Fleming, John A. "Geomagnetism: World-Wide and Cosmic Aspects with Especial Reference to Early Research in America." *American Philosophical Society*, Proceedings 84 (1941): 263–98.

Fleming, James Rodger. *Meteorology in America, 1800–1870.* Baltimore: Johns Hopkins Press, 1990.

Franzmann, Andreas. *Der Intellektuelle als Protagonist der Öffentlichkeit: Krise und Räsonnement in der Affäre Dreyfus.* Humanities Online, 2004.

— "Die Krise Frankreichs von 1870 und ihre Ausdeutung durch den Wissenschaftler Louis Pasteur – Eine Deutungsmusteranalyse." In *Wissen in der Krise*, edited by Carsten Kretschmann, Henning Pahl, and Peter Scholz, 117–156. Berlin: Akademie Verlag, 2004.

— *Disziplin der Neugierde. Eine empirische Untersuchung zur Professionalisierung von Erfahrungswissenschaftlern auf der Basis nicht-standardisierter Interviews mit Naturwissenschaftlern.* Frankfurt, forthcoming.

Frazier, Arthur H. "United States Standards of Weights and Measures: Their Creation and Creators," *Smithsonian Studies in History and Technology* 40 (1978), http://www.sil.si.edu/smithsoniancontributions/HistoryTechnology/sc_RecordSingle.cfm?filename=SSHT-0040.

Freidel, Frank. "A Plan for Modern Education in Early Philadelphia." *Pennsylvania History* 14, no. 3 (July 1947): 175–84.

— *Francis Lieber, Nineteenth-Century Liberal.* Baton Rouge: Louisiana State Univ. Press, 1947.

Freidson, Eliot. *Professional Powers: A Study of the Institutionalization of Formal Knowledge.* Chicago: Univ. of Chicago Press, 1986.

Friedman, Lawrence Meir. *A History of American Law.* 2nd ed. New York: Touchstone, 1985.

Garbi, Antonello, *La disputa del Nuovo Mondo: Storia di una polemia, 1750–1900.* Milano: Riccardo Ricciardi Editore, 1955. Translated by Jeremy Moyle, *The Dispute of the New World: The History of a Polemic, 1750–1900.* Pittsburg: Univ. of Pittsburg Press, 2010.

Geiger, Roger, ed. *The American College in the Nineteenth Century.* Nashville: Vanderbilt Univ. Press, 2000.

Geitz, Henry, Jürgen Heideking, and Jurgen Herbst. *German Influences on Education in the United States to 1917.* Cambridge: Cambridge Univ. Press, 1995.

George V. Fagan. "Alexander Dallas Bache. Educator." *The Barnwell Bulletin* 18, no. 75 (April 1941): 4–44.

Gilliss, James Melville. *The U.S. Naval Astronomical Expedition to the Southern Hemisphere.* Washington: A. O. P. Nicholson, 1856.

Goetzmann, William H. *Army Explorations in the American West, 1803–1863.* New Haven: Yale Univ. Press, 1959.

— *New Lands, New Men: America and the Second Great Age of Discovery.* New York: Viking, 1986.

Goode, William J. "Community within the Community: The Professions." *American Sociological Review* 22 (1957): 194–200.

Gould, Benjamin Apthorp. "An Address in Commemoration of Alexander Dallas Bache," American Association for the Advancement of Science, *Proceedings* (1868).

Gouldner, Alvin W. *The Future of Intellectuals and the Rise of the New Class: A Frame of Reference, Theses, Conjectures, Arguments, and an Historical Perspective on the Role of Intellectuals and Intelligentsia in the International Class Contest of the Modern Era.* New York: Seabury Press, 1979.

Grattan-Guinness, Ivor. "Benjamin Peirce's Linear Associative Algebra (1870): New Light on Its Preparation and 'Publication'," *Annals of Science* 54, no. 6 (November 1997): 597–606.

Greene, John C. *American Science in the Age of Jefferson.* Ames: Iowa State Univ. Press, 1984.

Greenfeld, Liah. *Nationalism: Five Roads to Modernity.* Cambridge: Harvard Univ. Press, 1992.

Haber, Samuel. *The Quest for Authority in the American Professions, 1750–1900.* Chicago: Univ. of Chicago Press, 1991.

Hahn, Roger. *The Anatomy of a Scientific Institution: The Paris Academy of Sciences, 1666–1803.* San Francisco and Los Angeles: Univ. of California Press, 1971.

Hall, Peter Dobkin. *The Organization of American Culture, 1700–1900: Private Institutions, Elites, and the Origins of American Nationality* (New York: NYU Press, 1984).

Haskell, Thomas. *Objectivity is not Neutrality. Explanatory Schemes in History.* Baltimore: Johns Hopkins Univ. Press, 1998.

— *The Emergence of Professional Social Science. The American Social Science Association and the Nineteenth-Century Crisis of Authority.* Urbana and Chicago: Univ. of Illinois Press, 1977.

Hatch, Nathan O. *The Professions in American History.* Notre Dame, Indiana: Univ. of Notre Dame Press, 1988.

Heck, N. H. "The Magnetic Survey of the United States." *Proceedings of the American Philosophical Society* 84, no. 2 (1941): 205–24.

Heitman, Francis. *Historical Register and Dictionary of the United States Army, From Its Organization, September 29, 1789, to March 2, 1903.* Washington D.C.: Government Printing Office, 1903.

Herrick, Cheesman Abiah. *History of Girard College*. Philadelphia: Girard College, 1935.

Hofstadter, Richard. *The Idea of a Party System: The Rise of Legitimate Opposition in the United States*. Berkeley and Los Angeles: Univ. of California Press, 1970.

Hogan, Edward R. *Of the Human Heart: A Biography of Benjamin Peirce*. Bethlehem, PA: Lehigh Univ. Press, 2008.

Holmfeld, John D. "From Amateurs to Professionals in American Science: The Controversy over the Proceedings of an 1853 Scientific Meeting." *Proceedings of the American Philosophical Society* 114, no. 1 (1970): 22–36.

Howe, Daniel Walker. *The Political Culture of the American Whigs*. Chicago: Univ. of Chicago Press, 1979.

— *What Hath God Wrought: The Transformation of America, 1815–1848*. New York: Oxford Univ. Press 2007.

Hunter, Louis C. *Steamboats on the Western Rivers: An Economic and Technological History*. New York: Dover, 1994.

Hutchinson, John. "Moral Innovators and the Politics of Regeneration: The Distinctive Role of Cultural Nationalists in Nation-Building." *International Journal of Comparative Sociology* 33, no. 1–2 (January 1, 1992): 101–17.

James, Mary Ann. *Elites in Conflict: The Antebellum Clash over the Dudley Observatory*. New Brunswick: Rutgers Univ. Press, 1987.

Jansen, Axel. "Die objektive Hermeneutik als Instrument der historischen Fallrekonstruktion." *traverse – Zeitschrift für Geschichte/Revue d'Histoire* 13, no. 2 (2006): 43–56.

— *Individuelle Bewährung im Krieg: Amerikaner in Europa 1914–1917*. Frankfurt and New York: Campus, 2003.

Jessen, Ralph, and Jakob Vogel, eds. *Wissenschaft und Nation in der europäischen Geschichte*. Frankfurt and New York: Campus, 2002.

Jewett, Frank B. "Alexander Dallas Bache. A Founder, First President and Benefactor of the National Academy of Sciences." *Proceedings of the American Philosophical Society* 84, no. 2: 181–86.

Johnson, Terence James. *Professions and Power*. London: Macmillan, 1972.

Judson, Lewis V. *Weights and Measures Standards of the United States: A Brief History*. Washington: U.S. Department of Commerce, 1963.

Kaestle, Carl F. *Pillars of the Republic: Common Schools and American Society, 1780–1860*. New York: Hill and Wang, 1983).

Karimi-Hakkak, Ahmad. "Revolutionary Posturing: Iranian Writers and the Iranian Revolution of 1979." *International Journal of Middle East Studies* 23, no. 4 (November 1991): 507–31.

Kelly, M. Ruth. *The Olmsted Case: Privateers, Property, and Politics in Pennsylvania, 1778–1810*. Pennsylvania History and Culture Series. Selinsgrove, PA: Susquehanna Univ. Press, 2005.

Kennedy, Roger G. *Greek Revival in America*. New York: Stewart, Tabori & Chang, 1989.

Kevles, Daniel J. "American Science." In *The Professions in American History*, edited by Nathan O. Hatch. Notre Dame, IN: Univ. of Notre Dame Press, 1988.

— *The Physicists: The History of a Scientific Community in Modern America.* Cambridge: Harvard Univ. Press, 1971.

Kimball, Bruce A. *The True Professional Ideal in America: A History.* Lanham, MD: Rowman & Littlefield, 1996.

Klein, Philip Shriver. *Pennsylvania Politics, 1817–1832: A Game without Rules.* Philadelphia: Historical Society of Pennsylvania, 1940.

Kohlstedt, Sally. "A Step toward Scientific Self-Identity in the United States: The Failure of the National Institute, 1844." *Isis* 62, no. 3 (Autumn 1971): 339–62.

— *The Formation of the American Scientific Community: The American Association for the Advancement of Science, 1848–1860.* Urbana: Univ. of Illinois Press, 1976.

Kohlstedt, Sally Gregory, Michael M. Sokal, and Bruce V. Lewenstein. *The Establishment of Science in America: 150 Years of the American Association for the Advancement of Science.* New Brunswick, NJ: Rutgers Univ. Press, 1999.

Koyré, Alexandre. *From the Closed World to the Infinite Universe.* New York: Harper & Row, 1957.

Labaree, David F. *Making of an American High School.* New Haven: Yale Univ. Press, 1992.

Langlois, Richard N., David J. Denault, and Samson M. Kimenyi. "Bursting Boilers and the Federal Power Redux: The Evolution of Safety on the Western Rivers." *University of Connecticut, Department of Economics, Working Paper Series* (May 1994). http://www.econ.uconn.edu/working/1994–01.pdf.

Larson, John Lauritz. *Internal Improvement: National Public Works and the Promise of Popular Government in the Early United States*, Chapel Hill: Univ. of North Carolina Press, 2001.

Larson, Magali Sarfatti. *The Rise of Professionalism: A Sociological Analysis.* Berkeley and Los Angeles: Univ. of California Press, 1977.

Laverty, Bruce, Michael J. Lewis, and Michele Taillon Taylor. *Monument to Philanthropy: The Design and Building of Girard College, 1832–1848.* Philadelphia: Girard College, 1998.

Lindstrom, Diane. *Economic Development in the Philadelphia Region, 1810–1850.* New York: Columbia Univ. Press, 1978.

Lippincott, Horace Mather. *Early Philadelphia: Its People, Life and Progress.* Philadelphia: Lippincott, 1917.

Littmann, Mark. *The Heavens on Fire: The Great Leonid Meteor Storms.* Cambridge, UK: Cambridge Univ. Press, 1998.

Livingood, James Weston. *The Philadelphia-Baltimore Trade Rivalry, 1780–1860.* Harrisburg: Pennsylvania Historical and Museum Commission, 1947.

Lurie, Edward. *Louis Agassiz: A Life in Science.* Univ. of Chicago Press, 1966.

Mahon, John K. *The War of 1812.* Gainesville: Univ. of Florida Press, 1972.

Manning, Thomas G. *The U.S. Coast Survey vs. Naval Hydrographic Office: A Nineteenth-Century Rivalry in Science and Politics.* Tuscaloosa: Univ. of Alabama Press, 1988.

Martin II, John L. "The Congressional Struggle over the Coast Survey, 1848–1851." M.A. thesis, Univ. of Hawaii, 1988.

McDonald, Robert M. S., ed. *Thomas Jefferson's Military Academy: Founding West Point.* Charlottesville: Univ. of Virginia Press, 2004.

McFaul, John M. *The Politics of Jacksonian Finance.* Ithaca: Cornell Univ. Press, 1972.

McKay, Ernest. *Henry Wilson: Practical Radical: A Portrait of a Politician* (Port Washington, NY: Kennikat Press, 1971).

McMahon, A. Michael. "'Bright Science' and the Mechanic Arts: The Franklin Institute and Science in Industrial America, 1824–1976." *Pennsylvania History* 47, no. 4 (October 1980): 351–68.

McMahon, A. Michael, and Stephanie A. Morris. *Technology in Industrial America: The Committee on Science and the Arts of the Franklin Institute, 1824–1900.* Wilmington, Delaware: Scholarly Resources, 1977.

McPherson, James M. *Battle Cry of Freedom: The Civil War Era.* New York: Oxford Univ. Press, 1988.

Merton, Robert K. *Science, Technology and Society in Seventeenth-Century England.* New York: Harper Torchbooks, 1970. First published 1938 as vol. IV, part two of *Osiris.*

Meyer, David R. *Networked Machinists: High-Technology Industries in Antebellum America.* Baltimore: Johns Hopkins Univ. Press, 2006.

Miller, Howard Smith. *Dollars for Research. Science and its Patrons in Nineteenth Century America.* Seattle, London: Univ. of Washington Press, 1970.

Miller, Lillian B. *The Lazzaroni: Science and Scientists in Mid-Nineteenth-Century America.* Washington: Published for the National Portrait Gallery, Smithsonian Institution, 1972.

Millett, Allan Reed, and Peter Maslowski. *For the Common Defense: A Military History of the United States of America.* New York: Free Press, 1984.

Morison, Samuel Eliot. *Three Centuries of Harvard, 1636–1936.* Cambridge: Harvard Univ. Press, 1936.

Morrell, Jack, and Arnold Thackray. *Gentlemen of Science: Early Years of the British Association for the Advancement of Science* (Oxford: Clarendon Press, 1981.

Moyer, Albert E. *Joseph Henry: The Rise of an American Scientist.* Washington: Smithsonian Institution Press, 1997

Münte, Peter. *Die Autonomisierung der Erfahrungswissenschaften im Kontext frühneuzeitlicher Herrschaft: Fallrekonstruktive Analysen zur Gründung der Royal Society.* 2 vols. Frankfurt: Humanities Online, 2004.

— "Institutionalisierung der Erfahrungswissenschaften in unterschiedlichen Herrschaftskontexten. Zur Erschließung historischer Konstellationen anhand bildlicher Darstellungen." *Sozialer Sinn* 1 (2005): 3–44.

Münte, Peter, and Ulrich Oevermann. "Die Institutionalisierung der Erfahrungswissenschaften und die Professionalisierung der Forschungspraxis im 17. Jahrhundert. Eine Fallstudie zur Gründung der Royal Society." In *Wissen und soziale Konstruktion*, edited by Claus Zittel, 165–230. Berlin: Akademie Verlag, 2002.

Murrin, John M. "A Roof Without Walls: The Dilemma of American National Identity." In *Beyond Confederation: Origins of the Constitution and American National Identity*, edited by Richard Beeman, Stephen Botein, and Edward C. Carter, 333–48. Chapel Hill: Univ. of North Carolina Press, 1987.

Numbers, Ronald L., and Charles E. Rosenberg, eds. "Science in American Society: A Generation of Historical Debate." In *The Scientific Enterprise in America: Readings from Isis*. Chicago: Univ. of Chicago Press, 1996.

Odgers, Merle Middleton. *Alexander Dallas Bache, Scientist and Educator, 1806–1867*. Philadelphia: Univ. of Pennsylvania Press, 1947.

Oevermann, Ulrich. "Die objektive Hermeneutik als unverzichtbare methodologische Grundlage für die Analyse von Subjektivität. Zugleich eine Kritik der Tiefenhermeneutik." In *'Wirklichkeit' im Deutungsprozeß: Verstehen und Methoden in den Kultur- und Sozialwissenschaften*, edited by Thomas Jung and Stefan Müller-Doohm, 106–89. Frankfurt: Suhrkamp, 1993.

— "Für ein neues Modell von Kulturpatronage." In *Die Kunst der Mächtigen und die Macht der Kunst*, edited by Ulrich Oevermann, Johannes Süßmann, and Christine Tauber, 13–23. Berlin: Akademie Verlag, 2007.

— "Objektivität des Protokolls und Subjektivität als Forschungsgegenstand." *Zeitschrift für qualitative Bildungs-, Beratungs- und Sozialforschung* 2 (2004): 311–36.

— "Regelgeleitetes Handeln, Normativität und Lebenspraxis. Zur Konstruktionstheorie der Sozialwissenschaften." In *'Normalität' im Diskursnetz soziologischer Begriffe*, edited by Jürgen Link, Hartmut Neuendorf, and Thomas Loer, 183–219. Heidelberg: Synchron, 2003.

— *Strukturprobleme supervisorischer Praxis. Eine objektiv hermeneutische Sequenzanalyse zur Überprüfung der Professionalisierungstheorie*. Frankfurt: Humanities Online, 2001.

— "Theoretische Skizze einer revidierten Theorie professionalisierten Handelns." In *Pädagogische Professionalität. Untersuchungen zum Typus pädagogischen Handelns*, edited by Arno Combe and Werner Helsper, 70–182. Frankfurt: Suhrkamp, 1996.

— "Wissenschaft als Beruf – Die Professionalisierung wissenschaftlichen Handelns und die gegenwärtige Universitätsentwicklung." *Die Hochschule - Journal für Wissenschaft und Bildung* 14, no. 1 (2005): 307–18.

O'Leary, Cecilia Elizabeth. *To Die for: The Paradox of American Patriotism*. Princeton: Princeton Univ. Press, 1999.

Oleson, Alexandra, and Sanborn C. Brown, eds. *The Pursuit of Knowledge in the Early American Republic: American Learned and Scientific Societies from Colonial Times to the Civil War*. Baltimore: Johns Hopkins Univ. Press, 1976

Onuf, Peter S. *Jefferson's Empire: The Language of American Nationhood*. Charlottesville: Univ. of Virginia Press.

Onuf, Peter S., and Leonard J. Sadosky. *Jeffersonian America.* Oxford: Blackwell Publishing, 2002.

Parish, Peter J. *The North and the Nation in the Era of the Civil War.* Edited by Adam I. P. Smith and Susan-Mary Grant. New York: Fordham Univ. Press, 2003.

Parsons, Talcott. "The Professions and Social Structure." *Social Forces*, no. 4 (May 1939): 457–61.

— *The Social System.* Glencoe, IL: Free Press, 1951.

Pasley, Jeffrey L. *The Tyranny of Printers: Newspaper Politics in the Early American Republic.* Charlottesville: Univ. of Virginia Press, 2001.

Peterson, Edward. *History of Rhode Island.* New York: J. S. Taylor, 1853.

Potter, David M. "The Historian's Use of Nationalism and Vice-Versa." In *The South and the Sectional Conflict*, 34–83. Originally published in 1963. Baton Rouge: Lousiana State Univ. Press, 1968.

Rao, Joe. "The Leonids: The Lion King of Meteor Showers." *WGN, Journal of the International Meteor Organization* 23 (August 1, 1995): 120–35.

Reese, William J. *The Origins of the American High School.* New Haven: Yale Univ. Press, 1999.

Reingold, Nathan. "Alexander Dallas Bache: Science and Technology in the American Idiom." In *Science American Style*, 110–26. New Brunswick: Rutgers Univ. Press, 1991.

— "American Indifference to Basic Research: A Reappraisal." In *Nineteenth-Century American Science: A Reappraisal*, edited by George H. Daniels, 38–62. Evanston: Northwestern Univ. Press, 1972.

— "Definitions and Speculations: The Professionalization of Science in America in the Nineteenth Century." In *The Pursuit of Knowledge in the Early American Republic: American Learned and Scientific Societies from Colonial Times to the Civil War*, edited by Alexandra Oleson and Sanborn C. Brown, 33–69. Baltimore: Johns Hopkins Univ. Press, 1976.

— *Science in Nineteenth-Century America, a Documentary History.* London, Melbourne, Toronto: Macmillan, 1966.

— "Science in the Civil War. The Permanent Commission of the Navy Department." *Isis*, no. 3 (1958): 307–18.

Remini, Robert V. *The Election of Andrew Jackson.* Philadelphia: Lippincott, 1963.

Riesman, David. *The Lonely Crowd: A Study of the Changing American Character.* New Haven: Yale Univ. Press, 1950.

Risjord, Norman K. *Jefferson's America, 1760–1815.* Madison: Madison House, 1991.

Romano, Louis A. *Manual and Industrial Education at Girard College, 1831–1965: An Era in American Educational Experimentation.* New York: Arno Press, 1980.

Rosenberg, Charles E. *No Other Gods: On Science and American Social Thought.* Baltimore: Johns Hopkins Univ. Press, 1978.

Ross, Dorothy. *The Origins of American Social Science.* Cambridge: Cambridge Univ. Press, 1991.

Rumpf, Lorenz. *Naturerkenntnis und Naturerfahrung: Zur Reflexion epikureischer Theorie bei Lukrez.* München: Beck, 2003.

Scholnick, Robert J. *American Literature and Science.* Lexington, Kentucky: Univ. Press of Kentucky, 1992.

Searle, John R. *Intentionality.* Cambridge: Cambridge Univ. Press, 1983.

— *Speech Acts: An Essay in the Philosophy of Language.* Cambridge: Cambridge Univ. Press, 1969.

— *The Construction of Social Reality.* New York: Free Press, 1995.

Sharp, James Roger. *American Politics in the Early Republic: The New Nation in Crisis.* New Haven: Yale Univ. Press, 1993.

Shryock, Richard H. "American Indifference to Basic Research." *Archives internationales d'histoire des sciences* XXVII (1948): 50–65.

Sinclair, Bruce. *Early Research at the Franklin Institute: The Investigation into the Causes of Steam-Boiler Explosions, 1830–1837.* Philadelphia: The Franklin Institute of the State of Pennsylvania, 1966.

— *Philadelphia's Philosopher Mechanics: A History of the Franklin Institute.* Baltimore: Johns Hopkins Univ. Press, 1974.

Slaughter, Thomas Paul. *The Whiskey Rebellion: Frontier Epilogue to the American Revolution.* New York: Oxford Univ. Press, 1986.

Slotten, Hugh R. *Patronage, Practice, and the Culture of American Science. Alexander Dallas Bache and the U.S. Coast Survey.* Cambridge: Cambridge Univ. Press, 1994.

— "Science, Education, and Antebellum Reform: The Case of Alexander Dallas Bache." *History of Education Quarterly* 31, no. 3 (Autumn 1991): 323–42.

— "The Dilemmas of Science in the Unites States. Alexander Dallas Bache and the U.S. Coast Survey." *Isis*, no. 84 (1993): 26–49.

Smith, Adam I. P., *No Party Now: Politics in the Civil War North* (New York: Oxford Univ. Press, 2006).

Smith, Henry Nash. *Virgin Land. The American West as Symbol and Myth.* Cambridge: Harvard Univ. Press, 1950.

Smith, Jeffery Alan. *Franklin and Bache: Envisioning the Enlightened Republic.* New York: Oxford University Press, 1990.

Stampp, Kenneth M. "The Concept of a Perpetual Union." In *The Imperiled Union: Essays on the Background of the Civil War*, 3–36. New York: Oxford Univ. Press, 1980.

Storr, Richard J. *Beginnings of Graduate Education in America.* Chicago: Univ. of Chicago Press, 1953.

Stratton, Julius Adams, and Loretta H. Mannix. *Mind and Hand: The Birth of MIT.* Boston: MIT Press, 2005.

Terhart, Ewald. "The Adventures of Interpretation: Approaches to Validity." *Curriculum Inquiry* 15, no. 4 (Winter 1985): 451–64.

Theberge, Albert E. "The Coast Survey, 1807–1867." http://www.lib.noaa.gov/noaainfo/heritage/coastsurveyvol1/CONTENTS.html.

Tyack, David B., and Elisabeth Hansot. *Managers of Virtue: Public School Leadership in America, 1820–1980*. New York: Basic Books, 1982.

Veysey, Laurence. *The Emergence of the American University*. Chicago: Univ. of Chicago Press, 1965.

— "Who's a Professional? Who Cares?" *Reviews in American History* 3 (December 1975): 419–23.

Walter, Thomas Ustick, and Jennifer A. Amundson. *Thomas U. Walter: The Lectures on Architecture, 1841–53*. Philadelphia: Athenaeum of Philadelphia, 2006.

Walters Jr., Raymond. *Alexander James Dallas, Lawyer, Politician, Financier, 1759–1817*. Philadelphia: Univ. of Pennsylvania Press, 1943.

Warner Jr., Sam Bass. *The Private City: Philadelphia in Three Periods of Its Growth*. Philadelphia: Univ. of Pennsylvania Press, 1987.

Weber, Gustavus Adolphus. *The Bureau of Standards: Its History, Activities and Organization*. Baltimore: Johns Hopkins Press, 1925.

Weber, Max. "Wissenschaft als Beruf." In *Gesammelte Aufsätze zur Wissenschaftslehre*, 582–613. Tübingen: J. C. B. Mohr (Paul Siebeck), 1988.

Wilentz, Sean. *The Rise of American Democracy: Jefferson to Lincoln*. New York: Norton, 2005.

Index

History

Nikolaus Katzer, Sandra Budy, Alexandra Köhring,
Manfred Zeller (eds.)
Euphoria and Exhaustion
Modern Sport in Soviet Culture and Society
2010, 363 pages, ISBN 978-3-593-39290-5

Sebastian Jobs, Alf Lüdtke (eds.)
Unsettling History
Archiving and Narrating in Historiography
2010, 253 pages, ISBN 978-3-593-38818-2

Angelika Epple, Angelika Schaser (eds.)
Gendering Historiography
Beyond National Canons
2009, 244 pages, ISBN 978-3-593-38960-8

Susanne Lachenicht, Kirsten Heinsohn (eds.)
Diaspora Identities
Exile, Nationalism and Cosmopolitanism in Past and Present
2009, 164 pages, ISBN 978-3-593-38819-9

Mathias Albert, Gesa Bluhm, Jan Helmig,
Andreas Leutzsch, Jochen Walter (eds.)
Transnational Political Spaces
Agents – Structures – Encounters
2009, 322 pages, ISBN 978-3-593-38945-5

Martin Baumeister, Stefanie Schüler-Springorum (eds.)
»If You Tolerate This ...«
The Spanish Civil War in the Age of Total War
2008, 300 pages, ISBN 978-3-593-38694-2